THE WHOLE KIDS CATALOG

For Adventure Seekers Of All Ages

THE WHOLE KIDS CATALOG

Created by

Peter Cardozo

Designed by

Ted Menten

BANTAM BOOKS

TORONTO / NEW YORK / LONDON

THE WHOLE KIDS CATALOG

A Bantam Book / October 1975

Published simultaneously in the United States and Canada.

Bantam Books are published by Bantam Books, Inc. Its trademark, consisting of the words "Bantam Books" and the portrayal of a bantam, is registered in the United States Patent Office and in other countries. Marca Registrada.

Bantam Books, Inc., 666 Fifth Avenue, New York, N.Y. 10019

Printed in the United States of America

0 9 8 7 6 5 4

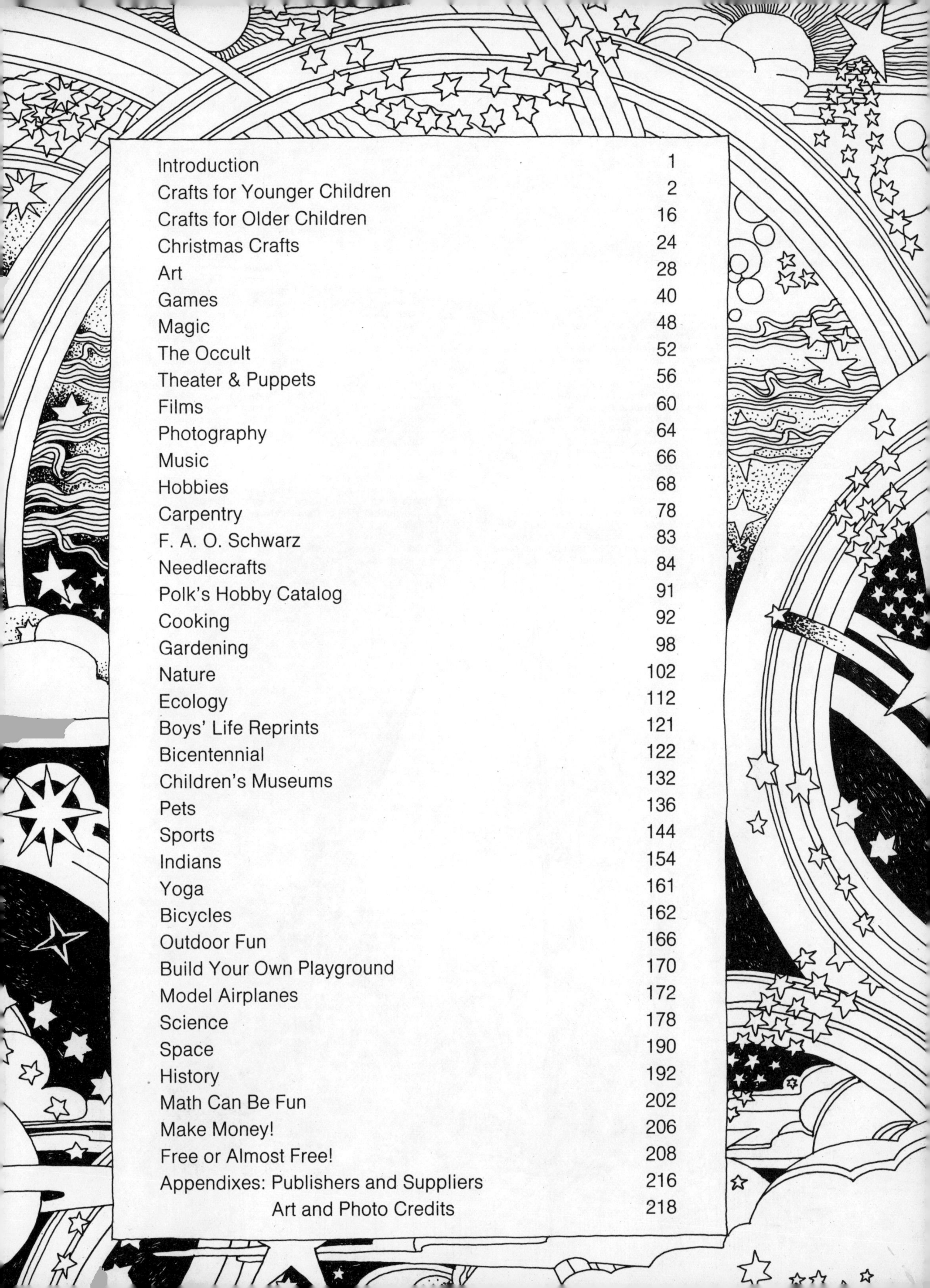

This is a magical book!

It will conjure up exciting adventures.
It has the power to turn you into an explorer.
It is an open sesame to discovering new ways to have fun . . .
Indoors, outdoors . . . quietly . . . or laughing your head off.

Use it as a **source book** and you'll learn that you can build everything from a model car to a tree house, grow plants in a terrarium, breed gerbils, fly kites, read Tarot cards, put on a magic show, train pets, create sandcastles, collect stamps, make puppets, knit and weave, produce a play or a movie . . . and a thousand other things that mean a learning experience without feeling you're being taught.

Use it as an **activity book.** Throughout the pages there are all sorts of things that you can actually do or make. You can color pictures of soldiers and animals, cut out historical paper dolls, solve puzzles, do mazes, create collages, fold paper airplanes, make a robot out of spools, turn an egg carton into a silly sea serpent, cook spaghetti from a special recipe, perform a mystifying mind reading trick, write in code, do scientific experiments. There's enough fun for a year.

Use it as a **guide book to free things.** Here are free samples, free posters, free pictures, free books, free recipes, free coloring books, free coins, plus a whole chapter of different things that you can get free. Wherever you see **FREE** ☞ there's something you can write away for.

This is a book for action. It happens all over the place . . . and in places you never thought of looking. It happens through a process of exploration which starts with this book and grows to include the whole world.

Only one thing is needed to make it all come alive . . . you!

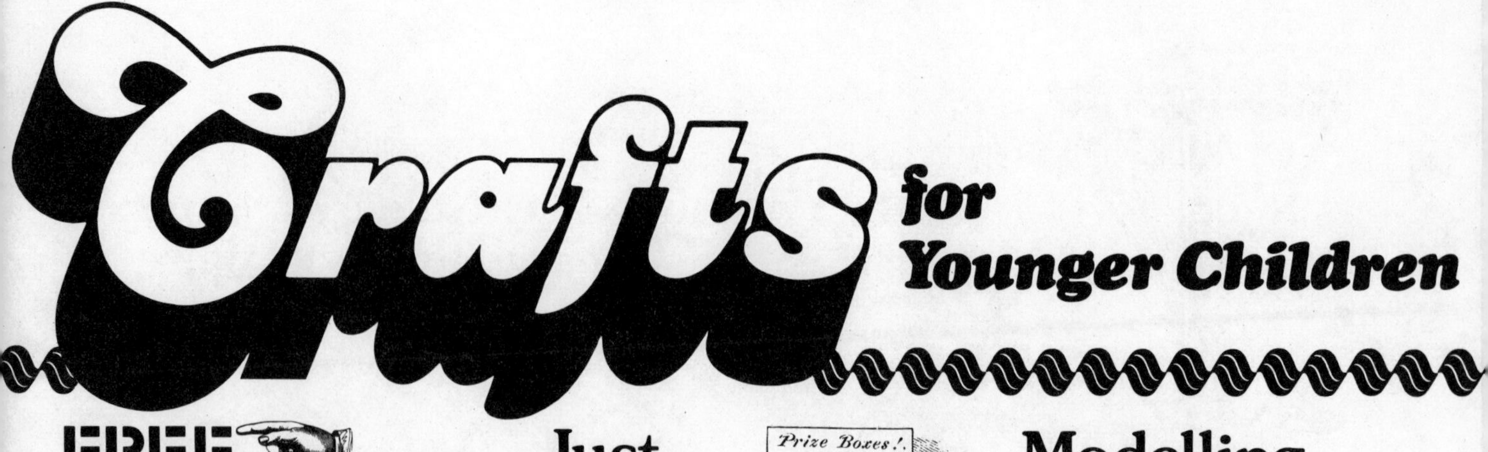

FREE Move Over Michelangelo

Put 2 cups of baking soda and 1 cup of cornstarch in a saucepan. Blend thoroughly with a fork. Add 1¼ cups of cold water. Mix until smooth. Boil 1 minute over medium heat until the mixture looks like moist mashed potatoes. Spoon into a plate. Cover with a damp cloth and let cool.

Now you have a batch of play clay! *Move Over Michelangelo* will show you what to make with your clay: vases, pencil holders, bookends, paperweights, storage containers, wall decorations, and plaques. The pamphlet will teach you how to mold, finish, and color the clay. For your free copy WRITE TO Church & Dwight Co., 2 Pennsylvania Plaza, New York, N.Y.

FREE Raft of Crafts

Everyone loves to receive a gift, and now you can increase your enjoyment of giving one by making it yourself. All you need are some old scraps found around the house to create unique gifts for your family and friends. How about a mini-garden planted in caps from empty aerosol cans, or a decoupaged box, or a wind chime made from empty tin cans, or a personalized wooden hanger? All these and more ideas are free in the booklet *A Raft of Crafts.* SEND your request on a postcard to Johnson Wax, Dept. WWC, Golden Rondelle, P.O. Box 567, Racine, Wis. 53403.

Just a Box?

Prize Boxes! Containing WHITE ELEPHANTS. ONE CENT EACH.

You can transform an ordinary box into a totem pole, a tepee, a zoo, a puppet, a stage, a store, a castle—boxes can even become trains, boats, and planes. You can decorate the things you make with crayons, felt-tip markers, poster paint, yarn, magazine ads, buttons, and beads. Want to learn how? Read *Just a Box?* written by Goldie Taub Chernoff and illustrated by Margaret Hartelius. Scholastic Book Services. 95 cents.

Modelling Is Easy

Modelling Is Easy—When You Know How, by Mary Dewing and Kathleen Douet, is the full title, and it's the "when you know how" that makes it easy. All the projects are splendidly illustrated by Ann Rees, with full-color photographs and instructions easy enough for 7-year-olds to follow. The materials you learn to work with are clay, papier-mâché, popsicle sticks, tissue, felt, fabric, and foil. The ingenious projects include a tribal

Make tin-can stilts, pasta jewelry, puppets, a totem pole. Fold, paste, cut, staple, paint, and hammer . . . and you're into 101 crafts. Spooncraft, tincraft, leathercraft, cartoncraft, papercraft, kitchencraft, scrapcraft . . . all the way to suppertime.

mask, "Wild Bill the Bandit," the Holy Family, a house and garden, a paper doll, a robot puppet with electric eyes, goldfish, soccer players, a safari park, and a pink elephant. Arco Publishing Co. $4.95.

Boden's Beasts

A charming craft book that will excite the creative and inventive talents of the young reader. *Boden's Beasts,* by John Woodside, is about making fun animals out of everyday household throwaway material. The text is in rhyme—actually it's a dialogue: Mr. Boden "speaks" in brown print to a little girl, Loren Jane, who "speaks" to Mr. Boden in gray print. The result? A menagerie of animals. Two Continents Publishing Group. $4.95.

Arts and Crafts You Can Eat

Now your kitchen can be an artist's studio with you as resident artist. You can transform ordinary foods into works of art that are as good to eat as they are to look at. The things you can create include inlaid pancakes, chocolate marshmallow scratchboards, stained-glass cookies, a kohlrabi monster and other vegetable beasts, and a gingerbread stick puppet.

Arts and Crafts You Can Eat was produced by the writer-illustrator team of Vicki Cobb and Peter Lippman. J. B. Lippincott Co. $1.95.

Kitchen Crafts

You can turn your kitchen into a creative workshop with *Kitchen Crafts,* by Linda and John Cross. This book shows you how to create dozens of useful and decorative objects without leaving your kitchen or investing in expensive tools and equipment. Every technique is illustrated and outlined, and "recipes" are provided for each project. You'll discover a wealth of hidden possibilities in flour, eggs, fruit and vegetables, nuts, beans, seeds, pasta, candles and wax, cans and containers, and leftovers. All you need is imagination and enthusiasm. Among the skills taught are macrame, tie-dying, batik, candlemaking, tin cutting, soap carving, and papier-mâché. Some projects are delectably edible, including a great cheese omelet, a gingerbread house, cookie castles, and marzipan! Macmillan Publishing Co. $8.95 cloth. Collier Books. $3.95 paper.

Scrapcraft

Look around the kitchen and you'll find 90 percent of the materials needed to make the projects in this book. You can be both ecologist and artist as paper bags become masks, puppets, and costumes. Instructions tell how to paint with string, cord, sponges, and fingers. You'll learn how to print with kitchen items like bottle caps, forks, and potato mashers, as well as vegetables and fruits. Other make-it-yourself craft items include a shadow box, tin-can stilts, tin-can bank, pasta jewelry, pasta-decorated mirrors, seed collages, an egg carton jewelry box, picture paperweights, and personalized switch plates. There are 50 easy-to-make handicraft projects. *Scrapcraft,* by Judith Choate and Jane Green (with illustrations by Steve Madison), is a good beginner's book. Doubleday & Co. $4.95.

Felt Craft

Felt is a wonderful craft medium, colorful and easy to handle. The projects in *Felt Craft,* by Florence Temko, are designed to be made by children without adult help; they are an introduction to felt craft for anyone of any age. Here are finger puppets, a hat, a poncho, a collage, bookmarks, a headband, and more. Felt is fun to work with, and you do not have to make the things exactly as they are shown in the book—you can vary them any way you like, make them larger; make them smaller; decorate them. Doubleday & Co. $4.95.

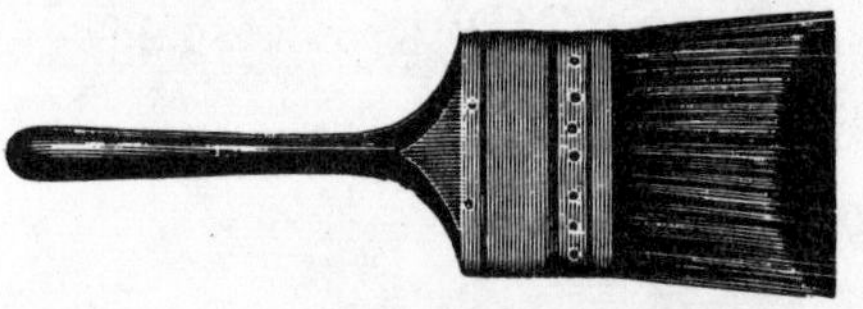

Fun with Next to Nothing

Here are some handicraft projects for you—with a difference. With the odds and ends from your boodle box you can make all sorts of interesting things. (A "boodle box" is where you keep those useless things you've saved in case you need them someday.) Wesley Arnold and Wayne Cardy show you how to make a model of a raft, a canoe, a cart, a train, a covered wagon, or a Viking ship. You learn about early man by making some of the houses he lived in—a cave, an igloo, a tepee, an adobe house, a grass hut, even a medieval castle. There are illustrations so you can "travel into space with rocket ships and a moon base." *Fun with Next to Nothing* is a Starline Edition from Scholastic Book Services. 75 cents.

Pebbles and Pods, a Book of Nature Crafts

This is not just another nature book or crafts book, but a unique combination of both. Here are easy-to-follow directions for a variety of imaginative activities that will show youngsters how to use natural materials like stones, bark, water, sand—even clouds and shadows—to make beautiful and original art projects. You will learn about spatter prints, wood and leaf rubbings, shell prints, stone figures, mosaics—and how to make beautiful sand drawings—from *Pebbles and Pods: A Book of Nature Crafts,* by Goldie Taub Chernoff (illustrated by Margaret Hartelius). Scholastic Book Services. 95 cents.

FREE Beautiful Junk

Material for building toys and games doesn't have to cost a lot of money. A lot of it is free, like this booklet that tells you about sources (moving companies, printshops, ice-cream stores) and what to make with what you get: puppet racks, flannel boards, wheel toys, and so on. For your free copy of *Beautiful Junk* WRITE TO U.S. Department of Health, Education, and Welfare, Office of Child Development, Washington, D.C. 20201. Be sure to ask for No. 73–1036.

Fat Cat's Craft and Coloring Book

You can have fun making the things in this book: everything from "Easel Squeezles" to "Masked Marvels," even "Stuffets and Zoobies." The introduction says: "The fanciful pictures by Donna Sloan are for coloring. They also give you a general idea of how the project could look. These designs are only models; make any changes you wish. Use your imagination. You are the craftsman." *Fat Cat's Craft and Coloring Book* is by Karen Whyte. Troubador Press. $1.50

Patterns for Miniature Furniture

With simple tools and the help of this catalog of patterns you can make miniature tables, chairs, beds, cupboards, cradles, fireplaces, barrels, and 50 other pieces of furniture. The catalog also lists metal miniature accessories, including a fireplace set, dishes, pots and pans, candlesticks, and picture frames, and there is a pageful of wallpaper and floor coverings. For *Patterns for Miniature Furniture* SEND 50 cents to Green Door Studio, Dept. W, 517 E. Annapolis St., St. Paul, Minn. 55118.

Fold, Paste, Whittle, Paint and Hammer

This is another "all kinds of things to make with simple materials" book and it's fun, illustrated with a nice sense of humor, and shows you how to make everything from peanut birds to a pushmobile. Robert Pierce put it all together. Golden Press. $1.50.

ZOOBIES

Turn household discards into whimsical, charming animals; whether a mythological beast to play with or an enchanting bank to give as a present.

Materials

plastic bottles
balloons
paper plates
wrapping paper, paper towel or toilet paper cores
cotton
clothes hangers
glue or paste
strips of newspaper
yarn
felt
spools
old toy wheels or tinker toys
acrylic or poster paints
shellac or plastic glaze

1. Use an unusual shaped plastic bottle (with or without handles) or a balloon for the body. Dip strips of newspaper in glue or paste and cover entire body. A paper or aluminum plate makes a lady bug.
2. For raised areas such as eyes and nose, dip cotton in paste and stick in place. Cover and shape the raised areas with strips of newspaper. This kind of padding can be used to shape legs and arms close to the body.
3. Cut clothes hangers in desired lengths for legs or use paper towel cores. Long necks are made in the same way. Glue on or tape to body, then, wrap the wires or cores with strips of newspaper pasted in place.
4. Wheels can be attached by threading heavy wire through the side of the core, then through the wheel and out the other side of the core. Bend down ½ inch of wire on each side. Cover wire with newspaper strips.
5. Paint with brilliant colors. When paint is dry, give your Zoobie a coat of shellac or plastic glaze. For banks, cut a slit in top of animal. You can add other decorations by gluing on felt, buttons, sequins, yarn or real feathers.

Spoolcraft

Wooden spools can be used to make all sorts of toys: clowns, peace pipes, mobiles, pincushions, airplanes, printing blocks, and games of all kinds. This book gives you clear and complete directions; only the simplest tools are needed. Chapter headings include "Spool People"; "Dolls and Doll Furniture"; "Spool Toys"; "Knitting and Printing with Spools"; "Spool Games"; "Indian Novelties"; and "Gifts, Gadgets and Decorations." Try your hand at making a robot, and if you like spoolcraft, get *Spoolcraft,* by Arden Newsome. Lothrop, Lee & Shepard Co. $2.50.

Robot

You Will Need:

1 large spool
1 small spool
7 tiny spools (the size found in mending kits)
cardboard
paints: silver, black, white, and red
2 thumbtacks
7-inch piece of thin, flexible wire
a piece of thin dowel stick (smaller than tiny spool hole)
tracing paper
glue, pencil, scissors, brushes

How to Make It:

1. Glue three tiny spools one on top of the other. Do the same with three other tiny spools. These will be the robot's legs. Stand the legs side by side and glue the large spool body on top of them. Glue the small spool head on the body and the last tiny spool on top of the head.
2. Trace the patterns of foot and arm onto tracing paper and transfer them to cardboard. Make two feet and two arms, then cut them out.
3. Glue a foot to the bottom of each leg. Fold back the tab of each arm and brush glue on the tabs, then stick an arm to each side of body.
4. Wind the piece of wire around the thin stick. Slide the wire off the stick and push one end into the tiny spool on top of the head. It will look like a spring.
5. Paint the robot with silver paint. Let dry. Paint square black eyes and a red rectangular mouth on the face spool. Paint a white square for panelboard on front of the body spool and let dry. Paint a small black rectangle on the white panelboard. All robots have dials. When paint is dry, push the thumbtacks into the panelboard for knobs to turn the dials.

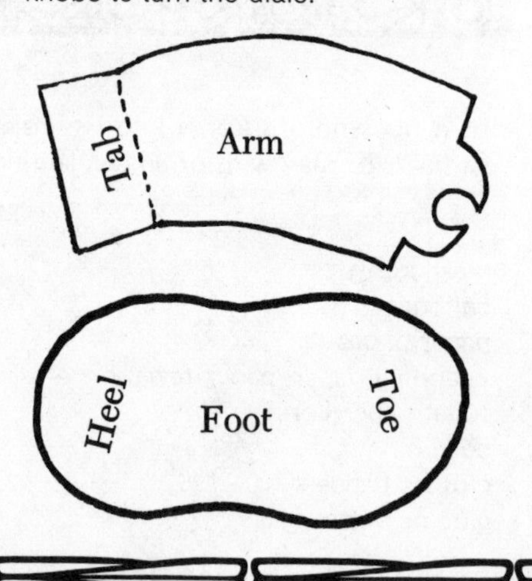

Folding Paper Toys

This is really a toy chest in book form, filled with playthings you can make out of plain paper: sailing and flying toys, puppets, magic tricks and noisemakers, and all sorts of furniture for a dollhouse. These paper-fold toys don't just stand there, they do something.

The puppets make a wonderful bathtub regatta, the puppets nod and "talk," the party noisemakers work, the flying objects glide and dive, and the tricks are really tricky.

Shari Lewis and Lillian Oppenheimer, authors of *Folding Paper Toys,* give you simple step-by-step diagrams to follow and there are large photographs of the finished toys. For boys and girls age 6 and up. Stein & Day. $4.50.

Tie-Dyed Paper

Tie-dying with everyday, round-the-house papers is an exciting new craft specially developed by author Anne Maile to introduce young children to the pleasures and fun of tie-dyeing. You will learn how to use only safe, inexpensive cold-water dyes to turn paper napkins and towels, and tissue, wrapping, and typing papers, into strikingly radiant designs. The techniques are easy to learn from step-by-step illustrated instructions. Inspired by the photographs (many of them in full color), you'll soon be making tie-dyed collages, posters, origami, mobiles, dolls, simple hand puppets, masks, and hats. Looking for a new and easy craft? Get *Tie-Dyed Paper*. Taplinger Publishing Co. $12.95.

The Gadget Book

The Gadget Book tells you how to make useful and useless gadgets. Here is everything from an alarm clock set off by the morning sun shining in the window to a burglar alarm for your door. They are all homemade figured-out-as-you-go devices. Instructions and diagrams give basic principles with clear helpful information about materials and how to put things together, but the precise details are up to you. Author-illustrator Harvey Weiss says: "The fun of a gadget comes not only from using it, but from putting it together in your own personal way, according to your own ideas." Age 8 and up. Thomas Y. Crowell Co. $4.50.

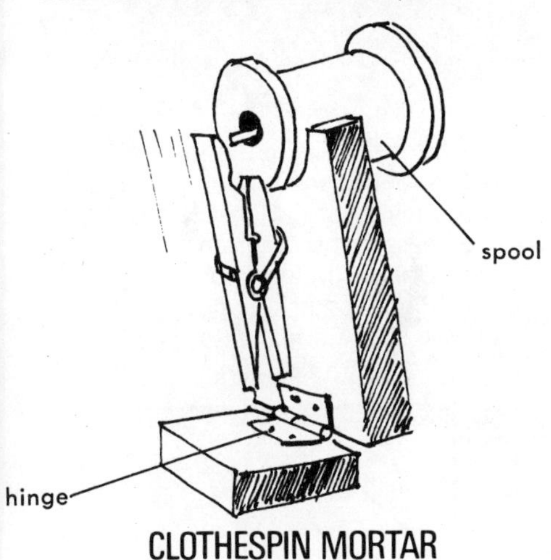

CLOTHESPIN MORTAR

International Folk Crafts

International folk crafts projects are easy and educational. Children in different parts of the world make these crafts, following instructions passed on from generation to generation. When you follow the directions in *The International Folk Crafts Book* (written by Ginger Scribner and illustrated by Gretchen Schields), your finished folk crafts will look as if they were made in a foreign country. The materials used can be found around the house or at a nearby hardware or sewing store. Troubador Press. $1.50.

The Fat Cat

A kangaroo with flowers in her pouch, a jeweled elephant carrying a splendid maharaja on his back, a charming fat cat purring on his luxurious tasseled cushion—all these and many more animals greet you from whirling, swirling scroll-like drawings. And there are limericks to make you smile.

With fishes in schools, you would guess,
They'd be smart to the point of excess,
But the way that they play
And frolic all day
Their schooling's a day long recess.

The Fat Cat Coloring & Limerick Book is a delightful coloring and limerick book with drawings by Donna Sloan and limericks by Mal Whyte. The book is in large format, printed on heavy vellum paper so you can cut out your colored drawings and use them for decoration. Troubador Press. $2.00.

Recipes for Art and Craft Materials

Instead of buying all the materials for your art and craft projects, here's an easy way to make a lot of them at home. Use the large collection of "recipes" in Helen Roney Sattler's book. You can learn to make your own pastes, inks, modeling compounds, paints and paint mediums, papier-mâché, and preservatives from *Recipes for Art and Craft Materials*. Lothrop, Lee & Shepard Co. $4.95.

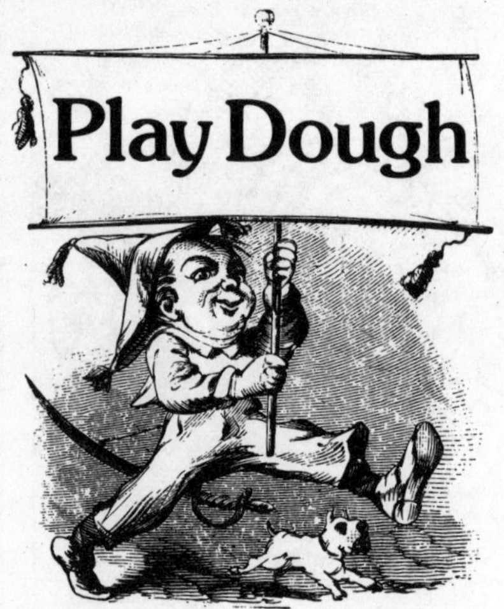

Play Dough

You Will Need:

2¼ cups non-self-rising wheat flour
1 cup salt
1 tablespoon powdered alum
4 tablespoons vegetable oil
1½ cups boiling water
food coloring or poster paints

How to Make It:

1. Mix flour, salt, and alum. Add vegetable oil.
2. Stir in boiling water. Stir vigorously with a large spoon until mixture holds together.
3. Knead the dough until it is smooth.
4. Divide the dough into several lumps. Add a few drops of food coloring or poster paint to each lump and knead to mix the color into the dough.

Makes about 3 cups.

How to Use It: Model as with clay. Objects will dry to a hard finish if left in the open air. Paint dried pieces with enamel, hyplar, or tempera.

Stored in an airtight container, Play Dough will keep a long time.

You Can Make Ragbag Treasures

Full of amusing, original ideas for making things out of everyday objects found just about anywhere, this book is translated from French, which may be what makes these creations so fresh. The book shows you how to make stick-on pictures, quick-knit creatures, puppet mascots, mitten marionettes, button banners, and many other unusual things. *You Can Make Ragbag Treasures* has an approach that will make you smile—and want to "do." Other Pinwheel Craft Books are *You Can Make Seaside Treasures, You Can Make Country Treasures,* and *You Can Make Tasty Treasures*. All are edited by Louis Beetschen. Pantheon Books. $1.25 each.

The Little Kid's Craft Book

When *The Little Kid's Craft Book* was published it was chosen as a selection by seven book clubs. Several impertant magazines acclaimed it as outstanding for its ingenious suggestions about craft projects made out of easily accessible materials. It includes sections on coloring, painting, clay modeling, building and manipulating puppets, and learning about nature with decorative leaf or seashell collections. Fully illustrated with actual craft projects of young children, *The Little Kid's Craft Book*, by Jackie Vermeer and Marian Lariviere, includes a stoplight made from bits of colored paper, and even an awesome array of cloth ghosts for Halloween. Taplinger Publishing Co. $8.95.

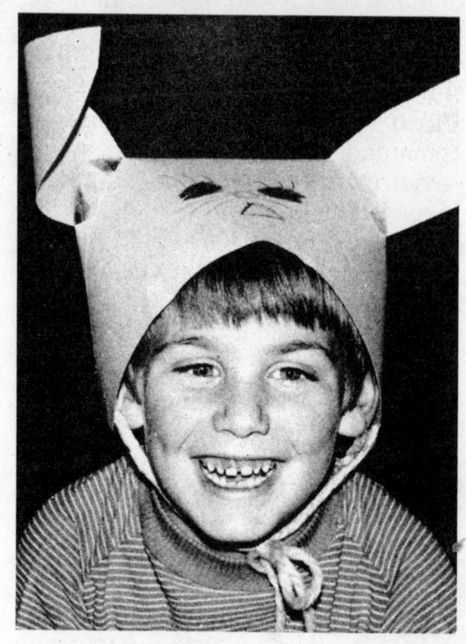

The bunny hat is made from a 12" by 18" sheet of construction paper. Draw on bunny's face, then cut and staple as shown in the diagram.

1. Cut on dotted lines.
2. Place point A on point B and staple. Repeat with C and D.
3. Add yarn ties.

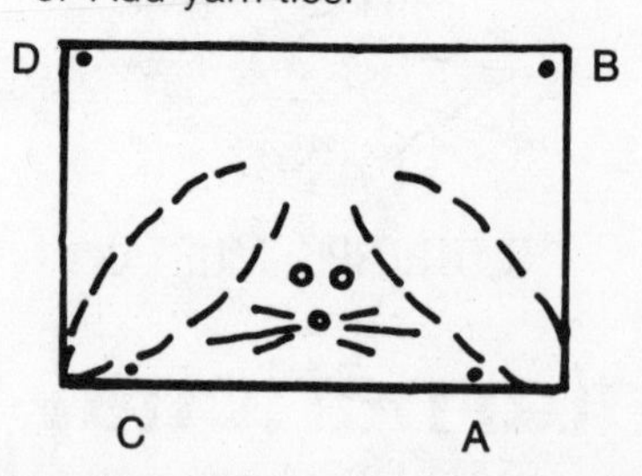

FREE Pit-Niks Make History

Contrary to popular opinion, avocado pit-niks did not come over on the *Mayflower*. The lost tribe of pit-nik was discovered only a few years ago by an art class in Pasadena, California. Some students were learning how to recycle scraps into craft items when the pit-niks found them!

You don't have to be artsy-craftsy to make a pit-nik. Several irresistible creatures are already ready, like the humpty-dumpty or the magic mouse: just follow the easy directions. There are animals, people, even out-of-space creatures hidden beneath every avocado's skin. For directions to make pit-niks and a lot of crafty things from avocado pits (paperweights, party favors, animals, toys, and so on), WRITE TO Pit-Niks, P.O. Box 2162, Costa Mesa, Calif. 92626, for your free copy of *Avocado Pit-Niks*.

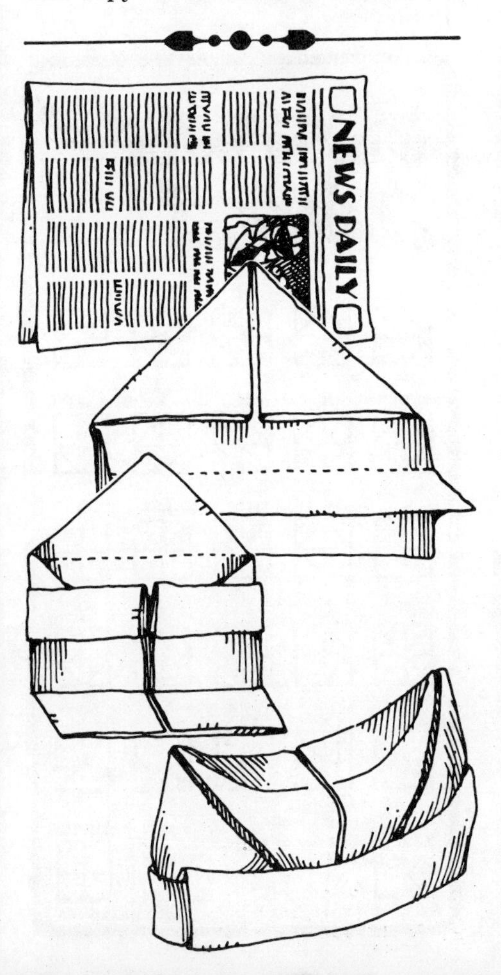

Steven Caney's Toy Book

Here are toys that parents and children make together; toys that work; toys that are inexpensive; toys that are fun. There are more than 50 of them in *Steven Caney's Toybook*. SEND $3.95 (paper) or $8.95 (cloth) plus 50 cents for mailing and handling to Workman Publishing Co.

Steven Caney's Play Book

A new collection of over 100 activities—each with photographs and diagrams. Projects from constructions to simple games require only discards and found materials. Things to make include an almost impossible card game, a geodesic dome doghouse, tire tread shoes, a paper clip and strip trick, fingerprints, room weaving (the whole room becomes a loom), and much, much more. For your copy of *Steven Caney's Playbook* SEND $4.95 (paper) or $9.95 (cloth) plus 50 cents for mailing and handling to Workman Publishing Co.

The Little Kid's Four Seasons Craft Book

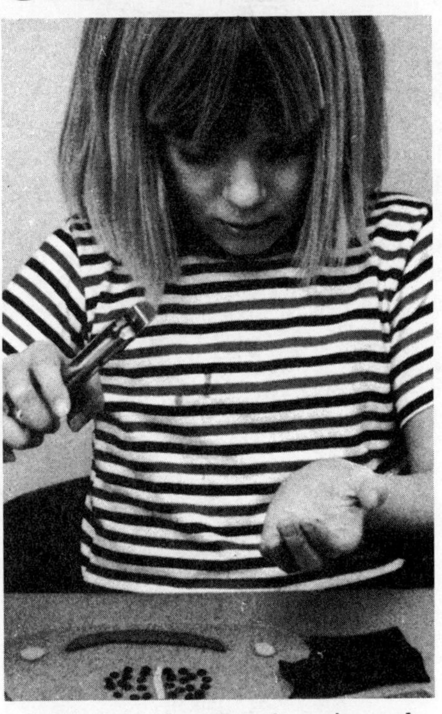

This new handcraft primer by the authors of *The Little Kid's Craft Book* shows you how to make a variety of imaginative objects to celebrate the year's changing seasons. For fall, make a nature collage out of leaves or a festive Thanksgiving turkey from pine cones. At Christmas, make your own cards. At Easter, create a basket made of milk cartons. For summer, build boats out of walnut shells and popsicle sticks. *The Little Kid's Four Season Craft Book* is published by Taplinger Publishing Co. $9.95.

Dough snowman

Snips and Snails and Walnut Whales

Pine cone birdhouses, seashell jewelry, and gumdrop trees are only three of over 100 natural craft projects Phyllis Fiarotta includes in her new book. She shows how to use and recycle flowers, twigs, leaves, pine cones, stones, seashells, nuts, even spider webs, into a variety of craft projects. Natural objects saved from the beach, gathered on hikes, or discovered in the backyard are combined with popular craftwork—stenciling, decoupage, sand-casting, batik, patchwork, drying flowers, sculpture—to create playthings, decorations, games, and gifts. A special three-act play on conservation is included with instructions on how to make the costumes and scenery. For your copy of *Snips and Snails and Walnut Whales* SEND $4.95 (paper) or $9.95 (cloth) plus 50 cents for mailing and handling to Workman Publishing Co.

Feathered Friends

If you look at a pine tree, you will see birds flying in and out. Many birds nest in these trees because they offer good shelter and a nice place to live. You won't mistake pine cones for birds, unless they are decorated with paper wings and faces. You can make these pine cone birds if you follow the directions below. With a little imagination they will look like common birds you have seen in the forest. Collect long, round, and oval pine cones for this craft.

Things You Need

different shaped pine cones
pencil
scissors
colored construction paper
liquid white glue
crayons
string

Let's Begin

HANGING BIRD
1. Choose a long pine cone.
2. Cut two paper wings from colored construction paper and glue them to the sides of the pine cone, see illustration.
3. Cut a strip of paper twice as long as it is wide.
4. Fold the paper in half.
5. Cut out a circle from the folded paper, leaving part of the fold uncut, see illustration.
6. Cut a very small slit into the folded side of the circle.
7. Cut a small beak shape from construction paper and push it into the cut slit. Draw an eye on both sides of the beak.
8. Open the paper head slightly, and glue over the wide end of the pine cone.
9. Tie string to a top petal and hang.

TURKEY
1. Lay a round pine cone on its side.
2. Cut several paper feathers from construction paper, see illustration.
3. Draw vein designs on the feathers with crayons.
4. Glue the feathers into the top petals of the pine cone.
5. Cut a head and neck out of construction paper, see illustration. Draw an eye on the head and glue on a red paper circle under the eye for a wattle (the fleshy piece of skin turkeys have under their necks.
6. Glue the head to the bottom of the pine cone.

OWL
1. Stand an oval pine cone on its bottom end.
2. Cut round paper eyes and a beak out of construction paper.
3. Glue the eyes and beak to the top of the pine cone.

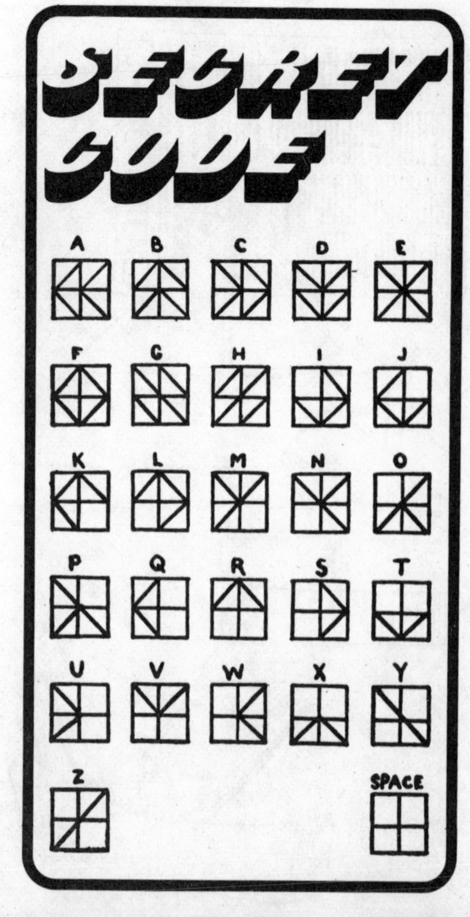

Cut and Color Toys and Decorations

This is not only a coloring book but a novel way to make greeting cards, mobiles, trapeze artists—even hand puppets! In a three-dimensional coloring book that combines the pleasure of making things with that of coloring them, Michael Grater has created a series of fanciful, frisky figures that are not only fun to make but can be played with after they're made. The projects in this book are simple enough for the youngest children and imaginative enough for older ones. For your copy of *Cut and Color Toys and Decorations* SEND $1.50 plus 35 cents for postage and handling to Dover Publications.

Paper People

This fine book explores some of the creative ways in which human figures of all kinds—clowns, witches, kings, queens, knights, acrobats, strong men, angels, and athletes, to name a few—can easily be cut, folded, or molded using inexpensive sheets of construction or wrapping paper, aluminum foil, or newspaper. Finishing touches may be added with crayons, paints, and odd scraps of brightly colored materials. The result is a delightful and amusing gallery of novel hand, finger, and glove puppets, marionettes, mobiles, jumping jacks, standing figures, ornaments, and action toys.

Says the author, Michael Grater: "Working in paper is a simple and fun activity. It is an unpretentious craft, but it can be creatively rewarding because so many things can be made from paper."

Particularly well illustrated, *Paper People* will show you how self-made toys can give you more fun than ones from a store. This book may well become one of the most useful paper-craft books in your library. Taplinger Publishing Co. $9.95.

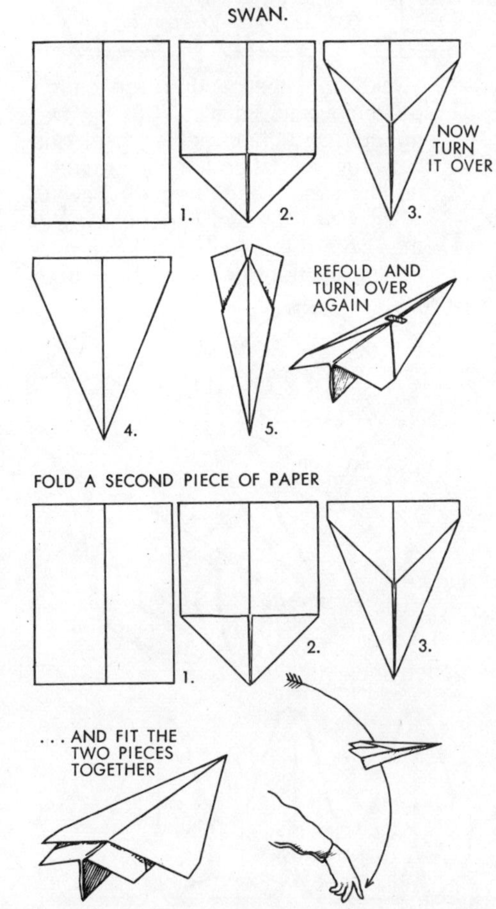

Trash Can Toys and Games

Author Leonard Todd says, "Let's face it: trash is here to stay." This book shows and tells you how to make toys from things you usually throw away: wood, paper, paperboard, cloth, glass, plastics, and metals. You don't need special tools. Out of paper you can make masks, hats, and puppets. From paperboard you can construct whole cities, pyramids, helmets. You can make a miniature garden in a bottle, or a water truck that has a nozzle that actually sprays water, or a walkie-talkie to send confidential messages.

You may not be able to solve the general problem of waste disposal, but you'll help a little, and have fun besides, following the suggestions in *Trash Can Toys and Games*. Viking Press. $6.95.

Kitchen Carton Crafts

This book has hours of fun in it: easy directions for over 45 toys, games, party favors, and gifts which you can make by yourself or with a friend. The materials needed are simple ones: empty cereal boxes, paper towels, waxed paper tubes, egg cartons, milk cartons, and ice-cream containers. Each project has clear diagrams showing the parts, the various steps, and the finished project. *Kitchen Carton Crafts* is by Helen Roney Sattler. Lothrop, Lee & Shepard Co. $4.95 cloth, $2.45 paper.

Silly Sea Serpent

You will need two pressed-cardboard egg cartons and two marbles to make this sea serpent.

Cut away the lid of the first carton. Then make a cut right through the middle of the eggcups on each side of the center section, as shown by the dotted lines in Figure 1.

Remove the triangular section at one end to form a tail (Figure 2).

Figure 1

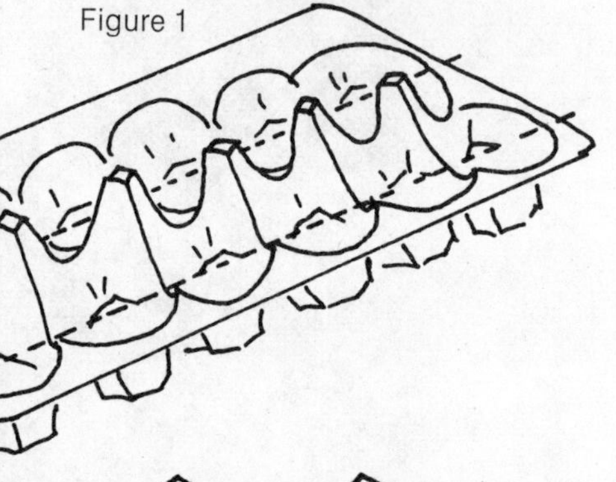

Figure 2

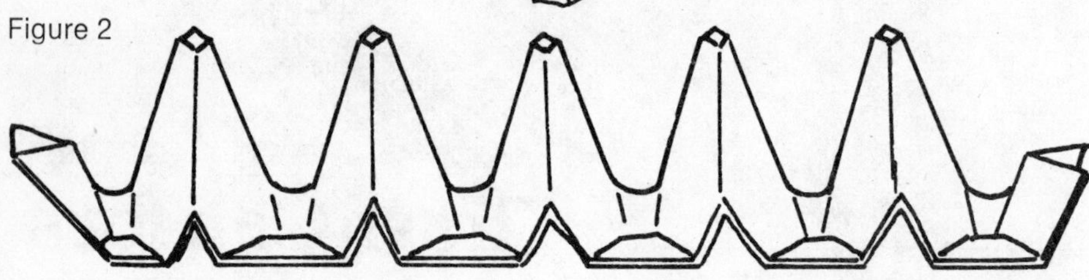

Next, take the second egg carton and remove its lid also. Cut out two eggcups and their center peak, as shown by the dotted lines in Figure 3, and throw the rest of the carton away.

Trim away the two eggcups on the other side of the peak, leaving a partitioning section for the neck (Figure 4).

Figure 3

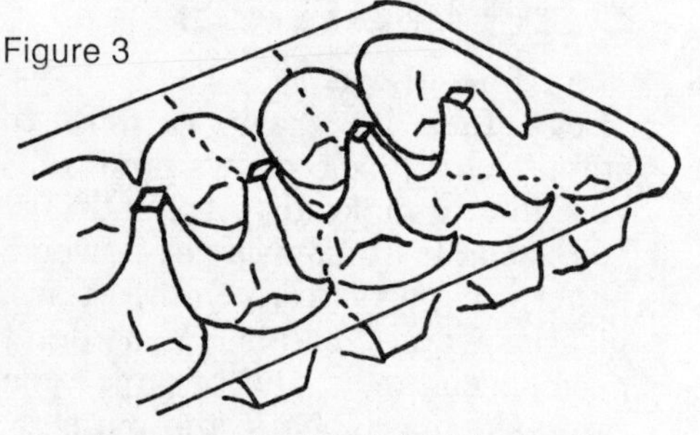

As shown by the dotted line in Figure 4, make a slit in the peak for the mouth. Glue in pieces of red crepe paper for fiery breath.

Then glue the head onto the front of the body section (Figure 5).

Paint the sea serpent green. Glue marbles in place (Figure 4) for eyes.

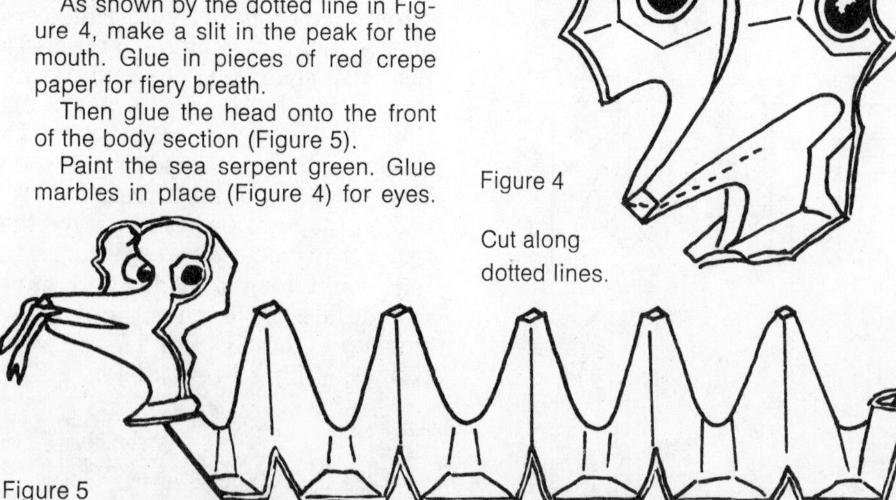

Figure 4

Cut along dotted lines.

Figure 5

Hobby Tray

This is one of those "Why doesn't someone make a . . . ?" And someone did! A Messy-Play & Hobby Tray made of sturdy plastic. Use it for playing: with finger paint, glue, chalk, papier-mâché, clay, sand, pennies, mud, oil. Make a miniature lake and sail boats on it. Make a play pond for a turtle. Use it for building: assemble model airplanes, model trains, radios, hi-fis, puzzles. Keep tiny parts from sneaking away. Open the box and build in the tray. You save space and time. Work on a project; put it away. Take it out to work on it again. A 23 x 14 inch tray costs $5.00. You can find it at your nearest toy or hobby shop. For more information WRITE TO Educational Resources Division, Educational Design, 47 W. 13th St., New York, N.Y. 10011.

The Doll House Book

Kristin Helberg has created a most unusual and charming book-that-becomes-a-doll-house. The spirally bound book opens into a Victorian doll house with doors that open and close, and 12 sheets of furniture for you to cut out and put together. The rooms (living room, dining room, kitchen, bedroom) are printed in black and white so you can decorate them with crayons, colored pencils, or watercolors. The furniture-to-make includes a sofa, chairs, tables, a fireplace, a stove, a kitchen cupboard, a bedroom dresser, and a canopy bed. The author says: "This little doll-house book is a glimpse of what life was like when your great-grandmother was a girl around 1890." For your copy of *The Doll House Book* and for hours of fun SEND $6.00 plus 75 cents for postage and handling to Kristin Helberg Toys, Town Square, Nicasio, Calif. 94946. Allow 3 to 4 weeks for delivery.

What Can I Do Today?

What Can I Do Today? by Joan Fincher Klimo, can be its own answer. It has sculpting, printing, pasting, stitching, decorating. It's a treasury of crafts for young children, with diagrams and drawings and pictures that show exactly how each project is done. Make a collage. Braid a rug. Sew up a pillow. Create a different kind of beanbag. Be artistic with papier-mâché. The directions for each project are in colorful pictures with a complete diagram of all the materials you need. Pantheon Books. $1.95.

Helena Hippo

Use spray can for body. Tape on four short paper tubes for legs. Shape body, ears, and tail with foil.

Monster Masks

Masks are for wearing at ceremonies, for decoration, and for wall hangings. They're all here in this beautifully illustrated book (100 photographs, 25 in full color). *Monster Masks* is written to delight monster lovers everywhere. Using the helpful instructions by Chester J. Alkema you will soon be able to make your own monster masks. The chapters include "Crayoned Masks"; "Drinking Straw Masks"; "Seed and Gravel Masks"; even "Sand-Cast Masks." Sterling Publishing Co. $3.50.

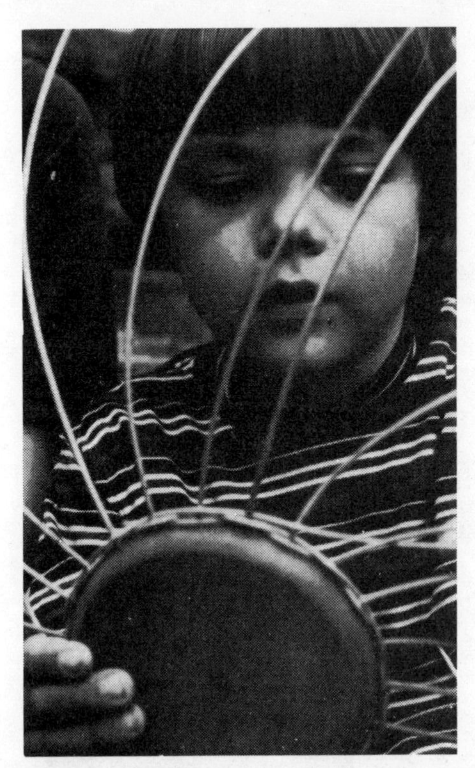

Woodstock Kid's Crafts

A great new Woodstock book for all you kids who want to get into crafts. You might decide to:

Do Your Own Macrame: All you do is tie a bunch of knots in different kinds and colors of string—then make headbands, chokers, hangings . . .

Make Your Own Terrarium: You'll find moss and ferns and other greenery, add some interesting bits of wood or stone, and relocate a frog or salamander into the new minihome you've created . . .

Make Beautiful Rocks: Find some rocks and let them tumble, go to sleep and hear them rumble. When they're finally polished, make rings, necklaces, charms . . .

Put On a Street Show: You don't need much money, any grown-ups, or a stage. Use real events, dreams, costumes, masks, or just your own funny face . . .

And there's more. Philosophy of this book: you can do anything you set your mind to. Lots of pictures and drawings show how to get on with it and how to get it on. The rest is up to you. *Woodstock Kid's Crafts* is by Jean Young. Bobbs-Merrill Co. $7.95 cloth, $3.95 paper.

Best Rainy Day Book Ever

Richard Scarry's *Best Rainy Day Book Ever* is really the best rainy day book ever! It has more than 500 things to color and make: trains to cut out, paper airplanes to fly, decorations to make, some special friends to color, games to play, and stories to read. You get your own 12-month calendar. You make these things right out of the book: all you need is some crayons, a pair of scissors, and paste. Age 5 and up. Random House. $3.50.

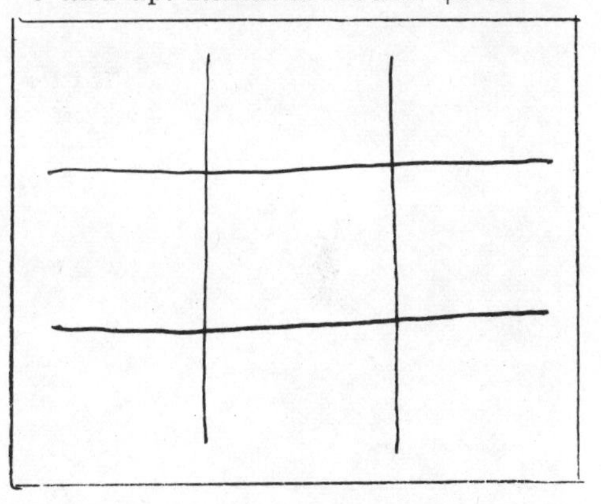

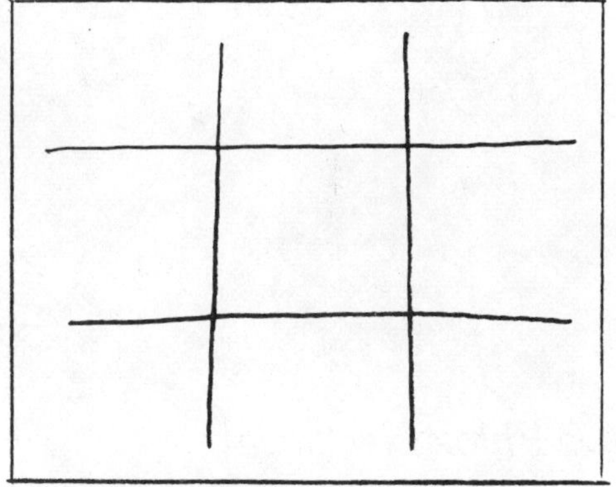

TICK-TACK-TOE

Would you like to play?

A Phlox Necklace

A necklace of phlox blossoms can be made by pushing the base of one flower into the center of the next until you have a chain as long as you would like. Be patient; if the flowers are not firmly pushed together, the chain will easily fall apart when you handle it.

Folk Toys Around the World and How to Make Them

If you're looking for a new project, how about building some folk toys? All of the toys (from 22 countries) in this colorful book can be made with inexpensive or scrap-size material. You'll have fun making and playing with them. Folk toys include a yo-yo from the Philippine Islands, a balancing fisherman from Portugal, a corncob donkey from Venezuela, and a pecking bird from Poland. *Folk Toys Around the World* is written by Joan Joseph and illustrated by Mel Furukawa, with working drawings and instructions by Glenn Wagner; the publisher is Parents' Magazine Press in cooperation with the U.S. Committee for UNICEF. For your copy SEND $3.50 plus 25 cents for mailing and handling to UNICEF, 331 E. 38th St., New York, N.Y. 10016.

A SHADOW PUPPET FROM INDONESIA (JAVA)

Green Fun

In every field and forest, lawn and garden there are free materials for an endless number of quick crafts. You don't need a knife, glue, string, tape, or anything except your hands, some maple leaves, an acorn, or perhaps some clover blossoms and a light heart.

Using lovely photographs and a very few words which even a beginning reader can follow, Maryanne Gjersvik shares some lively examples from old and favorite pastimes: dandelion curls, a plantain violin, snapdragon puppets, burr baskets, a daisy wreath, and many more. If you know some nature tricks that are not in *Green Fun,* Mrs. Gjersvik wants to hear from you. Chatham Press. $1.95.

Creating with Styrofam and Related Materials

The possibilities for ingenious creations with styrofoam, in its many shapes and sizes, are unlimited. The first step is to become a collector, to save everything you have been discarding: meat trays, cups, spools, leftover bits of fabric and yarn. Then you follow the simple instructions and make an exciting city that looks three-dimensional. Or perhaps a clown. Or try a sculpture that can be suspended from the ceiling or hung on the wall. You can mount your sculpture on heavy cardboard or wood to make it look professional. Have the fun of making figures from styrofoam cups, then designing costumes that transform them into a thousand and one characters. Age 6 and up. *Creating with Styrofoam and Related Materials* is by Tom Harris and Elsie Harley. J. G. Ferguson Publishing Co. $4.95.

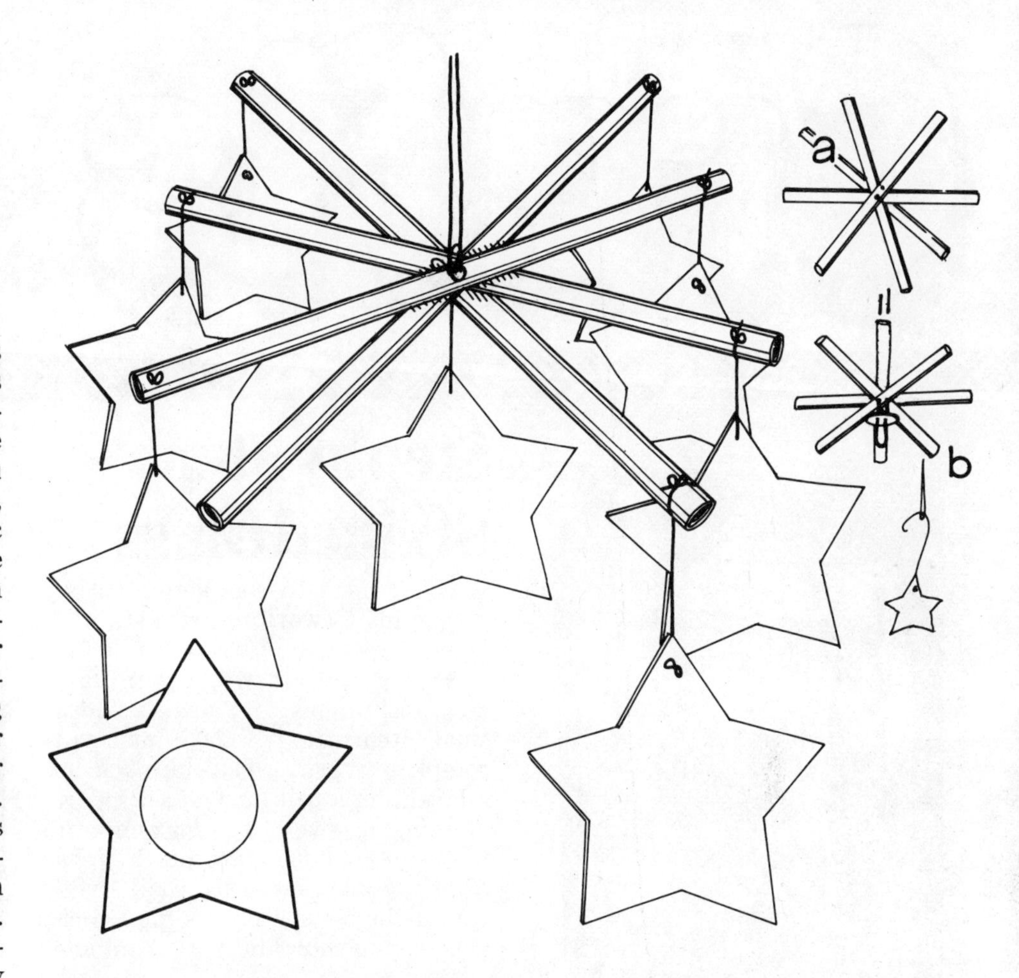

Sticks and Stones and Ice Cream Cones

Over 125 craft projects for children! A book so complete that *Harper's Magazine* called it "a 300-page treasure house that appeals to a very wide age range." *Sticks & Stones & Ice Cream Cones,* by Phyllis Fiarotta, has every craft from macrame, embroidery, and weaving to puppet making, egg blowing, patchwork, stenciling, painting, and papier-mâché. There are gifts to make, things to keep and a full-length puppet show to stage. For your copy SEND $4.95 (paper) or $9.95 (cloth) plus 50 cents for mailing and handling to Workman Publishing Co.

The ancient sailors used stars to guide them across the ocean. It is difficult to see stars, much less be a navigator, in your bedroom. What about a star mobile, then? With it, a galaxy of stars will orbit inside your room. They will lead you to whatever shores you wish to dream . . .

Things You Need

4 drinking straws
1 sheet of tracing paper
pencil
scissors
colored construction paper
needle and thread
1 small two-hole button

Let's Begin

1. Trace the bottom star (with the circle on it) from the book onto a sheet of tracing paper.
2. Cut out the star.
3. Use the cutout star to trace nine stars onto yellow or blue construction paper.
4. Cut out the nine stars.
5. Thread a needle, knot the thread, and sew through one point of a star. Now sew the star to one end of one of the straws, Fig. b, leaving enough thread between star and straw to allow star to dangle. Knot thread around straw. Repeat this operation for eight of the stars. Every straw should have two stars dangling from it, one at each end.
6. Place the four straws on top of each other. Form an evenly spaced star by crossing the straws at their centers, Fig. a.
7. Crush the straws with your finger at the place they meet.
8. Thread a needle with a long length of thread.
9. Sew down through the centers of the straws.
10. Pass the needle through one hole of the button placed underneath the juncture of the straws.
11. Sew up through the other hole of the button and back through the center of the straws.
12. Knot the thread.
13. Sew the last star through the center of the mobile.
14. Hang the mobile from the ceiling with tape.

Crafts for Older Children

Step by Step Beadcraft

If beadcraft leads you to think only of seed beads, take a look in this book for a pleasant surprise. The beads and materials used in its projects are as varied as your imagination. The basics—necklaces, chokers, bracelets, belt—are included, along with projects as diverse as a window hanging, a beaded knit purse, and melodic wind chimes. Ideas are borrowed from ancient Ecuadorian, African, Mexican, Egyptian, and American Indian beading patterns and techniques to form striking contemporary pieces. Complete instructions, detailed diagrams, and full-color photography combine to make *Step by Step Beadcraft,* by Judith Glassman, a unique guide for the beginner in the ancient and modern art of creative beadcraft. Golden Press. $2.95.

Step by Step Stained Glass

This book is a complete step by step guide to working with stained glass. It gives the fundamentals for executing a large window in various techniques; it also includes small items that a beginner can complete over a weekend, such as a lapel pin, a belt buckle, a window hanging, a jewelry box, or a mirror frame. A unique part of *Step by Step Stained Glass,* by Erik Erikson, is the section on design, which shows you how best to combine geometric shapes for the most pleasing design. There is a glossary of terms, and at the end of the book, a list of suppliers for tools and materials, a list of books for further reading, and a directory of schools and workshops offering courses in stained glass. Golden Press. $2.95.

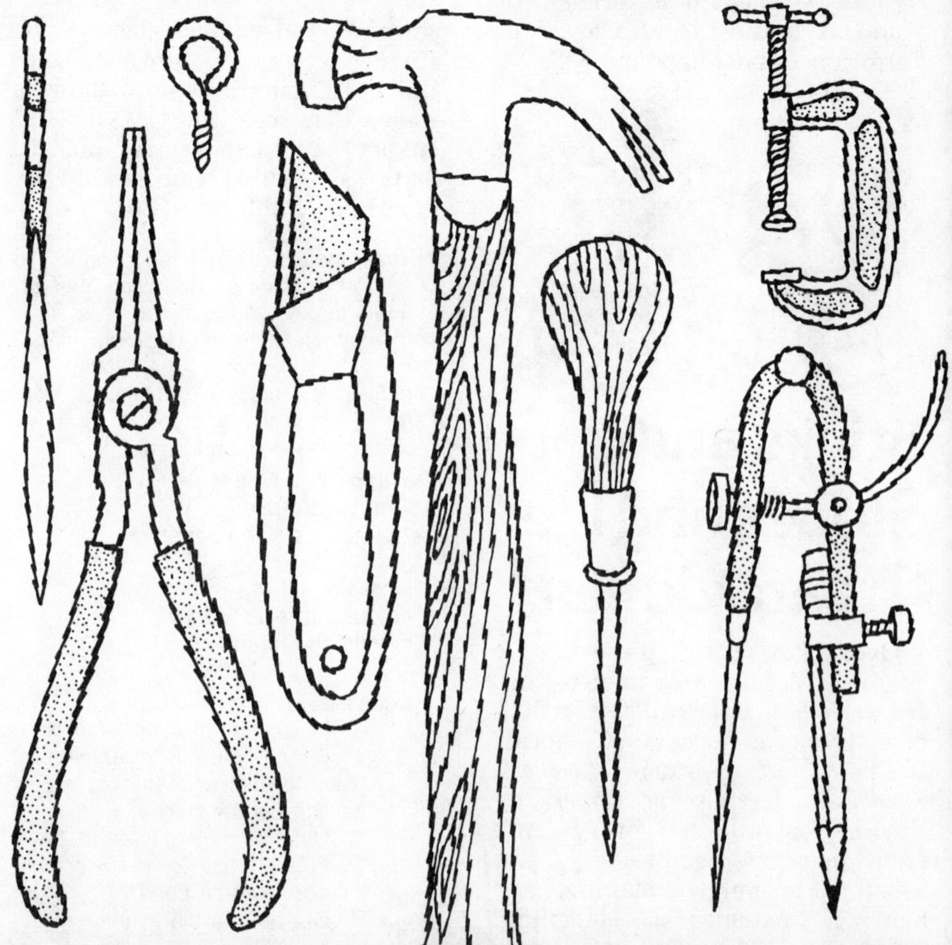

Jewelrycraft & pottery making & paper sculpture. Leathercraft & origami & stained glass. Scissorcraft & paper modeling & string art. Tincraft & beadcraft & shellcraft & enameling & candlemaking & picture framing & crafts & crafts & crafts!

Step by Step Enameling

Ideas plus projects are presented in this introduction to enameling by William Harper. Easy-to-follow, step by step directions guide you through the basic techniques—cloisonné, basse-taille, grisaille, champlevé, plique-à-jour, repoussé, and Limoges—also new techniques, and information on stenciling, foil application, and sgraffito. There are full-color photographs of examples by Mr. Harper and other leading enamelists. *Step by Step Enameling* has a glossary of enameling terms, a list of suppliers of the materials you'll need, and if you want to carry your talents farther there is a directory of schools and workshops where you can get more instruction. Golden Press. $2.95.

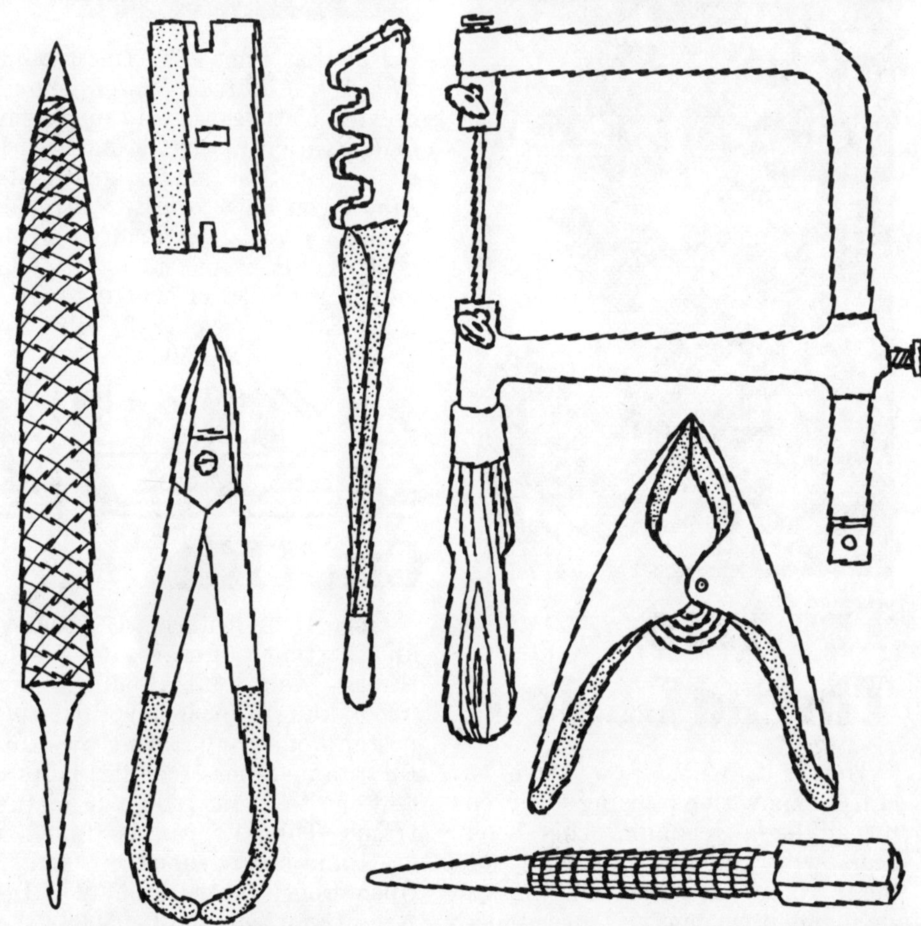

Folding Paper Masks

If you want to make 21 of the most original and most amusing masks you've ever seen, this is your book: *Folding Paper Masks,* written by television star Shari Lewis and by Lillian Oppenheimer, founder of the American Origami Center. The original masks were created by Giuseppe Baggi. You'll discover how with nothing more than a piece of paper you can become a clown or Cleopatra, a tiger, hobo, rabbit, or if the spirit moves you, a devil or witch.

Even beginners can follow the step-by-step illustrations and instruction that take the reader from the first fold to the last, and include a large photograph of every completed mask. If you're already into origami you will be pleased that most of the masks are original creations and do not appear in other books. E. P. Dutton & Co. $4.95.

Step by Step Candlemaking

This guide to creative candlemaking, by Mary Carey, covers all the essentials—materials and equipment; melting, coloring, scenting, and pouring wax; wicks; milds; safety; cleanup—plus ideas and projects. *Step by Step Candlemaking* guides you in producing a variety of decorative, unusual candles, whether layered, stacked, lace, freeform, hurricane, sand-cast, or floating. You will be inspired by the beautiful photographs of artistic projects that can be made by using basic candle shapes and a little imagination. Golden Press. $2.95.

Introducing Candlemaking

This is a simple but comprehensive introduction to home candlemaking. You can discover the thrill of making beautiful, unique candles right in your own kitchen. *Introducing Candlemaking*, by Paul Collins, shows how to make all kinds of candles from the simplest castings to the advanced poured and dipped varieties. Besides detailed step-by-step instructions, this book contains many useful suggestions about materials and designs, and about creating unusual candles for special events and holidays. Taplinger Publishing Co. $6.95.

How to Make Candles

This book has step-by-step instructions plus many photographs by Barbara Weakley that show how to make mold tapers, pillars, floating candles, container candles, and dipped bayberry candles. It answers the inevitable questions of the beginning candlemaker: what materials are needed, where to get supplies, what waxes to use, how to mix dyes and add fragrances, and most important, what mistakes to avoid. *How to Make Candles* was written by candlemaker Tom Weakley; to get your copy SEND $1.95 to Candle Mill, East Arlington, Vt. 05252.

The Art of Thread Design

There's a whole new wave of craft innovations using colored thread as a medium. This book deals with the most dramatic of these recent applications, known as mandala geometric thread design. It is an exciting new art form and all you need is a hammer, a board, a ruler, a compass, and some colored thread and you can create these designs.

But that's not all. After reading *The Art of Thread Design* you can create pictures of a suspension bridge, a locomotive, a stagecoach, a ship. Author Mark Jansen also shows you how to create three-dimensional art. If you can't find the book in your bookstore, try your local craft shop or WRITE TO Open Door Enterprises, 1249 Dell Ave., Campbell, Calif. 95008.

Craft Kits

Open Door Enterprises produces an assortment of craft kits. Besides thread design kit products, they make kits for coloring your own posters, one for creating a beautiful stained-glass candle lantern, and one to make your own picture frame. The kits can be found at craft and hobby shops or WRITE TO Open Door Enterprises, 1249 Dell Ave., Campbell, Calif. 95008.

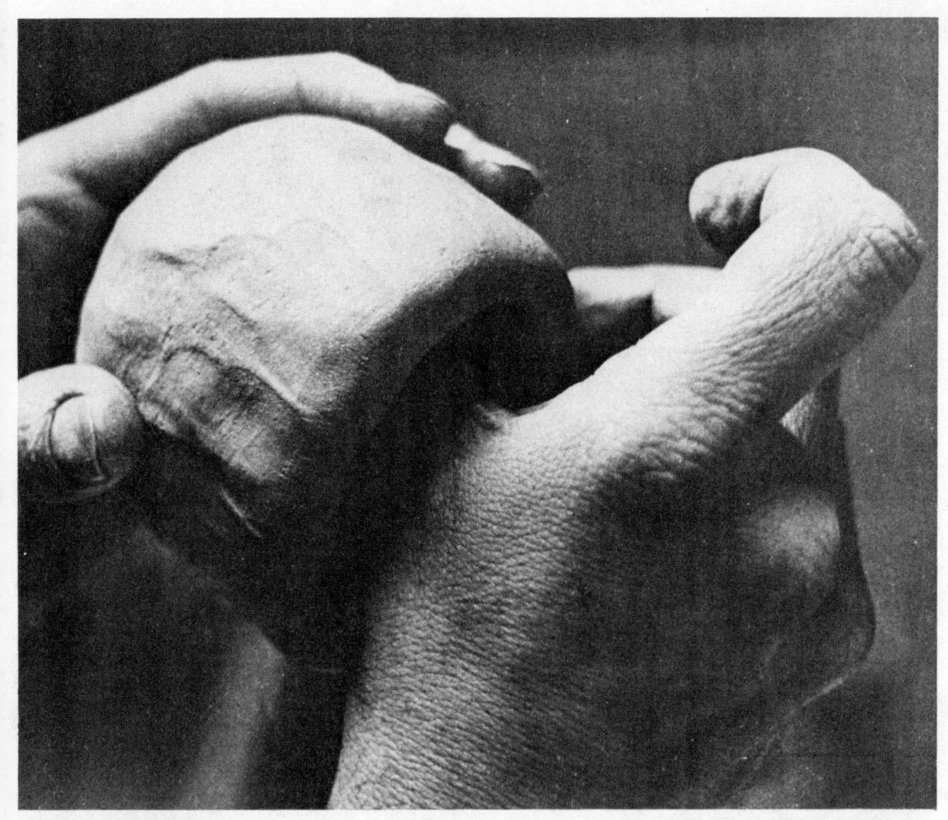

PHOTO: TRUE KELLY

Finding One's Way with Clay

A unique craft book, *Finding One's Way with Clay* offers the beginner as well as the experienced potter a new approach to making pots without a potter's wheel. Paulus Berensohn begins with the simple resources of clay and water—and the human imagination, which he feels is present in all of us—then he squeezes, presses, and pinches his clay into a bowl. There is a wealth of detailed instruction with hundreds of photographs by True Kelly, on making all kinds of pots: bowls, bottles, sculptural pieces, large pots, crooked pots, straight pots. There are exercises in shape, texture, and color, and a whole section on firing your creations, including directions for building your own sawdust kiln in the backyard. You can make the kiln out of common house brick or cinder block, or even from an old metal garbage pail. This book gives you the aesthetics as well as the craft of clay. Simon and Schuster. $9.95.

Exercise One
Making a Small, Symmetrical, Thin-Walled Pinch Pot

Opening

Hold a small ball of clay in your left palm. Rotate the ball slowly as you press slowly into the center of the ball with your right thumb. Start with the lower joint of your thumb bent forward horizontally. As you go deeper into the heart of the clay, straighten your thumb out and down into a vertical position. Keep going until you are almost one-fourth of an inch from the bottom of the ball. All we are trying to accomplish here is making the opening. Try not to make the opening wider than the size of your thumb. If you combine the act of opening with the act of widening, you will more easily lose control. To hold to symmetry, be sure to have a rhythmic relationship between your pressing thumb and the rotating ball. If you do not rotate, the shape of the thumb will immediately set up variations in the thickness of the clay around your opening. Just how much clay you leave at the bottom of the ball will affect the final shape of your pot; leave one-fourth inch of clay this time and experiment with more or less in future practice.

Potterymaking

This is a good introduction to pottery for children. Almost all the tools suggested in this book are household tools; by using oven-baked clay you can become a potter without leaving home. The most important tool is your own hands. It will feel good to get your hands in the clay. You'll mold it and shape it, and before long you'll make something that never existed before *you* made it. *Potterymaking*, by Virginie Fowler Elbert, shows you how to use the tools, glazes, wheel, and kiln. Things to make include a straight-side pot, a castle tower, a heart-shaped box, and a hollow bird. Doubleday & Co. $4.95.

Adventures with Clay

Adventures with Clay, by George C. Payne, offers you a variety of approaches to modeling and pottery making. There is an excellent section on making tiles. You'll learn about modeling from balls, cones, slabs, and rolls. Or you can try your own mosaic. Many photographs and drawings help you each step of the way. Frederick Warne & Co. $3.50.

FREE

American Handicrafts Catalog

If you're looking for the latest in craft ideas and all the supplies you'll need, no matter what craft you're working in, send for this free American Handicrafts Catalog. In full color you'll see ecology kits, string art kits, cathedral glass lamp kits, decoupage kits, over 200 prints, paper tole kits, craft books, molding mediums, candle molds, casting molds, copper tooling—supplies for every craft. For your free copy of the catalog WRITE TO American Handicrafts Co., Dept. AM, Advertising Dept., 1330 E. 4th St., Fort Worth, Tex. 76102.

FREE

Arts and Crafts Catalog

This catalog is truly an encyclopedia of creative art materials; it takes you from grade-school handicrafts to the most advanced art activities. It offers supplies for hundreds of art mediums: ceramics and sculpture; mosaics and stained glass; painting, watercolor, oil, and acrylic; drawing and silk screen; block printing; etching; weaving; leathercraft; and jewelry. There is a section on other art books to send for. There are 250 pages of items, each one with a detailed description and picture. You have the art world at your fingertips if you have this catalog. It's free. WRITE TO MacMillan Arts & Crafts, 9520 Baltimore Ave., College Park, Md. 20740.

Flower Making for Beginners

This book was written by Priscilla Lobley especially for young people who want to learn the craft of flower making. Her unique book shows you how to make a veritable garden of the most festive, decorative, colorful paper flowers imaginable—giant party roses, strikingly vivid morning glories, delicate sweet peas—all fashioned from crepe paper, wire, tape, and other inexpensive materials. While making these fabulous blooms you will learn much about the structure and function of various parts of flowers and plants. Clearly illustrated with excellent photographs, drawings, and diagrams, *Flower Making for Beginners* will get you started on a brand-new pastime. Taplinger Publishing Co. $6.50.

Origami

These books written by Tokyo-born Toshie Takahama are just right for beginners in origami; they are colorful, easy to understand, and filled with a variety of things to make.

Origami Toys shows 15 simple models: animals, puppets, tops, pinwheels, a jumping frog, and others.

Origami for Displays. These are beautiful if folded with colored paper. You can make fantastic table decorations or wall hangings, and ornaments to hang from a tree, Christmas or otherwise.

Japan Publications Trading Co. $2.50 each.

Jewelry Craft for Beginners

This is the first book the amateur needs to start making beautiful jewelry. With hundreds of clear drawings and photographs, the authors teach the simple techniques and basics of design along with all the how-tos for making necklaces, bracelets, cuff links, and pins as handsome as any you can buy.

This is an important new book for older children who want to make their own jewelry. Creating beautiful, useful jewelry does not require costly and complex equipment and materials. As the book's first projects illustrate, you can achieve wonderful results with common everyday tools and inexpensive, easily obtainable decorative materials.

Jewelry Craft for Beginners is written by Gloria R. Mosesson and Virginie Fowler Elbert. Bobbs-Merrill Co. $9.95. (You'll make that back with the first piece of handmade jewelry you sell.)

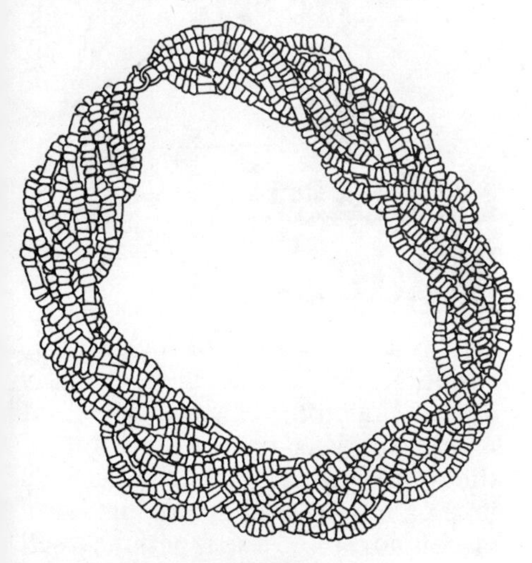

Step by Step Framing

In this handsome, fully illustrated book, Eamon Toscano, a professional picture framer, guides you step by step in making and finishing a basic picture frame. You learn how to frame photographs, prints, watercolors, and canvases. *Step by Step Framing* helps you find out about style of molding, type of finish, nature and purpose of mat, and suitability of mounts and liners, as well as other techniques: baguette, shadow box, glass trap, mirror framing, and passe-partout. There is a list of books for further reading and a list of suppliers. Golden Press. $2.95

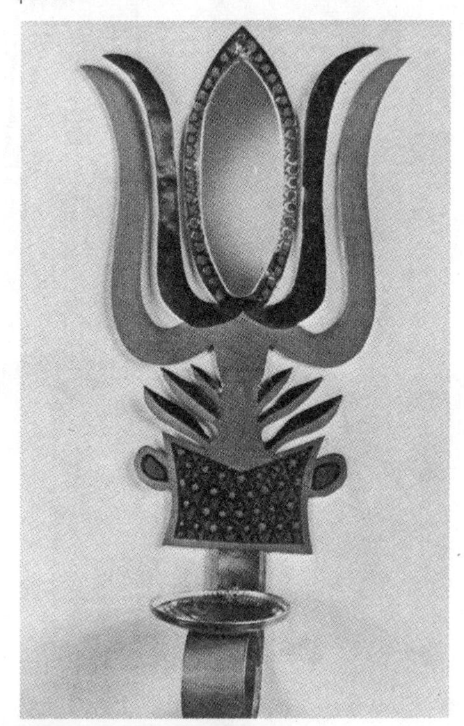

Tincraft

Tincraft provides an exciting solution to the ecological problem of what to do with useless, empty tin cans. They can become valuable craft supplies, to be recycled into beautiful decorations for your home, ornaments for your garden, or handsome pieces of jewelry. Complete with patterns, diagrams, and photographs, the book offers detailed instructions for over 85 creations. Imagine lustrous objects that glow and shine like pewter and silver and gold—all made with tin cans that come right from your kitchen shelf.

You can begin with simple projects and then you are led through a spectacular array of things to make. *Tincraft*, by Lacy Sargent, is a timely, appealing book that gives you the enjoyment of making something out of nothing, of turning waste materials into a treasured source of creativity. Simon and Schuster. $9.95.

The "Start Off" Books

All the books in the Basic Craft Series have been specially designed with an "easy easel" that makes each book stand up on its own, and the spiral binding on top allows you to flip over the pages. The texts are clear and concise, and following the project instructions should be a breeze. They are generously illustrated, including color photos of the finished projects. Start off each craft book by flipping to the first section, where you'll find information on the supplies and tools you'll need. Then flip to the first project . . .

Start Off in Shell Craft, by Cleo M. Stephens, shows the endless variety of attractive and amusing objects that can be made from shells: a basket filled with flowers, dolls, little animals, decorations, pins and earrings, hats and bags. The shapes, colors, and sizes of shells are so varied that almost any flower can be duplicated. Your creativity is only limited by your imagination.

Here are more of Chilton's Basic Craft Series: *Start Off in Boutique Quilting; Start Off in Cooking Crystal Craft; Start Off in Dough Craft; Start Off in Furniture Refinishing; Start Off in Leather Craft; Start Off in Making Christmas Creations; Start Off in Making Cloth Handbags; Start Off in Making Cloth Toys; Start Off in Nature Crafts; Start Off in Needlepoint; Start Off in Pop-Top Craft; Start Off in Stained Glass;* and *Start Off in String Craft.* Chilton Book Co. $1.95 each.

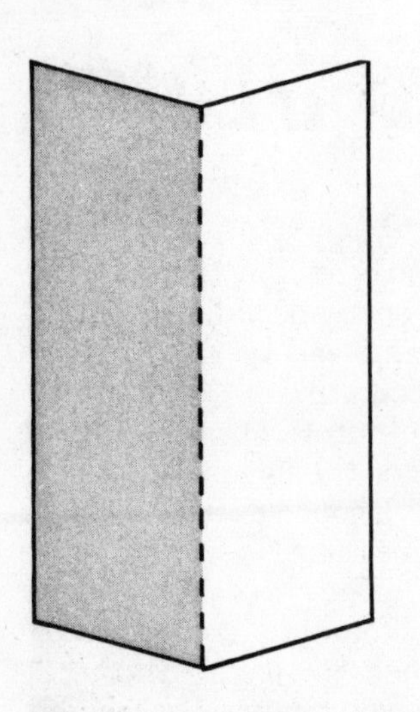

A rectangle of paper scored down its centre will stand upright, and can be cut and decorated to make a simple shape.

Make It in Paper

Paper sculpture is the art of raising flat paper into three-dimensional shapes. The tools and materials required are simple and inexpensive, and creators of all ages find the work thoroughly enjoyable.

There are certain basic techniques in paper sculpture. These are described by Michael Grater, stage by stage, from the most elementary beginning to a point of creative craftsmanship. You will learn how to make a curved score and how to curl your paper. *Make It in Paper* also has a section on decorating ideas for parties. Age 10 and up. Taplinger Publishing Co. $5.95.

Paper Folding for Beginners

This book anticipates all the difficulties you might have with origami. It has more than 275 carefully labeled diagrams that show the important stages of folding. William D. Murray and Francis J. Rigney have explained everything in detail so you can make a dog, a ship, a pagoda, trees, or a fish. For your copy of *Paper Folding for Beginners* SEND $1.25 plus 35 cents for mailing and handling to Dover Publications.

Adventures in Paper Modelling

Paper masks, models of animals and birds, decorated balloons—here are step-by-step instructions by George C. Payne on how to make these from a wide range of inexpensive materials. You will have hours of fun in learning to construct exciting masks and models from newspaper, cardboard, wire, and papier-mâché. The models you make can be taken to school for history, geography, and other projects. *Adventures in Paper Modelling* leads you toward self-expression and experimentation with various kinds of materials. The methods are carefully explained and accompanied by drawings and photographs. Frederick Warne & Co. $3.50.

Origami: The Art of Paperfolding

This is an illustrated guide to a fascinating pastime—the ancient Japanese art of folding a plain piece of paper into intricate and charming figures. The paperback book is easy to stick in your pocket with a couple sheets of paper—and carry your hobby with you. There are 50 different projects for you to try, among them a flapping bird, a frog, two penguins, Mother Hubbard's dog and Mother Hubbard, and a squirrel and a pink elephant. *Origami: The Art of Paperfolding,* by Robert Harbin, is easy to understand, just right for the beginner. For older crafts. Funk & Wagnalls. $1.25.

Origami Made Easy

The true beauty of origami is simplicity. This book contains many of the charming origami works of Kunihiko Kasahara, and all of them are filled with the artist's love of his art and sparkling sense of humor. The instructions will teach you to reproduce origami with ease.

The three pleasures of origami are the pleasure of folding, the pleasure of seeing what you have folded, and the pleasure of using your imagination to create. The latter is the main point of this book, which helps you develop origami of your own. *Origami Made Easy* has 128 fully illustrated pages of delightful figures, flowers, planes, decorations, and funny faces. Japan Publications Trading Co. $2.95.

owl

The cuts on the Owl were done freehand, but precise patterns have to be measured and drawn in with a soft pencil before cutting. This technique adds a great deal of light-and-dark contrast.

Paper: Folded, Cut, Sculpted

This book is the most complete guide to paper craft that has been written. From animal mobiles to sculptured lampshades, the possibilities of paper have never been so thoroughly explored. There are over 350 line drawings and 190 companion photographs to help you produce complex and fascinating objects. Just a few minutes and a sheet of paper can yield a delightful product. This book guides you step by step from the simplest to the most intricate operations. You will learn origami (the art of folding paper), kirigami (the art of cutting paper), and how to create three-dimensional paper sculpture. There is a section on Chinese paper craft that illustrates traditional yet unusual techniques, and a geographic exploration of the history of paper as a craft material. If you have ever had an interest in the art of paper craft, *Paper: Folded, Cut, Sculpted,* by Florence Temko, is a must for you. Macmillan Publishing Co. $12.95 cloth. Collier Books. $5.95 paper.

Christmas Crafts

The Meanings of Christmas

Do you know how the idea of the Christmas tree started? Who are the Three Kings we sing about in carols? Was there really a Santa Claus? Why do we exchange gifts on Christmas? These and many other questions are answered for you by Malcolm Whyte, with the added attraction of drawings by Varian Mace. You'll meet a funny. stocking-capped, white-bearded little man hammering away in his toy shop and learn that Santa Claus first appeared in the 4th century as a bishop of Myra named Nicholas. You'll discover that the earliest known holiday greeting card was printed in Germany in the mid-15th century and that in the United States we exchange 5 billion cards each season! To find out more about the most colorful of all holidays, get *The Meanings of Christmas*. Troubador Press. $1.50.

Christmas Creations

This book is packed with over 75 fun and easy-to-make projects for the holiday season. There are fabulous ideas for centerpieces, door and wall hangings, wreaths, ornaments, stockings and tree skirts, and doorknob decorations. Each project contains a detailed list of materials and step-by-step instructions that are clear and simple to follow. *Christmas Creations* is by Jane Berry. Chilton Book Co. $5.95.

A Child's Christmas Cookbook

This timeless collection of Christmas goodies for young people to make includes all kinds of recipes and ideas for the holidays. There are directions for an old-fashioned taffy pull, gifts such as pomander balls and potpourri, decorations made from popcorn and cranberries, tea parties, and recipes for yuletide supper dishes: serendipity salads, long-ago leftovers, Christmas cookies, and sugar plums. Gathered for today from an earlier era, *A Child's Christmas Cookbook*, by Betty Chancellor, is illustrated with nostalgic drawings from the old *St. Nicholas* magazine; many are by the famous Thomas Nast. Harvey House. $3.00.

Farm Journal Christmas Idea Book

This book from the editors of *Farm Journal* is brimful of things-to-do and things-to-make to help celebrate the Christmas season, with ideas that came from families all over the United States. This is actually three books in one. The first section has recipes for good things to eat: traditional foods, snacks and lunches, edible decorations, and festive foods. The second section is a veritable gold mine of handcraft ideas, complete with instructions for Christmas gift giving. And the third section, whether you have ten big rooms or one little one, will show you how to decorate your home in the spirit of Christmas. The *Farm Journal Christmas Idea Book*, edited by Kathryn Larson, will give you years of pleasure. Countryside Press. $6.95.

'Twas the week before Christmas and all through the house . . . hands were making Yule logs, hand-painted egg ornaments, Mexican piñatas, angels, and gifts galore. Make Christmas cards Tell Christmas tales. Sing Christmas carols. Merry Christmas!

Want to Make Your Own Christmas Cards?

Here are two very good Christmas card books:

Christmas Card Coloring Book, edited by Edmund V. Gillon, Jr., shows elegant but inexpensive cards. Designs include Noah's dove, Nast's St. Nick, and Christmas toys of 1892. You can create 16 personalized and unique cards and you hand-color them yourself. Envelopes are included.

Victorian Christmas Cards for Hand Tinting, edited by Theodore Menten, reproduces authentic Victorian designs, perfect for hand tinting—a light wash of delicate color which was a favorite method of the Victorians. Designs include detail from *Skating at Boston,* by Winslow Homer, Christmas toys from Marshall Field's illustrated catalog, *Scrooge's Third Visitor,* and other charming, hard-to-find Victorian scenes. You can make 16 cards. Envelopes are included.

To order SEND $1.50 plus 35 cents for postage and handling for each book to Dover Publications.

Christmas Stained Glass Coloring Book

This is not just a coloring book, but a way for you to make your own stained glass window decorations: 16 original designs by artist Theodore Menten have been printed on translucent paper: your favorite Christmas scenes, including snowmen, candles, a partridge in a pear tree, an angel, and a nativity scene. The pages can be colored with crayon, felt-tip pen, acrylics, watercolor, tempera, even oil paint. You can color one side or both, mix media—and when you mount the pages on your windows they look like stained glass. For your copy of this exciting, imaginative book SEND $1.50 plus 35 cents for postage and handling to Dover Publications.

Crafts for

Imagine using clamshells to make a lovely Christmas crèche or learning how to use seedpods and nuts to make a wreath. You will be inspired by the idea of spraying fruits gold or making a pinecone and nut tree. You'll learn to use materials you would usually discard. Tin cans, bottles, and cardboard cups can become extravagant decorations that will glitter

Parents' Magazine's Christmas Holiday Book

This is a one-volume source of Christmas lore, music, cookery, selections for reading, and family activities. In *Parents' Magazine's Christmas Holiday Book* Yorke Henderson and others provide a historical review of the development of Christmas including the origins of Santa Claus and the Christmas tree, and all the other ingredients that make Christmas a festival of joy. There are Christmas tales and Christmas carols, and recipes for the Christmas feast: roast goose, plum pudding, gingerbread cookies. This book will help your family do things together and there is "how-to" information on subjects ranging from shopping for gifts to decorating the tree and your home. Parents' Magazine Press. $4.95.

Christmas

on your tree or your Christmas packages. You can make Christmas lanterns, drums, and shiny angels: and with a little balsa wood and enamel paint will learn how to make some colorful wooden soldier ornaments. All these holiday preparations are easy to follow in *Crafts for Christmas*, by Katherine N. Cutler and Kate Bogle. Lothrop, Lee & Shepard Co. $5.50.

Christmas Crafts

The subtitle is *Things to Make the 24 Days Before Christmas*. If you have the time, there are some great projects here: maybe one of them could be your Christmas present. You will learn how to make 24 exciting things, everything from hand-painted egg ornaments to a bread-dough Christmas manger, from a Yule log to a Mexican piñata. Simple instructions by Carolyn Meyer: fun illustrations by Anita Lobel. Harper & Row. $4.95.

POMANDER BALL

What You Will Need

Apple, 1 box of whole cloves, cinnamon, ginger, small paper bag, yarn or ribbon, newspapers.

Making a Pomander Ball

Spread out the newspapers and put the apple on them. Push the stems of the cloves all the way into the apple, fitting them as close together as you can, until the whole apple is completely covered. Put a little cinnamon and ginger into the paper bag with the apple, close the top of the bag, and shake it until the apple is coated with the spices. Tie a piece of yarn or ribbon around the apple and make a loop at the top for hanging.

Pomander balls last indefinitely without rotting or molding, because the cloves help draw all the moisture out of the apple. But they do lose their spicy smell after a while.

Make Your Own World of Christmas

Use a pair of scissors, some glue, and four drinking straws to create a beautiful decoration for Christmas. There are dozens of other things you will learn how to make in this delightful book, *Make Your Own World of Christmas*, by Rosemary Lowndes and Claude Kaïler. You can cut out and glue Christmas ornaments such as the Christmas angel, the three wise men, St. Nicholas, and Santa Claus: large, three-dimensional figures you can hang on the tree. You can create fabled scenes with cutout people and a colorful background which will make Christmas tableaus from many lands including Germany, Italy, France, Holland, England, Czechoslovakia, Norway, Russia, Mexico, and the Holy Land. That's still not all; in among the pages you will find and cut out the months of a calendar—each with a colorful picture of one of the twelve days of Christmas. Bobbs-Merrill Co. $5.95.

The Young Designer

Tony Hart, the author of this interesting book of design, says in his introduction:

> We are all designers. Often we doodle away on a telephone pad or exercise book, sometimes unaware of the graphic value of what we do. Abstract shapes and patterns can and do lead to other things; notably in graphic design.
>
> In this book you will see work by designers in advertising, television, jewelry, textiles, and many other industrial products. Where possible I have included the idea which inspired the design and made it successful.
>
> As you turn from page to page, you will find practical exercises in design. They will help you look at nature and observe the simple way to use shape and structure to form your own designs, whether realistic, abstract, or semi-abstract.
>
> You do not have to be a technical artist to be a designer. You just have to be enthusiastic.

Be enthusiastic and read *The Young Designer*. Even if you're not an artist, this book is stimulating. It will help you to look—and see. Frederick Warne & Co. $4.50.

The Young Calligrapher

Calligraphy has a long history—and has helped preserve our history. *The Young Calligrapher* gives examples of fine illustrated manuscripts together with the development of writing in the Western world. The author, William C. Cartner, is a practicing designer and teacher who believes that calligraphy is a craft for anyone who can write his ABC's; he deals with the basic forms of writing as a special kind of drawing. To most of us letter writing is a drudgery; to the calligrapher it is a pleasure and an opportunity to practice his skill, which is reflected in his handwriting. Frederick Warne & Co. $4.50.

abcdefg

Cut paper into crazy patterns. Make collages and posters. Color a medieval alphabet. Painted sugar cubes glued to an old board. Snow sculpture. Sand castles. Mobiles. Finger painting. Twist a wire coat hanger into a shape. It's art. It's fun.

The Young Letterer

Tony Hart is a well-known artist and designer who has written this book to help the young letterer produce well-designed lettering using simple materials. Two styles of lettering are described: with a pen and with a brush. The lettered can soon draw serif and italic, simple roman letters, Gothic, Versal, and decorative styles, block letters, even the classic Roman alphabet. *The Young Letterer* has a section showing different forms of printed lettering today and an explanation of the best uses of each style of type. The book is illustrated with drawings and photographs and shows examples of great art and fine craftsmanship to open the way to appreciating the art of lettering. Frederick Warne & Co. $4.50.

Fine Art Reproductions

Now you can build your own inexpensive art collection with Artext Juniors. These authentic color miniatures average 3 x 4 inches and are specially designed for notebooks and art study projects. The cost is only 8 cents a print. Artext Juniors are printed on fine-art presses from color plates made in Europe by master engravers. A special technique has been developed to produce these small reproductions with the high degree of fidelity needed for serious art study. There are over 300 paintings to choose from; the schools include early Italian, Flemish, early French, early English, Impressionists, and modern American, while the artists include Botticelli, Dufy, Michelangelo, Rubens, El Greco, Corot, Hogarth, Whistler, Homer, and Wood. To get the Artext Juniors Catalog, which lists all the pictures available, SEND 50 cents to Artext Prints, P.O. Box 70, Westport, Conn. 06880. If you want the Folio Collection Catalog, which lists their 7 x 9 inch color prints (costs range from 35 cents to $1.50), SEND an additional 50 cents.

100 Ways to Have Fun with an Alligator & 100 Other Involving Art Projects

Here is an art text, by Alex Mogelon (with visuals by Anne Raymo), to aid you in the most important discovery imaginable—yourself. It will heighten the sensitivity of your senses and sharpen your ability to communicate visually. The mind-expanding projects in this unconventional book are "Room Transformation"; "List of Wet Things"; "Love Day"; "One Cent Day"; "Trivia"; "100 Things to Do with an Alligator"; and 95 other projects.

Read what a young teacher said about this book:

> *100 Ways to Have Fun with an Alligator* helped turn 560 loud 7th & 8th graders into 560 giggling loud creators of absurd fantastic drawings, constructions and discussions. Thanks, thanks, thanks.

Here is part of the introduction to *100 Ways to Have Fun with an Alligator,* by Albert Bush-Brown, former president of the Rhode Island School of Design:

> Alligators, strange lockers, Christmas trees, and a sidewalk; cookies, tattoos, masks, and a few moustaches; suitcases, old x-rays, mechanical toys, some snow, and one funeral director: ART?
>
> This book is playful. It is an invitation to a party. Imaginative and beguiling, it is also deceptive: not one of its pages preaches about art; yet none teaches anything else.
>
> Celebration enjoys hats and costumes and beards and masks. Everyone is invited. There should be banners and flags, shadows and lights, beacons and fountains, with lots of color and pictures on the walls. There should be games to play, poems to read, surprises, toys and musical instruments to bang and blow. It should last a week, a carnival with singing and dancing, a festival with flowers and films and acting. You would be invited. The invitations would be works of art; your acceptance would decorate the hall.
>
> That is really what this book is about. Somewhere, the text tries to tell you differently; it says something technical about getting new forms by breaking the conventions of size and logic. I do not understand that part. I came for the party, to sing songs, to act parts, to give gifts: Did you save a mask for me?

It sounds like a commercial, but . . . write for it today, you'll be glad you did and your life will be more exciting. For your copy SEND $3.75 to Art Education, Blauvelt, N.Y. 10913.

EDITION: HANSJORG MAYER

NEWSPAPER DRAWINGS

materials: *newspapers and glue.*

The object is to make a drawing or a composition by ripping or tearing out pieces of varying sizes, shapes and patterns from old newspaper. Do not use a scissors. Let the torn line of the rip-out, the shape of the individual pieces and what is printed on their surface, and speed determine the character of the composition.

variation: apply details to your newspaper drawing with brush and ink.

'I'M LOOKING FOR A KEY I LOST OVER THERE.' WHY NOT LOOK THEN, WHERE YOU LOST IT? 'IT'S TOO DARK OVER THERE. I LOOK FOR IT HERE WHERE THERE'S LIGHT.'

FREE How to Use Poster Color

When King Tutankhamen's tomb was opened in 1922, one of the most precious finds was a wooden chest, painted in tempera nearly 3400 years ago. The 20th-century adaptation of this painting medium is poster paint, which holds first place in today's art world. You will love working with it. It's fun, and can be used on almost anything—paper, cardboard, wood, cork, glass, and other materials. *How to Use Poster Color* is a free booklet that will tell and show you how. WRITE TO Milton Bradley Co., Springfield, Mass. 01101.

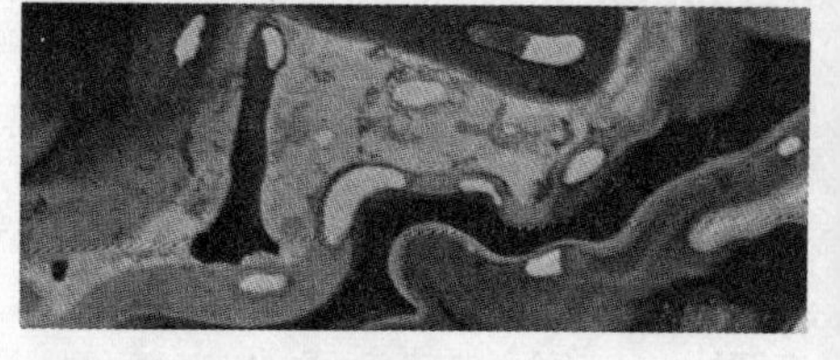

It's Fun to Make Pictures

This book with its extravagant and bright illustrations shows how simple it is to get your painting materials together, mix your paints, and then learn to look and create paintings. Your first lesson is landscape painting: a few small houses with trees and flowers and a small wood fence. You'll learn how to get the sun's light right on everything you see in the landscape. Then you can go on to make life-size portraits of your friends, or even a portrait of yourself, like those Rembrandt did so often. You will learn how to paint on material to make delightful hangings, how to make finger paintings and leaf paintings; and there's a mysterious way to use crayons to make "magic paintings." You'll learn to make collages, and to make pictures out of string and colorful mosaics out of colored bits of paper or colored eggshells. You'll get exciting, inspiring lessons from *It's Fun to Make Pictures,* by Eileen Deacon. Grosset & Dunlap. $3.95.

The Book of Posters

If you are into posters—just looking at them, being delighted by them, or if you want to find out how to make your own—then this imaginative, different, simply written, excitingly designed work, *The Book of Posters,* must be yours. It's giant size, 16 x 11 inches, with 80 pages of text and illustrations (28 in full color). Here is a fresh contemporary look at the art of the poster: its history and its modern technique covering the art of the poster from pre-Christian times to the present, and illustrating the work of outstanding poster artists including Toulouse-Lautrec, Henri Matisse, Ben Shahn, Roy Lichtenstein, Robert Rauschenberg, Frank Stella, Milton Glaser, and Peter Max. Most important: the text, by Norman Laliberte and Alex Mogelon, provides you with 30 unique, fully illustrated poster projects, to challenge your imagination and sharpen your ability to produce exciting and individual poster art. For a copy of this fabulous book SEND $5.95 to Art Education, Blauvelt, N.Y. 10913.

Damn Everything But The Circus
a poster with an e. e. cummings quote,
reprinted with the permission of the
Botolph Group, Inc., Boston, Mass.

The type face used in this poster is reminiscent of the early Barnum and Bailey posters which each year were plastered everywhere in town, proclaiming the forthcoming arrival of the circus. The size and character of the type is simply overwhelming—how could a small child not be impressed with the importance of the circus when the individual letters are human being in size!

The artist, Corita Kent, has taken a quote from a very contemporary poet and through both choice of type and size has made the phrase immediately recognizable and, at the same time, monumental and towering. The colors of the original—red, white and blue, the colors of the flag—almost brings a sense of patriotism to the poster and is highly effective on the eye.

What does it all mean? Perhaps this quote from the April 1969 issue of "Theology Today" and attributed to a commentary by Sr. Helen Kelley, explains it:

"... it means
damn everything
that is grim, dull, motionless, unrisking
inward turning,
damn everything
that won't get into the circle,
that won't enjoy,
that won't throw its heart
into the tension,
surprise,
fear
and delight of the circus,
the round world,
the full existence."

DAMN
EVERY
THING
BUT THE
CIRCUS
EE CUMMINGS

FREE

Magic with Water Color

Have you worked up to water colors? Here's a free booklet with instruction in this exciting medium. Among the topics covered are "How to Paint a Water Color Wash" and "How to Paint with Water Colors" (step-by-step instruction). For your free copy of *Magic with Water Color* WRITE TO Milton Bradley Co., Springfield, Mass. 01101.

Adventures with Sculpture

G. C. Payne, author of *Adventures with Sculpture* and principal of a primary school, has included ideas and methods which have been used successfully by children and which can provide useful starting points for many activities which can give a sense of achievement even to those without special artistic talent.

The book demonstrates by means of drawings, photographs, and step-by-step instructions how to make an exciting variety of sculptures, constructions, and models. Materials used include wood, plaster, wire, paper, scrap material, and the modern plastic foam and polyester resin. An introduction to these materials and creative ways of working with them is the aim of the book. Frederick Warne & Co. $4.50.

Altair Design and More Altair Design

About *Altair Design* the *New York Times* wrote:

> This provocative book is hard to describe. At first glance it is a series of pages of geometric patterns traced in lines. At second glance, it is a challenge to the imagination, for with colored pens or pencils, the black and white pages can be transformed into any pattern or design—simple or elaborate. This book has wings.

Try your hand at this fascinating art form. Can you see a shape in the pattern? Or would a geometric design please you? Anything goes—you are the artist, choose your color and your shape.

Over 50 different patterns have been programmed by a computer. The author, Ensor Holiday, started these designs as a hobby, but it soon became a consuming fascination. The variations were endless. *Altair Design* and *More Altair Design* are published by Pantheon Books. $1.95 each.

All-Around-The-House Art and Craft Book

Want to be inspired, take an art course, and feel your creativity expanding—all with one book? Get the *All-Around-the-House Art and Craft Book,* a very special book which should have stars all around the title so you'll know it's quite different from other books that it may sound something like. Just remember, this is not a scrap craft book; it is a book that uses scrap to create art.

Enjoy some of the preface:

> It can be easygoing, eye-opening and taste-expanding as you begin to see old things in a new light, as you begin to arrange and relate ordinary things in an extraordinary way.
>
> Just look around you! That's the first and most important step in creating an artist's environment for yourself. Imagination, ideas, a searching attitude and the material of ordinary life—those are the things an artist really works with.

A piece of board, sugar cubes, three poster paint colors, and Patricia Z. Wirtenberg's instructions are the ingredients for making a creative and attractive three-dimensional wall decoration. This is just one of 50 stimulating recipes for making something special from commonplace items found in every household: the projects include painting, printing, collage, assemblage, and sculpture.

The real fun of the book is that it sparks the imagination and opens new ways of looking at everyday things. Wire coat hangers, adhesive tape, laundry starch, newspapers, and spray paint become an airy place of sculpture. Screws and glue, straws and wire, milk cartons and plaster, magazines and cardboard, indeed almost anything can be combined with imagination to project a pleasing artistic result. Houghton Mifflin Co. $6.95 cloth, $3.95 paper.

PHOTO: ADRIAN BOUCHARD

Snow Sculpture and Ice Carving

This may be the only guide to this cool subject. If you have ever thrown a snowball you can learn how to pack snow together until you have a mouse and and a cat; Santa Claus and his sleigh; a duck family; three astronauts—even an igloo. You'll learn how to chip away at ice and make a sailboat; bears; turtles; elephants; whales; or a totem pole. Snow sculpture and ice carving are great crafts for kids age 8 and up.

Says author James S. Haskins:

> Ice and snow are free! All you have to do is go outdoors and get started. You will find great satisfaction and a new awareness in creating with this wonderful medium. If you have ever had any secret leanings toward the arts, and yet have felt it is something that can only be done by "experts," snow and ice sculpture is where you should begin. If your creation goes wrong, you knock it down and start over! The important thing is to begin, and to keep on going.

Snow Sculpture and Ice Carving is published by Macmillan Publishing Co. $8.95 (cloth) and Collier Books. $4.95 (paper).

The Art and Industry of Sandcastles

What is a sandcastle? "It's a private legend, a fortress made of sand and imagination for your own heroes, a stronghold for your daydreams. It is the strong desire to build, to create forms with your own meanings. It is most of all your own, out of your head, made by your hands." This book has lovely drawings that show you how to make castles by the water's edge; it also has exciting drawings showing the real castles of long ago. You will learn how they were made, who lived in them, how the castles changed in shape through the years. You will read about how some people spent their time in castles: for example, the austringer, who trained falcons, the master of hounds, and the young page who was learning to be a knight. This fascinating world will help make make you a master sandcastle builder. *The Art and Industry of Sandcastles* is written and illustrated by Jan Adkins. For your copy SEND $5.95 plus 25 cents for postage to Walker & Co., 720 Fifth Ave., New York, N.Y. 10019.

Adventures with Collage

Collage is a medium used by world-famous artists, and you can make collages, too. Collage is a picture or pattern built up from pieces of paper and/or other materials stuck on a background. Most children delight in collecting things; part of the fun is the amazing assortment of oddments which can be easily and cheaply obtained in the home or out of doors and applied to collage. Every aspect of the work is covered step by step, accompanied by illustrations of collages created by one child or by children in group projects. *Adventures with Collage* has been put together by Jan Beaney. She makes you want to create a collage. An Adventures in Learning book. Frederick Warne & Co. $4.50.

Drawing

How often have you said, "I wish I could draw!" especially when you wanted to keep in mind a scene that pleased you, or to illustrate with a sketch something that you couldn't make plain in writing? But you *can* learn to draw and here is a Teach Yourself Book, *Teach Yourself to Draw*, by Ronald Smith, from which you can acquire a knowledge of good draftsmanship, of materials, and of the various media in which the artist works. Chapters include "Drawing Materials"; "Object Drawing—Beginning"; "Perspective"; "Light and Shade"; "Drawing Outdoors"; and "Figures and Animals." Whatever you may achieve, one result is certain: if you learn something about drawing, you will forever after see the work of great artists with quickened pleasure and understanding. For this alone it is worth trying. For your copy SEND $2.00 plus 35 cents for postage and handling to Dover Publications.

The Children's Book of Painting

Alexander and Katinka, two delightful, friendly puppets, teach each other—and you—how to make paintings using some of these techniques:

How to paint and blow and make splotches and sprinkles with watercolors
How to print with a brush, a potato, and vegetables
How to make pictures with a paint rag, the fingers, even the whole hand
How to make scratch pictures
How to mix different shades of red, green, blue, and gray and how to paint with them
How to make paste-paper paintings

Alexander and Katinka talk to each other—and to you—and show you how to make fireworks, rainbows, castles, peacocks, and many other bright, gay pictures. You'll want to try them out yourself; your parents and teachers will want to look over your shoulder and try too. *The Children's Book of Painting* is by Lothar Kampmann. Van Nostrand Reinhold Co. $4.95.

Posters

Five huge full-size posters the exact size and color of the originals. These handsome, decorative posters, ranging in size from 26 x 20 inches to a giant 46 x 34 inches, were created by five of today's foremost poster artists: "100 Ways to Have Fun with an Alligator," by Milton Glaser; "The Book of Posters," by Peter Max; "Tempo of Today," by Norman Laliberte; "Art and the Future," by Bill Greer; and "Damn Everything But the Circus," by Corita Kent.

For the package of five exciting posters SEND $5.00 to Art Education, Blauvelt, N.Y. 10913. Ordered singly each poster is $2.00.

Art from Found Objects

Don't throw away that box of odds and ends you've been saving: a broken toy, a piece of driftwood, or some leftover nuts and bolts may be just what you need to create your own art from found objects. The projects in this excellent book are designed to show some inventive ways in which you can use everyday objects. Sculptures and mobiles, greeting cards and jewelry are just a few of the creative possibilities.

In clear, easy-to-follow directions, Jeremy Comins explains the procedures for each project. Many of the illustrations (there are dozens and dozens of them) show works made by the author's students, and these provide a source of ideas and inspiration. Just looking at them makes you want to create something of your own. You are continually encouraged to experiment and express your own creativity when you read *Art from Found Objects.* Lothrop, Lee & Shepard Co. $5.50.

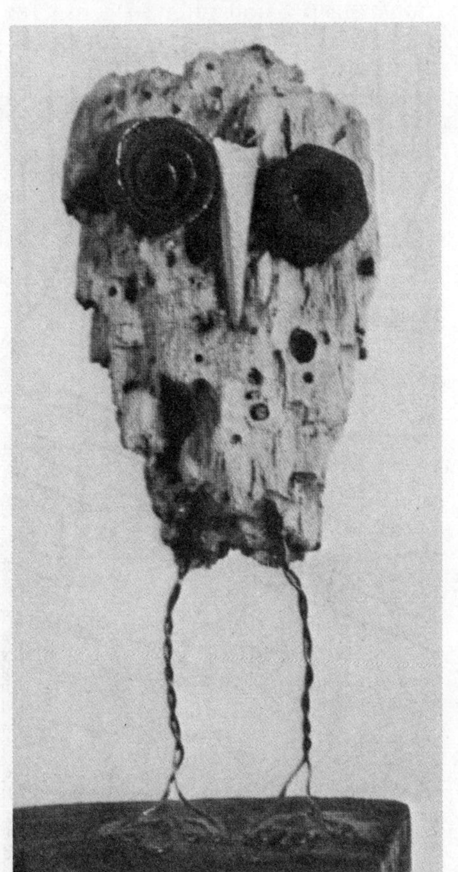

Draw 50 Animals

Did you ever wish you could draw a bird, or a dog, or a lion, or a monkey, and then decide you didn't have enough talent to bother? Did you ever sit down to draw an animal, but not know how to begin? If the answers are yes, then this book, *Draw 50 Animals,* is a must for you. With the guidance of graphic instructions and a little serious effort on your part, you will be able to draw. Lee J. Ames has compiled step-by-step instructions on how to draw each of 50 assorted animals. Start with a circle, then overlap an oval, and you have the beginnings of a body. The circles and lines you add now will give dimension and form and—presto—you have an animal before you! With a little practice you will be able to draw animals without the help of the book. Doubleday & Co. $5.95.

A Flower with Love

Here's a book which might be thumbed through too fast in a bookshop and not really seen. For those who find it and use it, it can add a new dimension to life. Says the publisher: "Do you want to know how to make someone a beautiful gift that costs nothing at all? Would you like to learn a new way of sending a message without words? This book tells you how to do this, and more. The secret is *ikebana*—a Japanese word which author-artist Bruno Munari interprets as 'a flower with love.' "

Clear, simple instructions and full-color photographs show and tell you how to create arrangements of natural objects that speak eloquently without words. Potatoes and pebbles, marigolds and moss—*ikebana* reminds us that these are a part of nature, as we ourselves are. And all things in nature contain hidden messages that can be used to help express our inner feelings: you can, for example, combine a wintry dry twig, a green leafy branch, and a single summer-bright blossom into a silent poem to celebrate the changing seasons; or you can make a composition of red radishes with green grass "whiskers" for a visual joke to share with someone you like.

The many examples in *A Flower with Love* are not meant to be copied, but to show you how to get started. The fun of *ikebana* is in creating your own designs, sending your own secret messages, arranging your own flowers—with love. Thomas Y. Crowell Co. $4.95.

Stained Glass Windows Coloring Book

The art of stained glass windows arose in Europe before the year 1000 and enjoyed a magnificent development throughout the Middle Ages. The flood of color these windows pour into the interiors of the great churches and cathedrals astonishes and delights visitors to this day.

Now you can make your own stained glass windows. The 16 illustrations in this book are renderings by Paul E. Kennedy of window designs from England, France, Germany, and Austria. You color the illustrations, which are printed on translucent paper; then you mount them on your windows and obtain an effect almost like that of the original stained glass windows. The pages can be colored with crayon, felt-tip pen, acrylics, water color, tempera, or even oil paint. It's a delightful book! For your copy of the *Stained Glass Windows Coloring Book* SEND $1.50 plus 35 cents for postage and handling to Dover Publications.

Art Instruction Series

These good beginner books by Walter Brooks are part of the Golden Press Art Instruction Series.

Creative Ways with Drawing. Directs the student toward meaningful observation: suggests proper tools and materials: provides examples and exercises. With this book anyone who is sincerely interested in learning to draw and is willing to work at it can develop a degree of proficiency and a strong foundation for other directions in art.

Creative Ways with Watercolor Painting. Watercolor is a fast medium. Broad washes, strong statements, delicate touches—all must be made quickly, with sureness and control. The student who is willing to study the properties of pigments and paper surfaces, to practice brush handling, and to pursue the exercises will be well on the way to self-expression in this versatile medium.

Creative Ways with Acrylic Painting. Acrylics are firmly established as a brilliant fine arts medium. It's a unique, versatile color system to satisfy the widest range of creativity and imagination. Even the beginner can rejoice in acrylic's convenience and opportunity for self-expression.

Other books in this series are *Creative Ways with Oil Painting*, *Creative Ways with Drawing Dogs and Cats*, and *Creative Ways with Flower Painting*. Golden Press. $1.25 each.

Step-by-Step Printmaking

This is a complete introduction to the craft of relief printing, beautifully illustrated in full color. Printmaking, with and without a press, is demonstrated in *Step by Step Printmaking,* by Erwin Schachner, who is known internationally for his prints. He guides you in the preparation of various relief surfaces, including linoleum blocks, wood blocks, and type, and in methods of printing from these surfaces by hand or on a proof or platen press. The book includes many examples of prints by the author and other printmakers. There is a glossary of printing terms and a list of suppliers from all over the country, plus a directory of schools where courses in printmaking are offered. Golden Press. $2.95.

FREE Adventures in Color

When it comes to painting, it's fingers before brushes—to help make finger painting a fun and an educational activity, here's a pamphlet complete with samples of third-, fifth-, and sixth-grade finger paintings. For your free copy of *Adventures in Color* WRITE TO Milton Bradley Co., Springfield, Mass. 01101.

A Medieval Alphabet Coloring Book

Do you know that the alphabet is named after the first two Greek letters, alpha and beta? Do you know that Bellerophon publishes a beautiful and exciting coloring book with letters from the medieval alphabet? There are 48 letters for you to color, 8½ x 11 inches. Between 500 and 1000 years ago many beautiful books were produced by hand. Only a few people could read, and books were precious, so great care was taken to make them beautiful. Colors were made with which to paint the handsome initial letters; even gold was sometimes used. Each page of *A Medieval Alphabet Coloring Book* is an initial letter that was made long ago, with all its swirls and designs, and you can work at coloring it just as those early artists did. You use crayons, colored pencils, or felt-tip pens. The book is available at your bookstore or SEND $1.95 plus 25 cents for mailing and handling to Bellerophon Books.

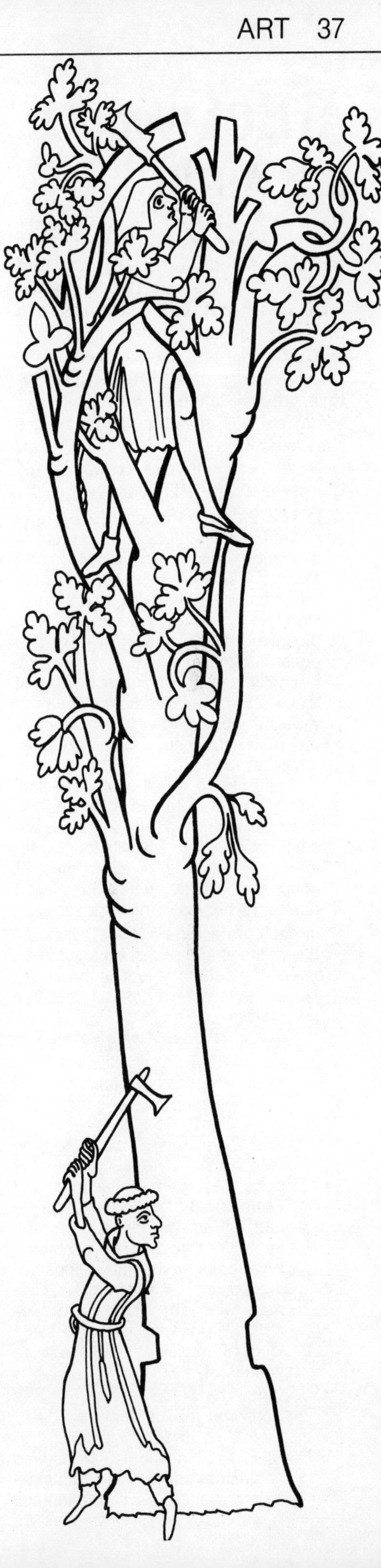

I is for insouciance

From Saint Gregory, Moralia in Job, from the Abbey of Cîteaux; Dijon, Bibliothèque Municipale MS. 170, f. 59; 12th Century.

Art: Tempo of Today

This is the sequel to *Art: Of Wonder & a World.* It will introduce you with exciting words and pictures to the language of art in relation to the contemporary scene. Most of all, this book is a stimulus to observe the world about you and truly *see.* For *Art: Tempo of Today,* by Jean Mary Morman and Norman Laliberte, SEND $3.50 to Art Education, Blauvelt, N.Y. 10913.

Art: Of Wonder & a World

Here is a truly exciting new approach to art (or is it to life?) written by Jean Mary Morman and Norman Laliberte. It's a wide-open door that will lead you to a new awareness of the world around you. There are 25 projects that make you stand on your mental tiptoes, and reach out, and look and seem to see farther than you've ever seen.

Here is a passage from one of the projects, "So This Is Art":

In our "phone and charge it" age, when even do-it-yourself kits are a challenge, what has happened to art? Two things! First, millions of people never know the thrill of getting a new idea and seeing it through, even though it is only making a cake that doesn't come in a box.

And second, we may have lost touch with the "stuff" of our world, with the wood and clay and stone—the materials men have used from cave days.

If you like to feel the wind, to dive under a wave, to walk barefoot in grass, or to make a snowman, you are close to art. If you like to cut paper into crazy patterns, or make kites, or soap box racers, you're in. If you don't, maybe you've never tried.

In each project there is a LOOK:

—at the shape of rust on a gutter.

—at the pattern of jets on the sky.

—at the design of a tiger-eye marble.

—at the colors in a patch of weeds.

. . . and a WHAT DO YOU THINK?

—is making things important to you? Building model cars? Making a dress? Cooking?

—Bob Dylan has a song like this: "Men make everything from toy guns that spark to flesh colored Christs that glow in the dark. It's easy to see without looking too far that not much is really sacred." What does he mean by things being sacred? Someone said: "The man who does not make, destroys." Does he mean the same kind of making that Bob Dylan speaks of?

. . . and a TRY THIS:

—See what you can make . . .

by folding and cutting paper.

by collecting pebbles or colored glass or match sticks and making a mosaic. You might try watching water movement for an idea.

by constructing a kite and flying it.

—Listen to music; watch a ballet on TV; sit quietly for five minutes.

There are 112 pages, 16 of them in color. This is a book that shouts "awareness." You can't be too young to use this book. It will make you see things you've never seen before. To get your copy of *Art: Of Wonder & a World* SEND $3.75 to Art Education, Blauvelt, N.Y. 10913.

Three Good Art Books

Taking Up Drawing and Painting, by Guy R. Williams. Designed through the simplicity of its approach to appeal to the young would-be artist, this book provides an introduction to all the pictorial artist's basic materials. Taplinger Publishing Co. $6.95.

Introducing Drawing Techniques, by Robin Capon. Extensively illustrated, this book is a guide to drawing materials and methods. It stresses the importance of art as a means of expression and offers the beginner instruction in line, point, tone, wash, and texture. Step-by-step advice is given for every kind of drawing technique. Taplinger Publishing Co. $7.50.

Taking Up Sculpture, by Sean Mullaney. In this illustrated primer, an experienced sculptor and teacher explains the fundamentals of the craft, detailing the necessary materials, tools, and equipment. Also included is information on clay modeling, wood and stone carving, and assemblage in metal. Taplinger Publishing Co. $6.95.

Optricks: A Book of Optical Illusions

You may have to look once—twice—or perhaps more times before you discover images hidden in a pattern of dots, paintings hidden inside other paintings, shimmering effects in a geometric pattern such as you might see in looking at screen doors, or overlapping picket fences. Hold an abstract pattern produced by a computer at a distance and lo and behold—you'll find a very well-known president's face instead! Some of these surprising optical illusions were done by artists such as Salvador Dali, René Magritte—and Maurits Cornelis Escher, who spent his life exploring visual illusions and mastering the art of black-and-white line rendering. You can enjoy Escher's sense of humor in his drawing of figures forever going up a staircase in a monastery. Train your eyes with this exciting visual puzzle book: *Optricks: A Book of Optical Illusions,* by Melinda Wentzell and D. K. Holland. Troubador Press. $1.50

The Love Bug

A coloring book full of wiggles, waves, and thick jungle scrolls, with chatty rhymes next to each one. On the first page wondrous bugs are on lush leaves winding and twining all over the page with a poem to go with it:

Though one has never been found
The Love Bug ought to be crowned
As prince of all things
For the love that he brings
As love makes the whole world go
'round.

There are decorative camels, crocodiles, koalas, and whales and, as the butterfly poem says: "Night and day they cheer up the sky." You will feel as happy as the bright butterfly as you color away with crayons, ink, watercolors, colored pencils, acrylic paints, and then hang up the result as a poster in your room. *The Love Bug Coloring & Limerick Book* has drawings by Donna Sloan and limericks by Mal White. Troubador Press. $2.00.

Scenic Sand Art

Scenic sand is a relatively new art form. It is the process of layering different colors of sand and shaping it into a design with a sand tool. It can be done in vases, bowls, bottles, anything that will let you see your sand art.

Here is a kit to help you start this new hobby. Just pour layers of different colors of sand. Design a pattern using the stylus in the kit, then add potting soil and plant your seeds. The bottom of your planter will be as beautiful as the top.

The kit contains five colors of fine-grade sand: red, blue, yellow, green, and black; a 4-inch high bubble ball planter; and potting soil. You get the seeds of a beautiful rainbow coleus, the sand tool, and, of course, complete instructions. You'll find the Scenic Sand Art Planter at your local hobby shop. Activa Products. Suggested retail price is $6.00.

Charm in Motion

Gently and ceaselessly turning, the mobile adds the elements of time and motion to sculpture, weaves a spell of charm and fascination. The outstanding artist and craftsman Takumi Shinagawa shows you how to use paper, wire, plastic, metal, wood, string, and so on, to create mobiles of animals, people, even Christmas angels. Then all you need is moving air and your butterflies, birds, and angels will whirl and twirl and fly! It's simple; just follow the simple and basic instructions in *Charm in Motion: A Collection of Mobiles*. There are 70 delightful constructions to inspire you. Japan Publications Trading Co. $9.95.

Do you see one vase or two faces?

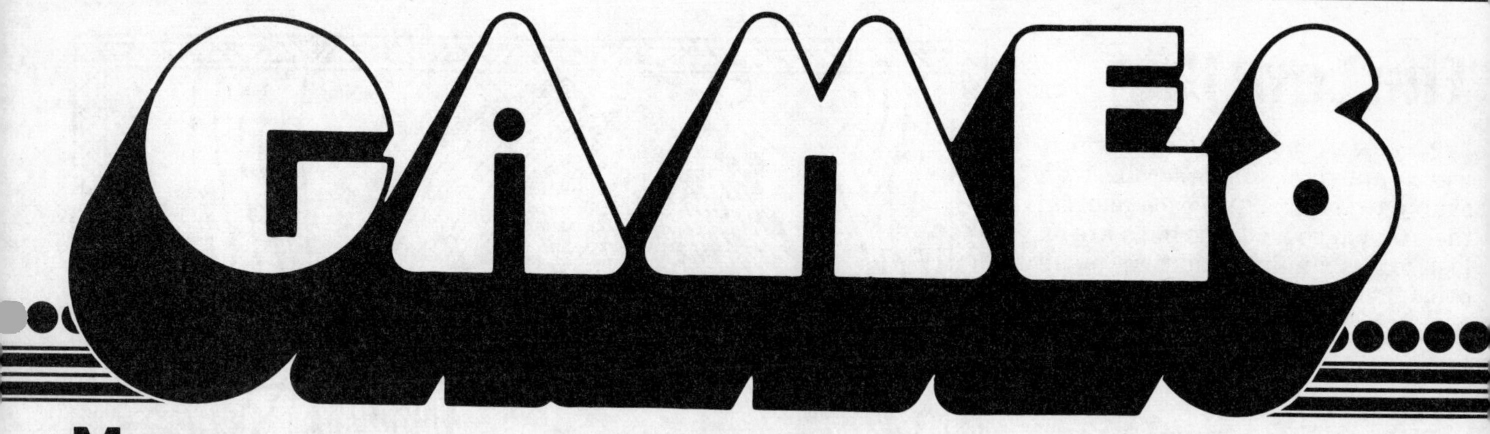

MAZES

"In ancient times, structures were built having many winding, twisting passages within them: they were called labyrinths. People unfamiliar with the pattern of the labyrinth would be confused at every turn and sometime these adventurers would become hopelessly lost in their wanderings.

"Today, labyrinths have become more fun. For one thing, they are generally known as *mazes,* and most of them take the form of graphic puzzles such as you will find in the books listed here. The challenge remains the same—to find a route that does not lead to a dead end, but rather, to a set destination. With a pencil you simply trace your way along the paths without crossing any solid lines. In some cases you will start at an entry point and meander to the exit. In others you will find that the correct path spells out a word or forms the outline of an object for which a clue has been given. Every maze has a rated time limit, and the solutions are in the back of each book. If you'd like to keep your mazes clean and reusable, simply put a sheet of tracing paper, or tissue paper, over the maze you select and trace your penciled path on the tissue paper.

"O.K., go to it—and try not to get lost!"

Those words were written by Vladimir Koziakin, who may well be the "king of mazes." He has designed hundreds of them, and there are at least ten books of his mazes which will frustrate and delight you at the same time.

Mazes for Fun 1, Mazes for Fun 2, Mazes for Fun 3. These mazes are Koziakin's easiest, and the best books for a beginner to try. The shapes of the mazes are fun: a frog, a subway, fish, creepy ivy, and ships. Grosset & Dunlap. $1.00 each.

Mazes, Mazes 2, Mazes 3, Mazes 4, Mazes 5. These mazes, also by Vladimir Koziakin, are challenging, sophisticated, and delightful mazes, ranging from easy 2-minute puzzles to intricate 45-minute labyrinths. There are 40 mazes of different shapes in each book. Grosset & Dunlap. $1.95 each.

MORE MAZES

Astro Mazes and *Mystery Mazes,* by Vladimir Koziakin. There are 50 mazes in each, with the solutions in the back. Tempo Books (an imprint of Grosset & Dunlap). 95 cents each.

ABC Mazes and *Dinosaur Mazes,* by Vladimir Koziakin. 31 mazes in each book, plus solutions. Tempo Books. 75 cents each.

Jacques Chazaud is another maze creator, with a number of books to his credit, including *Monster Mazes, More Monster Mazes, Strange and Amazing Mazes,* and *Storybook Mazes.* Each book contains 31 mazes. Tempo Books. 75 cents each.

There are three very interesting maze books by John Hull, with 19 mazes in each: *Maze Craze,* and *Maze Craze 2* and *3.* Troubadour Press. $1.25 each.

And now for you to ponder, peruse, and be perplexed by . . . mazes, codes, rebuses, crosswords, mathematical teasers, hide-and-seek words, optical illusions, riddles, acrostics, and tongue twisters. Then puzzles, puzzles, and more puzzles!

A Maze Jigsaw Puzzle

Are you a maze fan and a jigsaw puzzler too? Now you have it made with "An a MAZE ing Picture Puzzle." There are 500 interlocking pieces; the assembled size is 16 x 20 inches. First put the puzzle together; then you start solving the maze. The rated time for the solution of the maze is a tough 64 minutes. A felt-tip marker and the solution to the maze are included in the box. The creator is Vladimir Koziakin, the maze king of the 1970s. You can buy "An a MAZE ing Picture Puzzle" wherever you buy puzzles. $5.00.

The Maze Book

In this maze book with a difference, besides giving you 20 interesting mazes to do, Judy Jean Goodman teaches you how to make your own mazes. Step by step she shows you how to start, how to draw the path, how to make it difficult or easy, how to make branches and pinwheels and ladders. If you love to solve mazes, here is where your hobby can take a turn to the creative. *The Maze Book* has a heart for you to fill in with your own maze, and a ghost to fill in with your own spooky paths. Have fun, and soon you'll be writing your own maze books! For your copy SEND $3.00 to Activity Resources Co., P.O. Box 4875, Hayward, Calif. 94540.

Find It!

Over 100 eye-baffling puzzles by Stan Fraydas. For example, can you guess how many squares of all sizes there are on a chessboard? The answer: 204. If you don't believe it, get out your chessboard and start counting. If you like puzzles, find *Find it!* Golden Press. $1.50.

Make Your Own Chess Set

Play chess on a set that you've made yourself. Here are instructions for making up to 25 different chess sets from inexpensive, easy-to-find materials. Instructions and photographs will guide you to a new dimension of chess fun, win or lose. Ten more "instant" sets are described for spur-of-the-moment games along with suggestions for six boards. *Make Your Own Chess Set,* by David Carroll, also has an introduction to the history of the game and a fascinating history of chessmen. A must for chess devotees. Prentice-Hall. $6.95.

United States Chess Federation

Junior membership (age 6 and up) costs $5.00 per year. For information and an application WRITE TO United States Chess Federation, 479 Broadway, Newburgh, N.Y. 12550.

Chess for Children

What better way for you to learn chess than from the world-famous master and teacher Fred Reinfeld? Starting with the essentials, he guides you with step-by-step demonstrations of moves, values of the men, rules of checkmate, openings and gambits, tactics, defense, sacrifices, and notation—everything you need to know during the first two years of chess play.

Chess for Children is a good first book of chess and offers a firm foundation for a lifetime hobby. Sterling Publishing Co. $3.95.

The Teenage Chess Book

Reuben Fine, one of the world's leading authorities on chess, has written many books on the game. In *The Teenage Chess Book* he has kept in mind the player who has little or no previous knowledge of chess. His aim is to provide a useful practical technique for handling the game. In addition to basic information about the board, the chessmen, and notation, Dr. Fine, in a special chapter on chess "movies," introduces you to some of the delights of more complex play. The last chapter describes a study scheme which is bound to improve your playing ability. David McKay Co. $5.95.

Bobby Fischer Teaches Chess

Here is the fastest, most efficient, most enjoyable book on chess ever written. You, as the student, start at the beginning and progressively develop your skills by observing the situation on each page. You decide what you would do; then you turn the page and find out what former World Champion Bobby Fischer would do and why. Step by step you learn to think as Bobby does. After facing increasingly complex situations, you will be a better player after you take Bobby Fischer's chess course. (Authors: the champion, Stuart Margolies, and Donn Mosenfelder.) *Bobby Fischer Teaches Chess* works! Bantam Books. $1.95.

What's the Next Move?

If you know the rules of chess and have some familiarity with the game, *What's the Next Move?* by George Francis Kane, is the next book to help you improve your game. There are over 80 chess problems presented in large, open diagrams, one to a page, which represent critical points in chess games. You are challenged to answer the question in each case: What's the next move? Solutions are given in notation form on the back of each diagram. Charles Scribner's Sons. $3.95.

PuzzlePatterns

The basic principle of PuzzlePatterns is that a simple family of pattern elements will fit together in an infinite number of repeating patterns. Each PuzzlePattern set (Lines, Dots, Shapes) consists of 180 simple geometric shapes in brilliant color configura-

tions. Pieces can be assembled in any number of fascinating, expanding patterns, which demonstrate congruence, symmetry, complementarity and color reversal, continuity, order and disorder. Combinations and permutations are beautifully illustrated by every solution. Pieces are plastic-coated, heavyweight paperboard. Each set comes in a storage container. There are three sets: Lines, Shapes, and Dots. Ages 5 to 16. WRITE TO Creative Publications. $4.95 each. No charge for postage and handling.

Chess and Children

If you would like to play chess but are too young to read and understand a learn-how-to-play-chess book, tell somebody to buy *Chess and Children,* by George Francis Kane. It is written so that anyone can help you to learn chess, whether he or she knows how to play or not. Your friend, with this book as a guide, can show you how to set up the board and how the men move, and can give you problems to solve that will teach you some of the basic concepts of chess. There are over 100 diagrams to help, too. You will learn to see threats from the opponent and how to escape danger, and you will learn tactical moves that may win a game for you someday. You may be too young to read—but you're not too young to play chess! Charles Scribner's Sons. $7.95.

FREE How Do You Play Chess?

This excellent free booklet of 73 questions and answers is a basic introduction about an ageless game. It lists books you can get if you are interested in advancing your game of chess. WRITE TO *How Do You Play Chess?* Dover Publications.

The Backgammon Book

Here is *the* basic, authoritative book on how to play backgammon. Easily played by a 10-year-old, backgammon has challenged great game players for over 2000 years, never more than today. Now two of the world's greatest players, Oswald Jacoby and John R. Crawford, have written the most complete and up-to-date guide yet, *The Backgammon Book.* It has over 100 diagrams and provides a step-by-step account of how to play, from setting up the board and the crucial opening moves to the finer points of the middle and the end game. The interesting introduction tells you about the ancient origins of backgammon and some of its history, with pictures of some very old, magnificent backgammon boards. If you are a beginner, this book will give you the proper start. If you already know how to play backgammon, this book will improve your strategy. Bantam Books. $2.25.

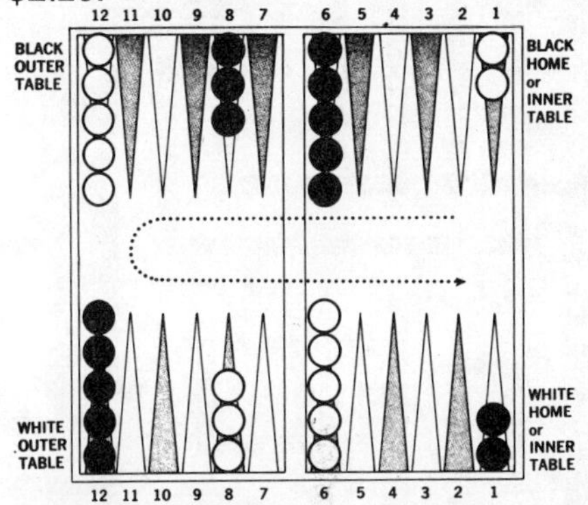

Word Puzzles

Hide & Seek Word Puzzles 1

All you need is a pencil to enjoy this collection of 50 word-search puzzles. Some are easy, some not-so-easy, but all are entertaining. The words you search for are words you know, and the subjects are up to date. You don't need a dictionary. Instructions and diagrammatic answers are included. Age 10 and up. *Hide & Seek Word Puzzles #1* is created by Jack Looney.

Hide & Seek Word Puzzles 2

Here are 48 brand-new puzzles for you: "Mystery Phrases"; "No Word Lists"; "Guess the Opposites"; "Name the Capitals"; and "Complete-the-Phrase" puzzles with themes such as favorite pop singers, sewing, and the Big Top. Age 10 and up. *Hide & Seek Word Puzzles #2* is created by Peter Tasler and Jack Dean. Bantam Books. 95 cents each.

FROM *PUZZLERS*

Are Mazes a Hang-up?

They can be, especially if you have "An a MAZE ing Hang Up." It's a 6 foot tall maze, 1½ feet wide. It's a maze for solving or for just looking. Roll it out. Hang it up—it makes beautiful wall decor. Since this is Vladimir Koziakin's most difficult maze, the manufacturer recommends that you use a soft pencil for your initial trip. You'll find "An a MAZE ing Hang Up" wherever puzzles are sold. $5.00. If you can't find either the picture puzzle or the wall hanging SEND $5.00 plus 80 cents for postage and handling to Gameophiles Unlimited, 16 Pine St., Morristown, N.J. 07960.

50 Card Games for Children

Want some new card games that are fun to play? Try menagerie, spade the gardener, frogs in the pond, and one called giggle a bit. There are 12 games of solitaire, 12 card tricks, some games that grown-ups play, and an easy lesson in contract bridge. These can be family card games. For a copy of *50 Card Games for Children,* by Vernon Quinn, SEND $1.00 to United States Playing Card Co., P.O. Box 12126, Cincinnati, Ohio 45212.

Junior Puzzle Book

Here is a book filled with all kinds of puzzles to ponder, peruse, and be perplexed by. They're great fun and make good company while you're traveling in the car, eating lunch, or baby-sitting. After you've challenged your brain with acrostics, crosswords, optical illusions, mathematical teasers, and more, be sure to share the book with your friends—there are plenty of puzzles to go around. To get *The Saturday Evening Post Junior Puzzle Book* SEND $1.00 plus 25 cents for postage and handling to Mary Alice Simpson, Saturday Evening Post, Youth Division, 1100 Waterway Blvd., Indianapolis, Ind. 46202.

Phuzzles

Phuzzles are mind-teasing photo puzzles. Carole Harper offers a variety of photographic puzzles that challenge you to recognize ordinary things shown in a new way. Close-ups make puzzling patterns by magnifying familiar objects. Cut-ups change well-known sights into bewildering collages of line and color. And photo-rebuses combine letters and photos to spell out amusing messages. Golden Fun & Games Book. $1.50.

Games (and How to Play Them)

Hopscotch . . . kick the can . . . Simon says . . . cat and rat . . . giant steps . . . musical chairs. Here are 43 wonderful games that will entertain you hour after hour. There are noisy games, quiet games, team games, traveling games, outdoor games, and rainy-day games, each clearly described and imaginatively illustrated with watercolor-and-ink drawings. They can all be played with little or no equipment and there are directions and rules to put an end forever to squabbles and confusion. *Games (and How to Play Them),* written and illustrated by Ann Rockwell, is great for ages 5 to 8. Thomas Y. Crowell Co. $6.95.

FROM *MAZE CRAZE*

The Domino Book

Muggins . . . Sebastopol . . . the magic pentagram . . . the cheater . . . a "cool" puzzle . . . sympathy . . . matador . . . a quadrille . . . Sir Tom . . . Fort Madison . . .

These are a few of the many dozens of games that can be played with the little black tiles called dominoes. The game as played in Europe and America is over 200 years old, and is played, by now, all over the world. Far older is the oriental form of the game, which can be traced back to the 2d century A.D. The author, Frederick Berndt, writes of the origins of the word "domino," the many different forms that domino tiles take, and the terms used in playing. In *The Domino Book* he tells you the rules for over 175 versions of the game, some simple, some complex. He has invented many solitaire games. The games are a challenge to the player's logic and mathematical prowess—and they can be enjoyed by the whole family. For your copy SEND $5.95 plus 50 cents for postage and handling to Thomas Nelson, Dept. BTM, 30 E. 42d St., New York, N.Y. 10017. Bantam Books. $1.95.

SnowCrystals

A beautiful pattern game for two, or a geometric perception puzzle for one. The object is to combine and match tiles so that symmetrical "snowflakes" are formed. There is no unique solution, so it doesn't matter if a few pieces get lost. A great game for mixed ages —the rules are simple enough for 5-year-olds, the strategy challenging enough for children up to 16. SnowCrystals demonstrates aspects of combinations, permutations, symmetry, repeating patterns, and shows how simple, low-symmetry systems combine into complex structures of higher symmetry. To order SnowCrystals SEND $4.95 to Creative Publications. No charge for postage or handling.

A Special Game Board

It's called the Write On–Wipe Off Game Board, because that's what you do. Write on the crystal-clear plastic surface and wipe off with a dampened tissue or towel. The board is always set up for you to play five games: crosswords, dots, hangman, dot-a-dart, and tic-tac-toe. There's a score board; complete instructions for each game are printed on the back. You can't lose the fiber tip pen, because it's tied on. Great fun for car trips. SEND $2.50 to Freelance, South Line and Prospect Aves., Lansdale, Pa. 19446.

The Complete Book of Solitaire and Patience Games

This book contains hours of do-it-yourself fun: over 225 of the most challenging and fascinating card games ever invented. In addition to the best and most frequently enjoyed games of solitaire like Canfield, Klondike, and patience, authors Albert H. Morehead and Geoffrey Mott-Smith have added a generous selection of brand-new games like boomerang, Bristol, and sudden death. There are complete instructions, illustrations, terminology, time requirements, and even odds against winning for each game in *The Complete Book of Solitaire and Patience Games*. Age 10 and up. Bantam Books. $1.00.

The sixth sheik's sixth sheep's sick.

A Twister of Twists, a Tangler of Tongues

Try saying "Peggy Babcock" five times as fast as you can. Chances are you can't do it, because your tongue won't cooperate. Tongue twisters are designed to tangle tongues, tickle funnybones, and lead to a gaggle of giggles.

Longer twisters are also good stories; one of the funniest in this collection concerns a seller of saddles named Sam Short, who's smitten with Sophia Sophronia Spriggs. The story runs to over 300 words, each of which starts with an *s*. "Peter Piper picked a peck of pickled peppers" is one of the oldest tongue twisters in folklore. It has been traced to an English grammar published in London in 1674. Alvin Schwartz, the author of *A Twister of Twists, a Tangler of Tongues,* deserves many thanks for this amusing and delightful compilation; so does Glen Rounds for his drawings. The collection includes twisters on rockets, biscuit mixers, rubber baby buggy bumpers, yo-yos, aluminum, preshrunk shirts, and so on. J. B. Lippincott Co. $1.95.

Brisk brave brigadiers brandished broad bright blades, blunderbusses, and bludgeons.

Some shun sunshine.

I need not your needles
They're needless to me,
For needing needles
Is needless, you see.
But did my neat trousers
But need to be kneed,
I then should have need
Of your needles indeed.

To complete a trilogy of wordplay from folklore (same author and artist, same publisher), get *Cross Your Fingers, Spit in Your Hat* ($2.25), a collection of superstitions and other beliefs; and *Tomfoolery: Trickery and Foolery with Words* ($1.95). Here's a sample from *Tomfoolery:*

One night a man paid his friend a visit and, much to his surprise, found him playing chess with his dog.

"This dog must be very intelligent to be able to play chess," he said.

"Oh, he's not so smart," his friend replied. "I just beat him three games out of four."

The Code & Cipher Book

0?&&!"(-?/# is a message in typewriter cipher, not something to mutter if you get stuck decoding a Morse code dispatch. *The Code & Cipher Book* unravels the riddle of encoding and decoding messages in a wide variety of boggling, baffling, bemusing, and amusing codes and ciphers. Secret ways of saying things range from very simple to extremely challenging. Some are traditional, some are modern—and they're all fun. With this book, written by Jane Sarnoff and illustrated by Reynold Ruffins (see *A Great Bicycle Book*), you'll learn how to crack the cipher of the ancient Celts, understand Cockney rhyming slang, make invisible inks, turn a telephone into a code machine, even make your own code wheel. Codes and ciphers can open up a whole new world for you. They're specially useful if you belong to a club or just want to communicate in secret. Charles Scribner's Sons. $6.95.

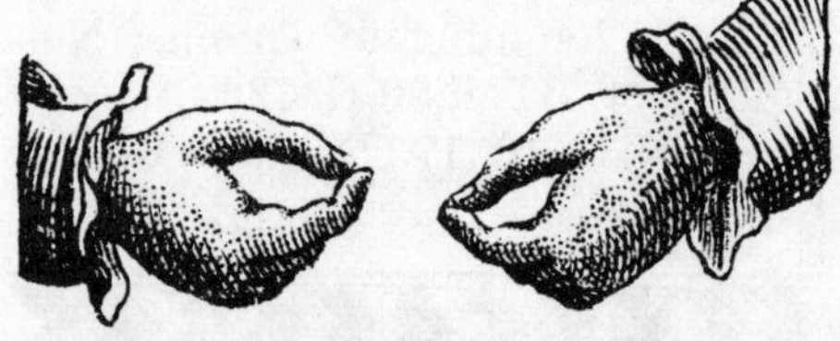

Like Riddles?

When did the Irish potato change its nationality? When it became French fried. And Jane Sarnoff has collected 499 other riddles, illustrated in full color (and full fun) by Reynold Ruffins in *What? A Riddle Book*. A variety of subjects are covered. Politics: In what month do politicians talk the least? Space: Where did the astronauts stand when they first landed on the moon? Economics: When can you have an empty pocket and still have something in it? Philosophy: When things are tough, what can you always count on? Natural science: Why do hummingbirds hum? Elephantology: How can you make an elephant float? If you like riddles, you'll love this book. Charles Scribner's Sons. $6.95.

Secret Codes and Ciphers

This is a fun book! It's hard to start reading it anywhere without wanting to go on....

OD UOY WONK TAHW SIHT SYAS?

If that sentence doesn't make any sense to you, read each word backwards. And if that was too easy for you, try this one:

DBO ZPV SFBE UIJT?

Of course you can. Just change each letter to the one that comes before it in the alphabet.

Both of the "secret" sentences above are in cipher. You don't have to look up anything in a book to find out what they mean. You just have to know the system.

The ciphers and the codes get more and more difficult, but you'll enjoy them all and the anecdotes that accompany them. You'll learn about concealment ciphers, transposition ciphers, cryptograms, substitution ciphers, the Vigenère tableau—enough to become a spy or maybe 007! Age 9 and up. *Secret Codes and Ciphers* is by Bernice Kohn (illustrated by Frank Aloise). Prentice-Hall (Treehouse Paperback). 95 cents.

How to Write Codes and Send Secret Messages

This is a beginner's code book. You'll learn about space codes, Greek code, alphabet codes, invisible ink, and how to deliver a code message. You'll learn how to create hidden word codes, and how to solve them with the special red transparent paper that comes with the book. Age 7 and up. *How to Write Codes and Secret Messages,* by John Peterson. Scholastic Book Services. $4.95 cloth, 75 cents paper.

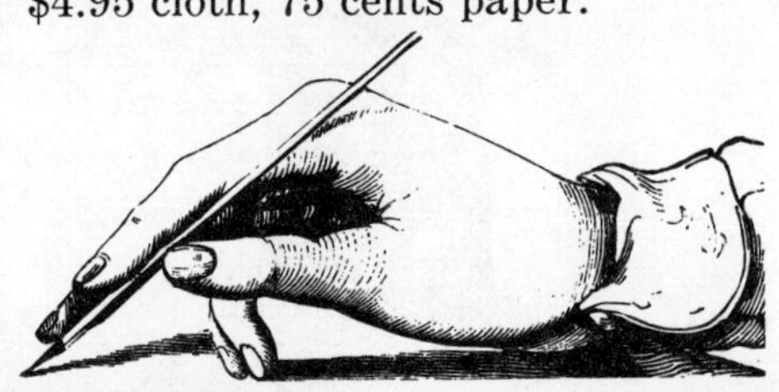

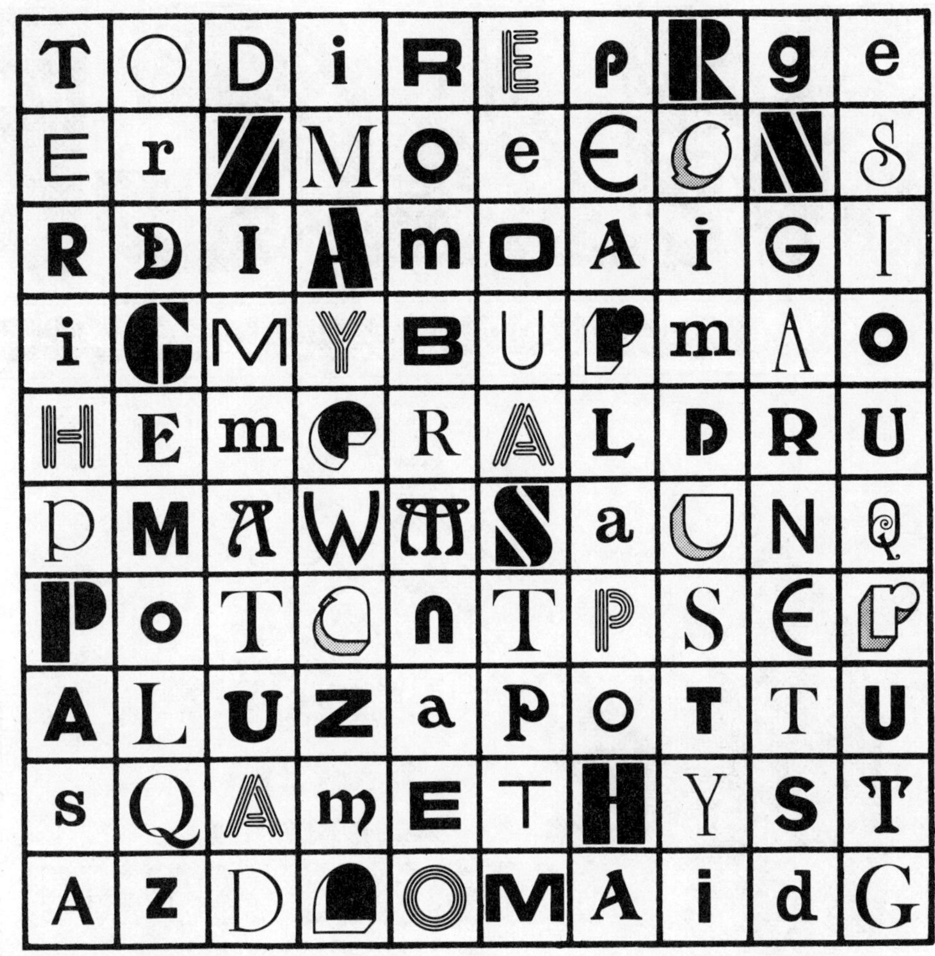

T	O	D	i	R	E	p	R	g	e
E	r	Z	M	O	e	E	O	N	S
R	D	I	A	m	O	A	i	G	I
i	G	M	Y	B	U	P	m	A	O
H	E	m	P	R	A	L	D	R	U
p	M	A	W	M	S	a	U	N	Q
P	o	T	O	n	T	P	S	E	P
A	L	U	Z	a	P	O	T	T	U
s	Q	A	m	E	T	H	Y	S	T
A	Z	D	D	O	M	A	i	d	G

Puzzlers

Puzzlers, by John Hull, contains a lot of puzzles, some hard, some easy. Here's an easy one for you. We put it in because the different type-faces for the letters of the alphabet make an interesting montage.

The instructions are: "Find your birthstone reading up, down, forward, backward, or diagonally."

January	Garnet
February	Amethyst
March	Aquamarine
April	Diamond
May	Emerald
June	Pearl
July	Ruby
August	Peridot
September	Sapphire
October	Opal
November	Topaz
December	Turquoise

Ages 7 and up. Troubador Press. $1.50.

Stop Dropping BreAd CrumBs on My YaCht Or: The Silent A B C

This is a treAsury of undouBted fasCination! 26 silly situations in which each letter has a chance to be seen but not heȧrd make up this original and engaging alphabet book. From the famous silent *e* to the elusive silent *v* the letters steal in and out of the book with wit and style. Graphic illustrations and distinctive calligraphy by the author, Cynthia Maris Dantzic, make the book even more fun. The silent ABC book—*Stop Dropping BreAd CrumBs on My YaCht*—is published by Prentice-Hall. $5.95.

MAGIC

Bill Severn's Magic Trunk

It's a library of magic: four books packed into a slipcase that looks like a magic trunk. The author, Bill Severn, gave his first stage performance of magic at the age of 12: before he was out of his teens he was entertaining in vaudeville and night clubs. Still an active performer, he is the author of many successful magic books, and a member of the Society of American Magicians, the International Brotherhood of Magicians, and London's Magic Circle.

Magic Shows You Can Give. A book that contains not only magic tricks but complete shows. The tricks will take some learning, of course, and the amount of applause you get will depend on how well you perform them. But part of the fun is learning. With the shows all planned for you, there is no wondering what to do or say next.

Magic in Your Pockets. A variety of tricks that can be performed anywhere with the ordinary contents of your pockets. This is closeup magic, the kind you do whenever a few friends get together. It is magic with small things, but the mystery is no less than if you perform with all the elaborate gadgets of a stage magician.

Magic with Paper. How to create tricks professional magicians often perform, without expensive props. The magic is done with paper or things made of paper. A wealth of material for all young magicians: the tricks in this book have been tested for you; the plots are given and the talk to use with them is suggested.

Magic Comedy. Magic with a fresh accent—tricks that will produce laughter as well as amazement. The tricks in this book are designed to be funny; there are none that require unusual skill or sleight of hand too difficult for the average person to master.

Bill Severn's Magic Trunk could —abracadabra!—turn you into a magician. David McKay Co. $9.95.

Wherein ladies are sawed in half, rabbits pop out of top hats, and silk handkerchiefs change color with the wave of a wand. *Plus* (don't tell how it's done!) most secret instructions so you can perform one of the most baffling of all card tricks!

Magic...Using Simple Things You Can Find Around the House

Mixed Bag of Magic Tricks, by Roz Abisch and Boche Kaplan. In this simple, easy-to-read book of magic fun you will learn to perform tricks which mystify and amuse, but do not require expensive or complicated props. Using ordinary household objects you can make an egg fly out of a cup, tie a knot in a piece of string without letting go of the ends, or make a piece of thread dance in the air. There are number schemes for guessing ages and dates, as well as card tricks, optical illusions, and mind-reading acts. Walker & Co. $4.95.

Spooky Magic, by Larry Kettlekamp (illustrated by William Meyerriecks). Ghostly spirits in old bottles; bodies floating in air; tables rising; water disappearing. This book teaches you how to do these tricks and others such as the Spirit Hand, Ghost Writing, the Spooky Silk, the Wandering Ring, the Mysterious Card, the Spirit in the Bottle, and the Floating Body. Everything you need can be found around the house. Scholastic Book Services. 75 cents.

John Fisher's Magic Book. Can you divide an object without separating it? Can you lift a friend off the ground with your fingertips? Can you make a man walk in two directions at once? Probably not. But with the help of this book and some simple equipment from around the house, you can do all these things and many more. There are 75 dazzling magic feats given; you will also learn showmanship to intrigue and mystify the most skeptical audience. Prentice-Hall. $4.95.

Card Tricks Anyone Can Do

Right on the cover of *Card Tricks Anyone Can Do,* by Temple C. Patton, it says: "No sleight of hand required." All you need to baffle your audience is to be be able to count; then this book will show you a unique approach to an intriguing branch of magic. The tricks are based on a regular deck of cards and simple mathematical principles. By following the easy instructions, you will soon be able to amaze your friends—and even yourself—with outstanding card tricks. Simon and Schuster (Cornerstone Library). $1.95.

More Magic Books

Your Book of Table Tricks, Your Book of Card Tricks, and *Your Book of Mental Magic Tricks,* all by Geoffrey Lamb. Thomas Nelson. $4.95 each.

Magic Across the Table, by Bill Severn. David McKay Co. $4.50.

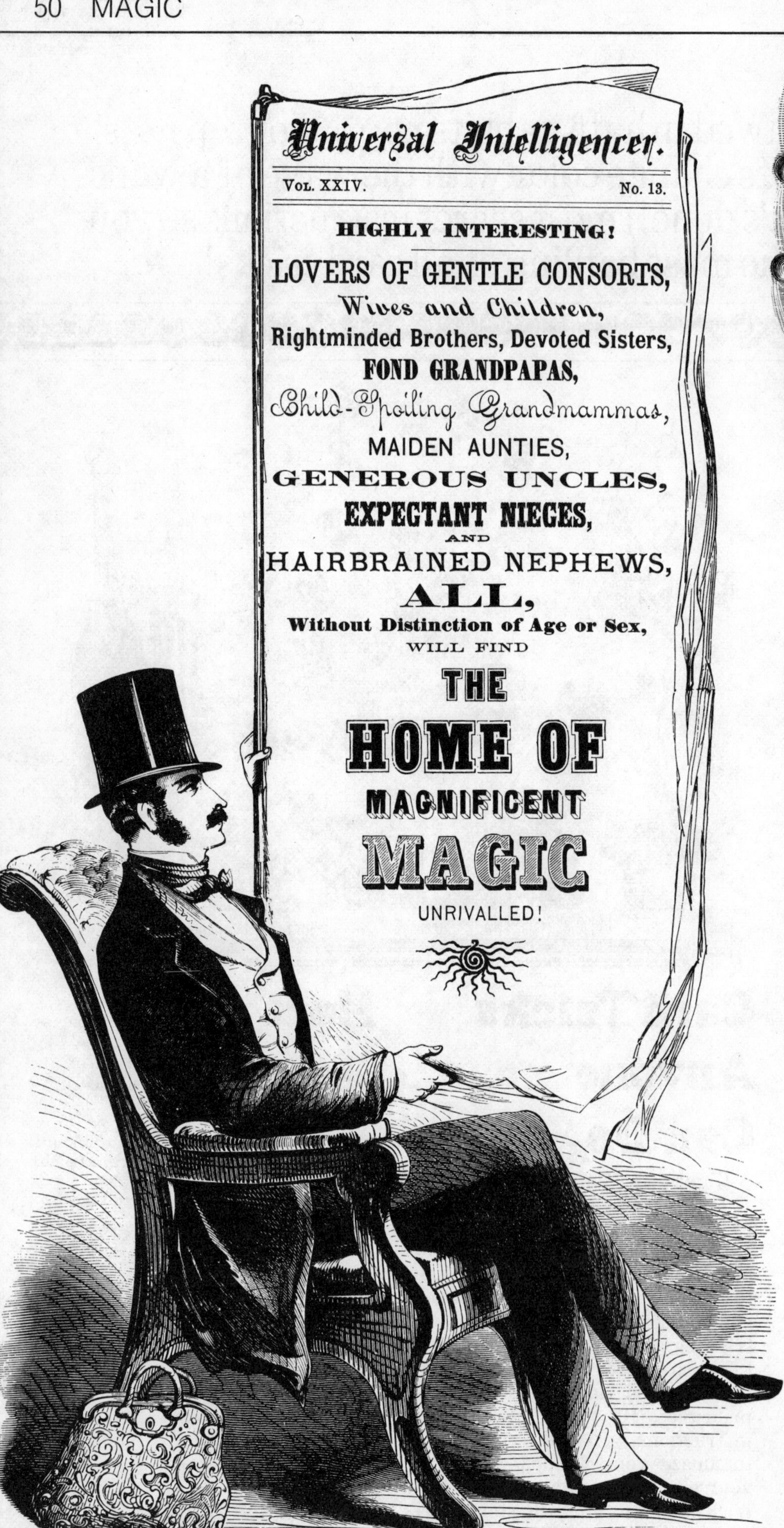

Be a Magician!

Would you like to become a magician and perform dazzling feats of legerdemain, mystify your friends, develop a new hobby, even put on a professional magic show? You can, and you are never too young (or too old) to start pulling rabbits out of silk hats, doing truly baffling card tricks, changing the colors of silk handkerchiefs, and pulling quarters out of the air. You can even learn how to saw a woman in half!

A good way to start is to visit a magic shop where professional magicians buy their equipment and their magic effects and illusions. One of the best is the Louis Tannen Magic Shop, 12 stories above Times Square in New York City. It's probably the world's most modern magic shop, and it has been catering to the needs of professional and amateur magicians for over 40 years. Many of the famous magicians you see on stage and television started their careers by buying their first simple magic tricks from Lou Tannen. And the men who work in his shop (all magicians) will gladly demonstrate any magic trick or effect you're interested in.

But you don't live in or near New York—and you still want to become a magician. It's easy: as Lou Tannen says, "Over 75 percent of our business is done by mail. We send out as many as five mail sacks a day filled with magic apparatus." One of his slogans is: "Lou Tannen's—Where Magicians Shop by Mail."

The Louis Tannen Magic Catalog

It's a hardbound book, 572 pages thick, the 1974 10th edition. It pictures and describes over 3000 tricks and feats of legerdemain that sell from $1.00 for a disappearing handkerchief to $197.50 for a buzz saw that "slices" harmlessly but realistically through a girl's midriff.

Here are the accessories, supplies, and gimmicks a magician needs; and hundreds of tricks that need no skill and can be performed immediately.

The catalog has card tricks: books on magic; coin and money magic; comedy magic; club and stage tricks; magic for youngsters; illusion and blueprints; liquid magic; mental and spirit effects; closeup magic; rope, paper, and flower tricks; silk effects; and many more. When you buy the Louis Tannen Magic Catalog, your name and address is placed on their mailing list, and four times a year you will receive a free copy of *Top Hat Topics,* which shows-and-tells about the latest in the wonderful world of magic.

Here Is a Mind-Reading Trick You Can Learn Right Now

This is a truly baffling trick!

You hand a spectactor in your audience a telephone book and a sealed envelope. You tell him that he has to guard the sealed envelope and make sure that nothing is disturbed inside.

A second spectator is given a piece of paper and a pencil. He is instructed to write down a three digit number.

A third spectator is asked to make certain calculations with the numbers and announce the result.

The first spectator who has been guarding the telephone book is now asked to look in the telephone book at a page and name indicated by the mathematical results and read aloud the name at that number.

He is now asked to open the sealed envelope . . . and lo and behold that same name appears on a piece of paper inside the envelope!

Here's how to do it:

Before you start, turn to page 108 in the telephone book and count down to the 9th name on the page. Write this name on a slip of paper and seal it in an envelope. This is the envelope you will later hand to the first spectator with the telephone book.

Now ask the second spectator to write down a three digit number.

Let's suppose that he writes down 653.

The third spectator is asked to reverse that number and subtract the lower from the higher:

$$\begin{array}{r} 653 \\ 356 \\ \hline 297 \end{array}$$

Now ask him to take the result (297), reverse it (792) and add the two together:

$$\begin{array}{r} 297 \\ 792 \\ \hline 1089 \end{array}$$

The answer will always be 1089, no matter what numbers are used.

Now ask the guardian of the telephone book to look on the page indicated by the first three numbers, 108, and to count down to the name indicated by the last number, 9. Ask him to read aloud the name that appears at that position. And then have him open the sealed envelope and read aloud the name that is printed on the paper inside. Of course, it will be the same name!

If you wish, you may use this as a telepathy stunt instead of a prediction. In that case do not write a prediction, but ask the spectator who is looking in the telephone book to concentrate on the name. With great concentration and drama, you are able to tell him the name he is looking at. Of course you have memorized it before the start of the trick. To vary the program, use a dictionary or a book instead of a telephone book.

This is just one trick from *101 Best Magic Tricks,* by Guy Frederik. There are 100 others: card tricks, coin tricks, rope and ribbon tricks, advanced tricks, and do-it-yourself magic. Sterling Publishing Co. $3.95.

The Magic Store That Comes to You

There is a seemingly never-ending list of books on magic. The Tannen catalog devotes over 100 pages to hundreds of magic books. If you want to be a magician, the best thing is to order the *Louis Tannen Magic Catalog* ($2.50, postage paid). WRITE TO Louis Tannen Magic Shop, 1540 Broadway, New York, N.Y. 10036.

Secrets of Magic

Do you want to know how the sensational trick of sawing a woman in half is done? Or how Houdini made an elephant vanish? Or the secret behind the daring and dangerous art of knife throwing? Or how a human being can be suspended in midair? Learn about the amazing and mysterious feats, not only of today, but of ancient times. A fascinating survey of the secrets of all the ages, starting with the secrets of the ancients; magic of the Middle Ages; modern mysteries such as the bulletproof man; magic of India (the Indian rope trick, the obedient cobra); oriental magic (fire walking, buried alive); secrets of stage magic (the living head, marching under water); and methods of fake mediums (spirit table lifting, spirit raps). From the dawn of time, the lure of magic has been felt by men of every race. *Secrets of Magic,* by Walter B. Gibson, illustrated by Kyuzo Tsugami, will open up the closely guarded secrets of the ages to you. Grosset & Dunlap. $2.95.

the soothsayer's handbook

Divination is the art of foretelling the future by interpretation of signs. *The Soothsayer's Handbook—a Guide to Bad Signs & Good Vibrations,* by Elinor Lander Horwitz, is a book for younger readers about divination and how to divine.

Techniques for forecasting the future by astrology, numerology, and palmistry go back to earliest times, and there has never been a country or age with no demand for the services of men and women gifted in predicting what is to come. Are you gifted? Do you have ESP (extrasensory perception)? The first chapter in this book will test your talents. It gives you projects to do with your friends, such as the coin-tossing test, the card test, and other forms of "mind reading." Other chapters include discussions of astrology, palmistry, numerology, telling fortunes with playing cards and tarot cards, tea-leaf reading, dice, dominoes, and crystal ball gazing. If you are curious about these subjects, this is a good basic book to start with. J. B. Lippincott Co. $1.95.

test your esp

Can you read the thoughts of others? Forecast events? Influence matter with your mind? *Don't scoff!* Believe it or not, you may have the power. Every day, thousands of people show uncanny ESP abilities—and don't even realize it. Find out the truth about your own mind. *Test Your ESP,* edited by Martin Ebon and written in collaboration with the Institute for Parapsychology directed by Dr. J. B. Rhine, offers a spellbinding array of provocative experiments designed to analyze the four major areas of ESP experience: telepathy, clairvoyance, precognition, and psychokinesis. Each experiment is simple enough to perform in your living room, yet when correctly performed and evaluated, it will give you a computer-accurate index of the psychic power you may very well possess. Signet (an imprint of New American Library). 95 cents.

Amaze your friends with the mysteries of the ancient seers! Learn how to be a fortune-teller, see the future in a crystal ball, read the secrets of life in tea leaves and Tarot cards. Amaze your friends by reading palms and astrological horoscopes.

Graphic astrology- the astrological home study course

Astrology is man's attempt to discover the unity of the universe, to learn how the planets in their motion react upon one another and influence a person's life. Ellen McCaffery, the author of this home study course, makes it easy for you to fully understand your horoscope. There is nothing mysterious about astrology; in fact, it is so logical that you will be amazed. There are hundreds of illustrations, many complete examples of charts of famous people, and diagrams to help you. If you want to learn astrology, start with *Graphic Astrology—the Astrological Home Study Course,* the simplest and most comprehensive textbook on the subject. Macoy Publishing Company. $5.50.

astrology: sense or nonsense?

What is astrology? What are its origins? How does it affect you? Or does it affect you at all? Does astrology have something real to offer?

These are some of the questions that Roy A. Gallant poses and answers in *Astrology: Sense or Nonsense?* A highly regarded astronomer and writer who has taught at the Hayden Planetarium, he takes a hard look at the roots of astrology in Babylonian civilization some 5000 years ago as an attempt to explain the universe. He shows how astrology gave rise to the science of astronomy, and what has happened to astrology and astrologers through the ages. Mr. Gallant discusses why, in this technologically advanced age, the popularity of astrology has reached a new high.

Liberally illustrated, this is a must for all students of the occult. Doubleday & Co. $5.95.

paths to inner power

This is a series of practical handbooks designed as an authoritative introduction to occult theory and practice. Each book is devoted to one facet of esoteric philosophy and teaching. New titles are added every year, ensuring a continuing and versatile coverage of occult themes. A few of the titles available are *Dreams: Their Mysteries Revealed* (G. A. Dudley); *First Steps to Astrology* (Preston Crowmarsh); *How to Develop Clairvoyance* (W. E. Butler); *How to Develop Psychometry* (W. E. Butler); *How to Understand the Tarot* (Frank Lind); *Karma Yoga* (Harvey Day); *Palmistry: Your Destiny in Your Hands* (Mary And); *Practice of Meditation* (Charles Bowness); *Precious Stones: Their Occult Power and Hidden Significance* (W. B. Crow); *Psychosomatic Yoga* (John Munford); *Secret Power of Numbers* (Mary Anderson); and *Your Character from the Stars* (T. Mawby Cole).

To get your copy of any of these books SEND $1.00 plus 35 cents for postage and handling to Samuel Weiser, Box S, 734 Broadway, New York, N.Y. 10003.

psychic sciences

This is a guide to becoming an amateur fortune-teller using all the well-known methods from casting horoscopes to reading tea leaves. The chapters on palmistry, fortune-telling with cards, numerology, and graphology are especially detailed and easy to master. Briefer treatments are given of other methods: dice, dominoes, colors, interpretation of dreams, "moleosophy," phrenology and physiognomy, telepathy, yoga, and other paranormal phenomena. The *Complete Illustrated Book of the Psychic Sciences* is by Walter B. and Litzka R. Gibson, with illustrations by Murray Keshner. Pocket Books. $1.50.

the tarot revealed

Tarot cards have been used for centuries to predict the future, shed light on the past, and reveal the mysterious patterns of character and destiny. *The Tarot Revealed* is a modern guide to reading the tarot cards which unlocks the secrets of the strange and beautiful symbols on them. With a certain amount of application anyone can learn to read the cards skillfully. This is an excellent beginner book to the reading of the tarot: each card is pictured, with its meaning and symbology. The author, Eden Gray, teaches you how to lay out the cards; first, using the ancient Celtic method: second, using the tree-of-life method. Signet (an imprint of New American Library). $1.50.

the tarot-a deck of cards

To embark on your study of the tarot you will need a good deck of tarot cards. The Rider Waite Tarot Deck is the original and only authorized edition of the famous 78-card tarot deck designed by Pamela Coleman Smith under the direction of the noted occultist Arthur Edward Waite. The pack consists of the Minor Arcana, 56 cards in four suits (the forerunners of modern playing cards), and the Major Arcana, 22 additional cards that contain the most ancient symbology. To obtain your cards SEND $6.00 to Tarot Deck, Samuel Weiser, 734 Broadway, Box S, New York, N.Y. 10003.

on astrology

You are a special person. When you were born the sun, moon, stars, and planets were in a unique position in the heavens. Astrologers believe that the location of these giant sky objects at the moment of your birth can help tell what kind of person you are. For thousands of years people have studied the heavens for clues to their souls and their fate. Today the ancient practice of astrology is kept alive by millions of believers.

On Astrology, by Peter Livingston (illustrated by Frederick Schneider), is a comprehensive first look at this popular subject. It explains how to make and interpret a simple horoscope and traces astrology from its emergence out of ancient thought to its place in the development of science and psychology in the 20th century. Prentice-Hall. $5.95.

personality and penmanship: a guide to handwriting analysis

With this book you can learn to analyze handwriting, gaining new insight into yourself and others. Each person's handwriting is as individual as fingerprints, and *Personality and Penmanship: A Guide to Handwriting Analysis* gives you the key to her or his personality profile. All the vital secrets of graphology (the science of handwriting analysis) are explained by Dorothy Sara in everyday language that anyone can understand and follow. Fully illustrated with many examples of handwriting. House of Collectibles. $1.00.

the book of prophecy

"Let me see your hand." "Look into the crystal ball." "What did you dream last night?" These are the ways of prophecy. In *The Book of Prophecy,* Edward Edelson takes a look at some methods of prophesying, including palmistry, dreams, the crystal ball, numerology, tea leaves, dominoes, dice, the tarot, playing cards, and the *I Ching*. You'll find out what a psychic "sees" in a crystal—what it should look like, what it means. There are explanations of how all these methods are used and of why there is an apparent resurgence of popularity nowadays. Doubleday & Co. $4.95.

zodiac coloring book

Do you know that the word "zodiac" comes from the Greek, meaning "a circle of animals"? Here they are in the *Zodiac Coloring Book,* by Dennis Redmond: the ram, the bull, the lion, the crab, the fish are all printed in decorative designs on heavy-duty vellum ready for you to color with ink, watercolor, felt-tip pen, pastel, or acrylic paint. Make your own posters, or frame your completed pictures for gifts. Troubador Press. $2.00.

If you are unable to find any of these books relating to the occult in your bookstore, WRITE TO Samuel Weiser, 734 Broadway, Box S, New York, N.Y. 10003. Be sure to include the title and cost of the book plus 35 cents for postage and handling. For a catalog of Samuel Weiser publications send 25 cents.

THEATER & PUPPETS

Making Puppets Come Alive

This is the best book on hand puppetry we've seen. Illustrated with unusually helpful demonstration photographs, it shows you (and all beginners) how to bring a hand puppet to life. You'll learn how to make your puppet wave, sneeze, yawn, point, clap, cry, walk, run, pick up things, and express grief, happiness, and lots of other emotions. Larry Engler and Carol Fijian, professional puppeteers, help you develop the skills needed to put on a good amateur puppet show complete with suitable staging, costumes, props, and special effects. Included are the fundamental elements of good theatrical technique: speech, voice use, and synchronization; stage deportment and interaction; improvisation, dramatic conflict, and role characterization. Every detail from making working puppets through the finishing touches of play giving is clearly explained and beautifully illustrated. *Making Puppets Come Alive* makes learning hand puppetry simple and fun. Taplinger Publishing Co. $9.95.

How to Make Puppets

The complete title is *How to Make Puppets and Teach Puppetry.* The book was written by Margaret Beresford. In her preface, she says: "The plays included in this book were created by children, and the manipulation is well within their capabilities. *Children, if allowed to act freely, will often surprise the adult by the high standards of their work.*" (Our italics.) It's a good book on puppetry, not only for teachers, but for young puppeteers. The chapters include "Elementary Puppet Making"; "How to Make Papier-Mâché"; "How to Make and Manipulate String Puppets"; "Shadow Puppetry"; and "Puppet Plays Written and Acted by Children." Taplinger Publishing Co. $3.95.

Theater is the wonderful world of make-believe... scenery and stages, lights and costumes, and putting on a show. Theater is also just you... pretending to be someone else. You're a monster! You make your audience laugh. Applause! Applause!

Felt Puppets

Detailed instructions for making 17 beautiful and colorful hand puppets out of felt. Basic patterns for cutting felt are included; full-color photographs show completed puppets. Otto Owl, Henry Hippo, Max the Monkey, and Loonie Lion are only a few of the puppets featured in the book. For your copy of *Felt Puppets* SEND $1.25 plus 25 cents for postage and handling to Mangelsen's, P.O. Box 3314, Omaha, Nebr. 68103.

Puppet Party

This unique book shows you how paper bags, boxes, tubes, old gloves and socks, rubber balls, and sponges can make all kinds of puppets: finger puppets, hand puppets, and stick puppets. Delightful illustrations by Margaret A. Hartelius and simple text and easy-to-follow directions by Goldie Taub Chernoff make *Puppet Party* a good book for the very young who want to learn to make puppets. Scholastic Book Services. 95 cents.

Easy Puppets

Hand puppets can be made at home! With simple directions and more than 100 pictures the author shows how a wide variety of puppets can be made from inexpensive materials in your house; apples, potatoes, eggshells, old gloves, buttons, even bottle tops. Using the simple patterns you can make turtle puppets, balloon puppets, potato-head puppets, and many more. You'll start by making simple puppets and progress step by step until you are making papier-mâché heads and complete stages with proscenium arches and curtains. The author, Gertrude Pels, and her artist-husband, Albert Pels, who illustrated the book, believe that any child can make a strong and sturdy puppet. If you like puppets, you'll love *Easy Puppets*. Thomas Y. Crowell Co. $4.50.

Making Easy Puppets

Shari Lewis is a familiar face as one of television's foremost puppeteers. In this book she shares with you some of her creative ideas and instructions for making more than 30 puppets. By following the simple, step-by-step directions, no matter how young you are you can have the fun of making your own puppet.

Using only materials available in almost every home—or easily and inexpensively obtained—Shari Lewis unfolds a world of exciting creativity by showing you how to turn a flip-top cigarette box into a jolly Santa Claus, a carrot into a sinister pirate, or a man's handkerchief into a long-eared rabbit.

There are directions for making four different kinds of stages and dozens of ideas for having fun with puppets: suggestions for plays and storytelling, what to put in a "rainy day box," songs to act out with favorite puppets; and there is a section on the history of puppets and marionettes from earliest times. The emphasis in *Making Easy Puppets* is on simplicity and imagination—a combination that will provide you with many hours of puppet pleasure. Illustrations by Larry Luria. E. P. Dutton & Co. $4.95.

Hand-some Johnny

Hand-some Johnny is a famous television star. Have you seen him?

You will need:

A tube of dark lipstick
Your hands

Here's How: With a lipstick, draw a face on the back of your hand. Follow the diagram for the position of the mouth. Carefully color the lower half of your pointer finger and the upper half of your thumb. But don't copy the features in the diagram. Draw any silly face that occurs to you. Now make a fist, wrapping the end of your pointer finger around the last joint of your thumb, so that your thumb becomes the lower lip, and your pointer finger becomes the upper lip. And that's all. As you move your thumb up and down, your puppet will talk, chew gum, eat, and sing.

Try sticking the thumb of your other hand through the mouth from inside the fist. Whoops! Looks like he stuck out his tongue at you, doesn't it? Tsk, tsk, tsk. Mustn't let this puppet get out of hand.

Want to Produce a Show?

You Can Put On a Show, by Lewy Olfson, is packed with ideas and information about how to put on any kind of show—variety or revue, puppet show, pageant, talent show, or play—anywhere, from your backyard to an auditorium. It describes each job, producer to curtain puller, tells how to coordinate talent, use blackout sketches and running gags, write your own one-act plays, create puppet theater, and put on one-man shows. The scenery, lights, and costumes can be created from materials around the house. Age 8 and up. Sterling Publishing Co. $4.95.

Costumes for You to Make

Would you like to be an Egyptian queen, an Aztec king, a Viking warrior, or a knight in full armor? With the right costume, you can transform yourself into all of these, even an astronaut, or a flower, to name just a few of the hundreds of costumes described in *Costumes for You to Make,* by Susan Purdy. No previous sewing skills are needed. Clear and simple directions tell you how to make hundreds of costumes with readily available, low-cost materials. There are masks and disguises which can be made in just a few minutes—some projects are more complex, but all are fun and exciting to make and wear. Where do you find costume ideas and materials? How do you use color and design in theatrical productions. How do you use make-up? The book answers these and many other questions. All ages. J. B. Lippincott Co. $5.95.

Found Theater

Found Theater is the name of the book—and we hope you will find it, because it is one of the few books on creative dramatics or children's theater that really puts it all together. We found it because Carolyn R. Fellman, the author, sent us a copy and wrote the following letter, which we'd like to share with you:

> . . . thought you might be interested in the enclosed idea book as a possibility for resources for children. The theme, as the title *Found Theater* suggests, is discovering the infinite possibilities for theater all around your environment and the wealth of techniques—invented and, as yet, uninvented—for utilizing them once you allow yourself a free mind and time enough to find them.
>
> It is particularly directed toward people who say, "Oh, well, I can't do anything like that. I'm not creative." At any rate it is my own way of trying to dispel some of the awful mystique shrouding the arts in general and encouraging some irreverent approaches to the arts in education in particular—and I would, of course, be most happy to be able to share these ideas with others.

Found Theater's Table of Contents

The book is 77 pages long (8½ x 11 inches), fully illustrated—and fully exciting! Here is the introduction:

> This is a collection of suggestions about doing theatre with kids and for kids. If you haven't seen very much of it, it's because there isn't very much to see. You and your kids may have to invent it. How do you do this? A lot of people think if you follow some magic, age-old formula of scripts and stages, lights and costumes, it will come out theatre. Perhaps this emphasis on product is why most theatre is so uninspiring.
>
> First of all you have to rethink what theatre is all about. It is about:
>
> people
> and dogs
> and monsters;
> familiar places
> and made-up places
> and places you've never been.
> It elicits response; it communicates; it moves.
>
> Sometimes it speaks more in action than in words because kids understand action better. It is music and movement and sounds and smells all around you every day, in fact and fantasy. It happens all over the place in spaces you never thought of looking for it. It happens through an exciting process of exploration which starts with yourself and grows to include the whole world, real and fantastic. And maybe, while you are exploring, you will find some moments along the way that are so exciting that you have to find a way of sharing them and the exploration takes on the discipline of an art form. Perhaps this book can help you find where to begin.

And here is part of the author's pep talk:

> You need a space to perform in.

> You do NOT need a stage and curtains!
>
> You need a set and costumes.

> In one ecstatic word: JUNK! ! !
> MOTTO:
> Never pass a garbage can too quickly.
>
> As for actors, technicians and so on, you've got star quality all around you. So, all in all, the resources at your disposal are tremendous. You have a theatre and an original one at that; you have some of the most imaginative, creative and original actors, artists and playwrights hanging around your room every day; you've all got energy and everybody wants to do it, so you've got super-high motivation; and the world is your garbage can—unlimited free materials. You have, in short, everything you need to do a production as good or better (albeit different) than any theatre in this city. If you believe this, you are ready to begin.

Where to Buy Found Theater

If creative dramatics is what you're looking for, you've found it in *Found Theater*. But you can't find it in a bookstore. SEND $3.50 to the author (and publisher): Carolyn R. Fellman, 313 S. Aurora St., Ithaca, N.Y. 14850, and she'll be happy to send you a copy.

FILMS

Young Filmmakers

Young Filmmakers tells you all you need to know to make an 8-mm or 16-mm movie, but it doesn't tell you how to, because Rodger Larson doesn't believe in telling young filmmakers how to. At the base of his teaching is the conviction that the creative directions of young filmmakers should determine the way they will go. The opening sections describe some of the films teenagers are making today. Other sections deal with the nature of visual language and the shooting of a movie beginning with the formulation of an idea and script, including the responsibilities of the filmmaker as his own director and cameraman.

But most important, Rodger Larson with Ellen Meade have translated into words the essentials of the filmmaking experience, helping young filmmakers understand how to use the technical means at their command to realize expressive ends. E. P. Dutton & Co. $6.95.

PHOTO: ALFONSO BARR

Be a Hollywood producer no matter where you live. Make a movie—live action, animation, or videotape. How do you plan a film? Choose a camera? Pan, tilt, dolly? How do you edit? Make titles? The answers are here. 8mm or video. Lights, camera, action!

Filmmaking for Beginners

Joan Horvath, who wrote this excellent book, is an award-winning film director, producer, writer, and reviewer. She produced, directed, and wrote a series of films for "Sesame Street" and has had her films shown at the White House Conference on Children.

Filmmaking for Beginners includes all the information you need to get started. It is an easy-to-understand introduction to the subject which particularly stresses the need to use imagination and self-expression in this fast-growing, fun-filled hobby and occupation.

How do you plan a film, write a script?
How do you choose a camera?
What film is best for a beginner?
How do you light a scene, choose an exposure?
What is a cutaway shot, a pan, a tilt, a dolly?
How do you use a telephoto or zoom lens?
How do you edit a film or add sound?
What are the tricks for animation or for special effects, such as slow, fast, or reverse motion, jump cutting, time-lapse photography, and superimposition?

These and many other questions about the techniques and creative aspects of movies are clearly answered in this book. The author tells how to compose your shots, direct your actors, and get weird effects. She points out the importance of matching direction of your action when shooting and of carefully blending your close-ups, medium shots, and long shots when editing. She brings to her subject a delight that is contagious, one that should win many new youthful devotees for filmmaking.

Here is part of her first chapter, "Movies by Amateurs":

Children, from the primary grades up, are learning how to create films both for their own pleasure and to express what they feel and want to communicate to others. All over the country, children and teachers are attending workshops where they learn the basic techniques of filmmaking.... Films by children have already been shown on network television and are being distributed to schools across the country.

It is reasonable to say that film is often the language of today's child. We have been in the midst of a communications revolution in this country for a number of years. The result has been a new kind of student, with new horizons, a new language, and new tools for the expression of that language. Five-year-old children are making movies. They edit, direct, operate cameras, act, write scripts, and create sound tracks for both 8 and 16mm films.

Learning to appreciate films, learning to make films can help the child in expressing himself, in developing his tastes and critical powers. It can train him to see with a perceptive eye. Making films can introduce the youngster to new experiences that are pleasing and enriching. His social growth can also be enhanced as he learns how to work in production units with others. Learning the basic skills of filmmaking and applying them in personal, creative experiments can be, if nothing else, a very special kind of fun.

For your copy SEND $5.95 plus 50 cents postage and handling to Thomas Nelson, 30 E. 42d St., New York, N.Y. 10017.

Want Your Own Film Distributed?

More and more people want to see and enjoy the motion pictures (live action and animated) that young people are making today. And now there is a distribution center for them: the Youth Film Distribution Center is a nonprofit agency and the major source for the rental and sales of youth-made films. Money made by the Center goes to workshops or individuals who produced the films. The basic purpose is to support new opportunities for young filmmakers. If you are interested in having your film distributed and want to see a catalog WRITE TO Youth Film Distribution Center, 43 W. 16th St., New York, N.Y. 10011.

The Yellow Ball Workshop

Yvonne Andersen, author of *Make Your Own Animated Movies,* has taught at Project, at the Newton Creative Arts Center, and at the Cellar Door Cinema. She and her husband, Dominic Falcone, now direct the Yellow Ball Workshop, where Miss Andersen has supervised the creation of over 200 films by children, some of which have won national film festival awards. Acclaimed for their striking simplicity, basic naturalness, humor, and poetry, these experimental films have been shown in schools and libraries all over the country. A 13-minute film about the Yellow Ball Workshop called *Let's Make a Film* is available. For further information about the Workshop and its films WRITE TO Yellow Ball Workshop, 62 Tarbell Ave., Lexington, Mass. 02173.

Making Films at the Yellow Ball Workshop

Some thoughts on children and filmmaking by Yvonne Andersen:

Film animation is an important new form of artistic expression. It combines the graphic arts of painting, drawing and sculpture with the art and techniques of theatre and film. The animated film is most often associated with humor, but it can also be used to express joy, terror, pathos, tenderness and social comment. Our students are encouraged to consider it first as a means of making their paintings and drawings come magically to life.

To achieve this, the film animator needs only his own artwork and some basic film equipment. He need not have actors, scenery, technical crews, makeup, or any of the cumbersome expensive things needed in live-action filming. He can sit down in a room all by himself and draw his actors and their scenes on small pieces of paper. Then, through his own efforts, he can make them come to life on the screen.

A large part of our work at the Yellow Ball Workshop has been with young people. This has been exciting. We have had fun, and we have found our students to be serious, hard-working artists who have important contributions to make. People of this age have special qualities. Because they are "new" people, they can see things in a new way, making interesting and important social comments and inventing marvelous new techniques in which to work.

Make Your Own Animated Movies

Since 1963 young people ages 5 to 19 have been making their own cartoon movies at the famous Yellow Ball Workshop in Lexington, Massachusetts. The director of this workshop, Yvonne Andersen, has created a book to show and tell children (or adults supervising filmmaking activities) all they need to know to make successful animated movies.

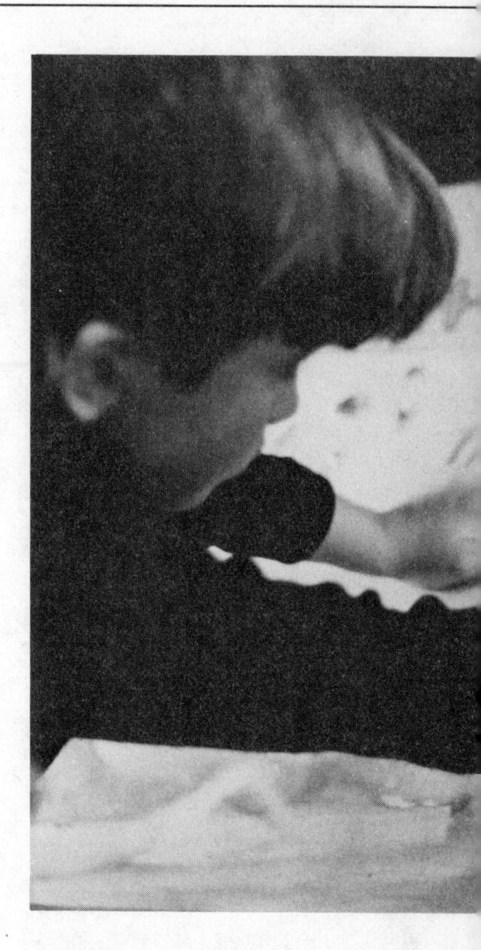

Information about equipment and supplies and hundreds of valuable tips tested by experience are packed into this fully illustrated handbook. How to set up and use camera and lights for animated films, the fine details of positioning, shooting, and film editing, how to make cutouts, clay figures, special effects, sound effects, and how to use the techniques of drawing on film or making "pixillated" films are included, with scenes from Yellow Ball Workshop films to illustrate the methods described. *Make Your Own Animated Movies* is a practical guide to an exciting and rewarding activity. Little, Brown and Co. $6.95.

Young Animators

This book is designed for young people who want to learn how animated films are made. There are instructive notes on how to make an animated movie with 8-mm film, and you will learn even more from the reports of 12 young filmmakers who share their discoveries and experiences with you. They tell you, in their own words, how they made their own animated movies.

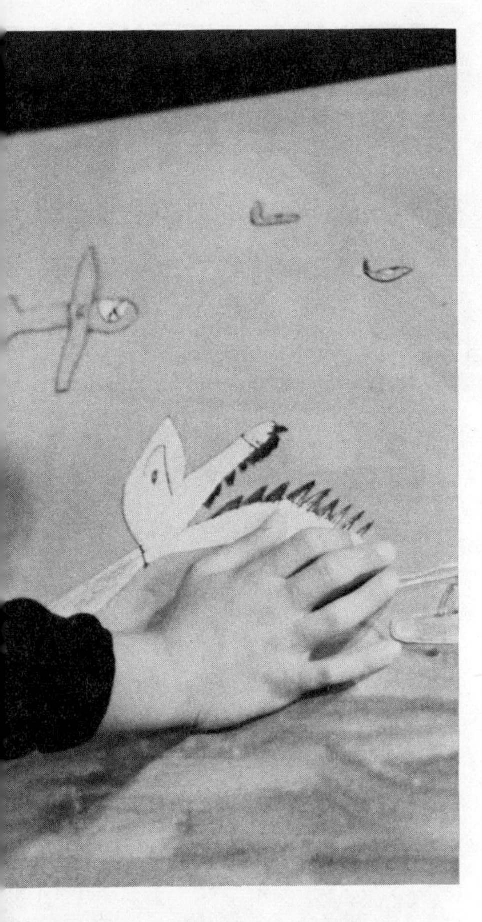

Many of them started making animated movies when they were 12 years old, often using their bedrooms and kitchens as their studios. You'll find out how you can communicate ideas and express yourself by using materials you can manipulate. Says Rodger Larson, executive director of the Young Filmmakers Foundation, the group that produced this book: "The phenomenon of young people in command of their own film production will, in the decade ahead, surely change the course of filmmaking." If you are interested in making an animated film, *Young Animators: And Their Discoveries* is an inspirational book. Praeger Publishers. $6.50.

Making Movies

With this book, a camera, and film, you can become a moviemaker as well as a moviegoer. *Making Movies: From Script to Screen* is designed to help fledgling moviemakers of all ages. In easily understandable language it explains the complete production process, broken down into manageable steps, spelling out the theory and practice of each step. Everything is here from concept to storyboard, from preparation for production (including budgeting) to direction, cinematography, sound recording, editing, and postproduction sound. You learn how distribution works and about careers in filmmaking and how to achieve one.

There are over 75 illustrations and a color insert illustrating innovations in color cinematography. Each chapter has been written with experts; technical points are reinforced by excerpts from successful scripts and interviews with leading directors.

Whether you aim for a feature-length film or a ten-minute home movie, an animated film, a documentary, or a television commercial, this is the book you need. By Lee R. Bobker with Louise Marinis. Harcourt Brace Jovanovich. $10.00 cloth, $6.95 paper.

Young Filmmakers of Westchester

Two young teachers of filmmaking in the New Rochelle, New York, elementary schools, Alice Mintzer and Michael Feller, have recently started an exciting filmmaking workshop for young people, ages 8 to 15. Each student completes his/her own super 8-mm animation or live action film. Classes are informal. No previous experience is necessary. For further information WRITE TO Young Filmmakers of Westchester, 201 S. Lexington Ave., White Plains, N.Y. 10601. The telephone number is (914) 428–2190.

To find a "filmmaking workshop for young people" where you live, ask your teachers at school, ask at your local camera store, or show these pages about filmmaking to some of your teachers and maybe they'll get interested and start one for you.

The Spaghetti City Video Manual

This well-organized, clearly written, cartoon-illustrated handbook gives advice on the use, repair, and maintenance of video systems and hardware. *The Spaghetti City Video Manual* includes practical discussions of the video camera, monitor, and videotape recorder (VTR), audio, synchronization, playback, editing and transfer, cables, termination, AC and DC power sources, preventive maintenance, and video technique. This handbook, which is invaluable to the nontechnical reader, is by Videofreex, an innovative group in Green County, New York, that is concerned with the educational uses of video. Praeger Publishers. $7.95 cloth, $4.95 paper.

The Electric Journalist

The author, Chuck Anderson, is a high-school teacher who has written extensively on media in education. *The Electric Journalist: An Introduction to Video* is an illustrated manual on television and videotape aimed at high-school students. The book discusses the basic principles of filmmaking; how the videotape recorder system works; editing tips; ideas for video productions; interviewing and street shooting; the video underground; the medium as a social conscience; how TV affects our lives; and the future possibilities of video, particularly cable TV and public access. Praeger Publishers. $6.50.

PHOTOGRAPHY

The Amateur Photographer's Handbook

When you reach the point of no longer being satisfied by little booklets of advice on photography . . .

When you reach the moment when you want the next pictures you take to be better pictures than the last ones you took . . .

When you say to yourself, "I want to learn about photography and the different things I can do with a camera" . . .

Then you're ready for *The Amateur Photographer's Handbook* (8th ed.), by Aaron Sussman, a book you can grow with and use all your life.

Some people call it the photographer's Bible. Others have commented: "An amateur's classic . . . one of the most practical guides around"; "From first chapter to last, this is one of the best written, best all-around books for the amateur to own."

This completely revised and enlarged book contains over 550 pages with 188 plates in color and monochrome; 140 diagrams and a chart; 53 tested formulas; and 33 tables.

The chapters include "What the Lens Does"; "The Mystery of 'f' "; "How the Shutter Works"; "What Camera Shall I Get"; "The Picture People and Closeups"; "Developers and Developing"; "Printing and Enlarging"; and "Fun with Your Camera."

Here are a few words from the foreword:

> Photography is more than a means of recording the obvious. It is a way of feeling, of touching, of loving. What you have caught on film is captured forever, whether it be a face or a flower, a place or a thing, a day or a moment. The camera is a perfect companion. It makes no demands, imposes no obligations. It becomes your notebook and your reference library, your microscope and your telescope. It sees what you are too lazy or too careless to notice, and it remembers little things, long after you have forgotten everything.

If photography is your hobby, this book should be on your bookshelf. Thomas Y. Crowell Co. $8.95.

Through the Seasons with a Camera

Here's what Paul Villiard says in the preface to *Through the Seasons with a Camera:*

> Not being able to see the things around you is nothing to be ashamed of. Most people have this blindness. But those who do *not* have a blind area and *do* see the things around them are never bored. They are never really alone or lonesome. Life to them is an ever changing panorama of wonderful adventures. They learn the answers to questions, and the more they learn the more they want to know, the more they become interested in life on this wonderful planet.

He tells about the wealth of photographs that can be taken through the seasons, and gives tips on how to arrange subjects, what cameras to buy in almost every price range and how to adjust them, focus them, and set them to take the best possible pictures. The book should help every young naturalist-photographer to see things outdoors as he or she has never seen them before. Doubleday & Co. $3.95.

It's So Simple, Click and Print

Says the catalog of the Workshop for Learning Things: "A book written for us by a sixth grade class who responded to our need for a guide to classroom photography by saying: 'We'll do it for you!' and they did . . . and here it is. Filled with their drawing and photographs, it's a real help to others getting started on their classroom camera adventure. All the information you need for picture taking, film developing, and printing . . . as described by a class that did it all." For your copy of *It's So Simple, Click and Print,* SEND $2.00 to Workshop for Learning Things, 5 Bridge St., Watertown, Mass. 02172.

See your world through a lens! A camera is a memory book that never forgets. It makes you see things you never saw before. Learn the mystery of "f," the secrets of the shutter, the right camera for you, how to develop your own pictures. Ready? Click!

How to Photograph Your World

Have you noticed the new tree on your block? Have you seen any big white clouds recently? Have you ever really looked at your family, your friends, your pets? Stop, for just a minute and look around—carefully. You will be surprised. You are surrounded by interesting people and things, all of them just waiting to be photographed.

In *How to Photograph Your World,* Viki Holland tells you in words and photographs how to select the best camera for you and how to use it. You learn what types of light produce certain effects, how you can plan ahead to get the shot you want, how to compose the picture, and how to use your camera to tell a story. But perhaps most important, you discover new ways to look and see and understand the world around you—your world. Charles Scribner's Sons. $5.95.

Writing with Light

This is a simple workshop in basic photography. It is a how-to book which makes you really want-to. The author, Paul Clement Czaja, claims photography is a "celebration of your seeing," and it is with seeing and with light that he is concerned. The first photographic experiments he recommends require no special equipment, but aim at perceiving light and its relation to objects as they are seen. After experimenting with shadowgraphs on slow photographic paper, you advance to the use of a camera—but one you make from a rubber ball.

Now you are able to produce photographic prints without having made any greater investment than a rubber ball, photographic paper, and a few chemicals, but you have a firm basic understanding of photography. The equipment and instruction become more sophisticated, but throughout, *Writing with Light* is concerned primarily with promoting understanding and appreciation. A fine selection of photographs illustrate the author's concept of photographic seeing. Chatham Press. $5.95.

A Manual for Shutterbugs

It's 64 pages short, but it covers everything from cameras and light to choice of film, from what's on the film to tricks with your camera. *Photography: A Manual for Shutterbugs,* by Eugene Kohn, is probably the fastest course on photography ever published. Prentice-Hall (Treehouse Paperback). $1.25.

Invitation to Music

"The best way to learn about music is to listen to it. But mere listening is not enough. It is *how* you listen that is important." With these opening words Elie Siegmeister, eminent composer, provides us with the "how" of listening to music to get the greatest degree of enjoyment. Supplying enough history to enrich our understanding of how music has evolved since early times, the author acquaints us with its development from early chants and primitive drumbeats to the complexities of classical and modern forms. Included is a thoughtful and appreciative chapter on the evolution of jazz as a significant part of modern expression.

Invitation to Music is a good, basic book for the young person just coming to an interest and curiosity about the world of music. Harvey House. $7.75.

Sing, Children, Sing

This UNICEF songbook (edited by Carl S. Miller) contains songs, dances, and singing games from 34 countries, with pictures of children from each country. The original words of the songs are given, where possible, along with a singable English translation—and, of course, the music. The introduction is by Leonard Bernstein. For all ages. (Published by Chappell & Co. by arrangement with UNICEF.) To get your copy SEND $3.50 plus 25 cents for postage and handling to UNICEF, 331 E. 38th St., New York, N.Y. 10016.

The 50 Cent Guitar Book

A wildly hilarious, way-out, down-to-earth comic book put together by Bob Davis, who teaches guitar at the San Francisco Conservatory, and cartoonist John Adams, a teacher at the University of Colorado. And as the cartoon at the beginning of the book says: ". . . just let me say that a guitar is man's best friend! . . . Well, at least a true friend!" Get *The 50 Cent Guitar Book* and liven up your talents. Flash Books (an associated company of Music Sales Corp.). $3.95.

Nonsense Songs

Sing Song Scuppernong is a collection of 16 nonsense songs. Jeanne B. Hardendorff, the author, learned some of them during her childhood in Mississippi and Tennessee; others she learned from friends. The songs are always fun to sing; all of them, from "The Tale of the Burglar Bold" and "Zizzy, Ze Zum, Zum" to the song about "the girl who said she wasn't hungry but who managed to put away, if not a horse, then everything short of it," will delight you.

The songs may be sung by a group, with one other person—even alone, to cheer yourself up. All the songs have withstood the test of time, providing fun and entertainment over the years. Jacqueline Chwast's paper-cut pictures add delightful fun to the songs. Holt, Rinehart and Winston. $5.95.

Play a guitar, a piano, a violin. Pick the musical instrument that's right for you. Learn all about music from early chants and savage drums to 18th-century classical and 20th-century jazz. Sing loud or soft, but sing, sing, sing!

Want to Play a Guitar?

With *Beginning the Folk Guitar* a bewildered beginner will be able to have a sound foundation in basic techniques and be able to strum, hum, and sing such verses as: "Queen Elizabeth fell dead in love with me, we were married in Milwaukee secretly, but I snuck around and shook her, to go with Hooker to fight mosquiters down in Tennessee." Well-known folk guitar instructor Jerry Silverman not only tells you how to play and accompany yourself with delightful songs such as "Abdullah Bulbul Amir," "So Long, It's Been Good to Know You," and "Frankie and Johnny," he also gives helpful information on buying a guitar. Oak Publications. $3.95.

Sing a Song!

Sing It Yourself: Folk Songs of All Nations is a delightful illustrated songbook edited by Dorothy Gordon, who says, "Sing and you'll always be happy!" Most of the songs in this collection are folk songs, simple songs that sometimes tell stories and sometimes just express the feelings of the people who sing them. They are passed down from generation to generation; the ones that last the longest are usually the most fun to sing. Have you ever sung "Frog Went a-Courtin'"? Do you know all the verses? Here they are: all 16 of them! There are other favorites from America, Russia, Norway, England, France, plus songs from the American Indians. E. P. Dutton & Co. $4.50.

Understanding Musical Instruments

This book will help you select the musical instrument that is right for you. Jack Levine and Takeru Iijima, who are music teachers, describe the orchestral instruments—as well as some not found in the orchestra—and tell how each one is played. There are many illustrations to show how each instrument is held and a list of recordings featuring the individual sound of instruments.

If you are a would-be musician, or if you would like to learn more about musical instruments, *Understanding Musical Instruments—How to Select Your Instrument* is an excellent basic book for your library. Frederick Warne & Co. $3.95.

The Fireside Book of Fun and Game Songs

Group singing is great fun. Fun and game songs bring about this feeling like no others: these are songs that have activity and humor, and you can sing them around a campfire or coming home from the beach.

The Fireside Book of Fun and Game Songs contains over 110 songs. In part I, "One-Thing-After-Another," are the cumulative songs, the ones that keep adding things till by the end of the song your memory is taxed. Follow-the-leader songs are when you echo a phrase. Clapping snapping and making peculiar noises songs are full of motion and wordplay; and there are tales and rounds, and hello good-bye and hooray for us songs.

These songs for singing, songs for fun were collected by Marie Winn, with musical arrangements by Allan Miller and illustrations by Whitney Darrow, Jr. Simon and Schuster. $12.50.

Baxter's Kids' Guitar Manual

Bob Baxter gives you all the details on self-teaching, easy chords, plenty of well-known songs, and encouraging words to keep you going until you can strum and sing "Aunt Rhody," "Clementine," and "Blow the Man Down" with gusto —and the right chords! *Baxter's Kids' Guitar Manual* is published by Amsco School Publications. $2.95.

HOBBIES

From Autographs to Toy Banks

What do the following have in common: autographs, buttons, calendars, cartoons, coins, greeting cards, menus, playing cards, road maps, and toy banks? Answer: They are all things that people collect. Turning a haphazard collection into an organized hobby is fun, and anyone can do it. This book is a complete and stimulating guide to collecting. Collector's clubs, reference works, and books for further exploration are given for each hobby.

If you want more information get a copy of *Hobby Collections A-Z*, by Roslyn W. Salny, with illustrations by Robert Galster. Thomas Y. Crowell Co. $4.50.

Hobbies for Here & Now

Designed for the teen-ager who wants to learn about contemporary hobbies, *The Teen-Ager's Guide to Hobbies for Here & Now,* by Norah Smaridge and Hilda Hunter (with illustrations by Charles F. Miller), covers a wide range of possibilities from fly-tying to bird watching, from woodcarving to candlemaking. Limited to action hobbies, it's an excellent book because it supplies practical information: cost of tools and materials; amount of work space necessary; degree of talent or ability needed for success; and moneymaking potentialities of the hobby. It also directs the reader to specialized sources of information. Dodd, Mead & Co. $4.95.

Rock of the Month Club

Are you a rock hobbyist? Join the Rock of the Month Club. The dues are $2.00 a month and you can drop out anytime. Each month members are given a choice of cutting materials such as lace, moss agates, petrified wood, and geodes, or of crystal specimens such as fluorite, calcite, and selenite. Every 2 months you receive a membership newsletter with news of member activities, rock shows around the country, materials wanted by members, and a list of rock specimens for sale. If you want to build a collection of rocks, learn about field trips, trade or swap, meet other people who share your interests, or purchase rocks at a discount, WRITE Art Tatum, President, Rock of the Month Club, 3901 Pershing Drive, El Paso, Tex. 79903, for further information.

Collect stamps, rocks, autographs, coins. Build a model plane. Run a model railroad. Construct a radio. Write a story. Make a book. Enjoy doing everything in this catalog.

Stamp Collecting

Stamp collecting has always been a popular hobby and it's more popular today than ever. Stamps have become more interesting: once they showed only presidents, kings, queens, national heroes, and monuments; today they picture many subjects: animals, insects, maps, flags, science, people, historic events—you name it. And stamps have value! Every genuine stamp ever printed is worth cash. And the rarer the stamp, the more it's worth. Stamps tell you about people and events that made history.

Different Kinds of Collecting

Worldwide Collecting: The worldwide collector doesn't specialize, but accumulates stamps from every country. The fun is getting as many stamps as possible from each country, and learning about it from the stamps it issues.

Single Country Collecting: Many collectors specialize in collecting stamps from one country: the U.S., Canada, Germany, England, Israel, and France are among the most popular, along with Russia, China, and Japan. These have all issued many stamps, which makes collecting them a long-term adventure. Interesting collections can come from smaller countries too.

Topical Collecting: Did you know that the countries of the world may have issued hundreds of different stamps on your favorite subject? There are sports stamps, animal stamps, flower stamps, astronaut and space stamps, all kinds of stamps relating to specific topics. The topical collector picks a subject, then collects all the stamps he can find relating to it.

Specialized Collecting: Some collectors buy only airmail stamps; others, stamps of a certain color; or plate blocks (four or more stamps which have not been torn apart, and which contain a plate number in the margin); or first day covers (envelopes whose stamps are canceled on the first day of issue by the local post office where something historical pertaining to the stamp subject took place); or postal stationery (envelopes, postal cards, and letter forms that have the postage printed on them).

Get into Collecting

When it comes to getting into collecting, hardly anything can beat starting with a stamp-collecting kit. You can get these kits from Scott Publishing Company or from your local U.S. post office.

Scott Stamp-Collecting Kits

Scott puts out 16 Stamp-Collecting Kits. Each one is a topical kit, containing stamps related to a specific subject plus everything you need to begin collecting.

Stamps: From 30 to 100 all different, all genuine postage stamps illustrating different aspects of the topic.

Album Pages: 16 colorful, illustrated album pages (punched to fit a three-hole binder) with spaces for your stamps and interesting facts about them.

Stamp Hinges: An ample supply of stamp hinges to use for mounting your stamps on the album pages, prefolded to make them easy to use.

Collecting Tips: A special page that tells you how to handle stamps, identify them properly, and assemble your collection.

A Special Bonus Stamp: A giant mystery stamp (some are as large as playing cards) from a wonderful, far-off place like Ras al Khaima or Abu Dhabi.

These are the Stamp-Collecting Kits available from Scott:

Wild Animals
Mysterious Africa
British Worldwide
U.S. Commemoratives
Art Treasures
Birds of Land and Sea
Undersea World
Championship Pets
Blast Off into Space
Sports Action
Exciting Olympics
Uniforms and Costumes
Colorful Flowers
Nations of the World No. 1
Nations of the World No. 2
Butterflies and Insects

They're available at many stores, or you can order them by mail; WRITE TO Scott Publishing Co., P.O. Box 208, Floral Park, N.Y. 11002. Enclose a check or money order for $6.00 per kit plus 50 cents for postage and handling.

Scott Standard Postage Stamp Catalog 1976 (4 vols.): Vol. I (stamps of U.S. and former possessions, U.N. member nations, Great Britain and the Commonwealth), $13.00; Vol. II (world stamps, alphabetically A-I, excluding those in Vol. I), $15.00; Vol. III (world stamps, alphabetically J-Z, excluding those in Vol. I), $15.00; Vol. IV, *1976 Scott Specialized Catalog of U.S. Stamps* (in-depth listing of all U.S. stamps), $13.00.

If you can't find these books at your local stamp dealer WRITE TO Scott Publishing Company, P.O. Box 208, Floral Park, N.Y. 11002.

Scottalk, the Stamp-Collecting Newspaper for Kids

Scottalk is a stamp-collecting newspaper that comes out six times a year and is written and edited by kids. They're experienced collectors, so they fill *Scottalk*'s pages with fascinating and informative articles, useful collecting tips, entertaining stamp games, puzzles, and lots more. It's a great newspaper, well worth subscribing to. If you want to get into stamp collecting and want to learn what's what in collecting from people your own age, subscribe to *Scottalk*. For a year's subscription SEND $1.00 plus 25 cents for postage and handling to Scottalk, P.O. Box 208, Floral Park, N.Y. 11002.

The Teen-Ager's Guide to Collecting Practically Anything

Aimed directly at the teen-ager who wants to make a worthwhile, long-term collection, this book keeps budget and capabilities in mind. It suggests only collectibles that can be obtained at little cost and that are likely to increase in value. Answering the question "Can a teen-ager collect genuine antiques?" *The Teen-Ager's Guide to Collecting Practically Anything,* by Norah Smaridge and Hilda Hunter (with illustrations by Charles F. Miller), points to exciting possibilities in Victoriana and "bygones," and covers fields in which teen-agers are especially enterprising—pop or junk, and salvage items.

All the tried and true collectibles are covered: stamps, coins, autographs, and items in metal, wood, ceramics, glass, paper, and so on, with emphasis on such new and fertile areas as ecology, the environment, woman's lib, and American arts and crafts. One section explains how to buy at sales; how to clean and store items; and to write up a collection and display it. The last chapter shows how the tenn-ager can put his collectibles to novel but practical use as dress accessories, or in his room, home, or yard. Dodd, Mead & Co. $4.50.

All States Hobby Club

The membership fee of $2.00 per year includes six issues of *All States Hobbyist,* the magazine that brings hobby news from coast to coast. For a membership application WRITE TO Tom Beam, All States Hobby Club, 101 Chestnut Hill Lane, E., Reisterstown, Md. 21136.

Your Local Post Office Is a Stamp Collectors' Treasure Trove!

You're familiar with your local post office as the place to buy stamps to mail a letter or package; it's also the place to visit if you're interested in stamp collecting. Uncle Sam's Postal Service runs a major chain of retail stamp-collecting centers all over the country—10,000 of them, in local post offices.

USPS Stamp-Collecting Kits

The post office has 11 Stamp-Collecting Kits available. They are mostly topical; each one is chock-full of genuine U.S. and foreign postage stamps. Each exciting Stamp-Collecting Kit contains the stamps; a colorful 20-page album with illustrated stamp spaces; plenty of prefolded stamp hinges to use for mounting your stamps; and a terrific 32-page booklet, *The ABC's of Stamp Collecting,* which tells you everything you need to know about how to start a fine stamp collection.

Here is the list of Stamp-Collecting Kits:

- Space (2d ed.)
- Flowers
- The World of Sports
- Travel Through the Ages
- The Animal Kingdom
- Birds and Butterflies
- Masterworks
- Diamonds and Triangles
- 50 Stamps from 50 Countries
- Flags, Maps, and Coats of Arms
- United States (2d ed.)

Available at your post office. $2.00 each.

Stamps & Stories

Stamps & Stories is the first book a stamp collector should get. It tells our nation's story in a most unusual way: through stamps. It contains full-color reproductions of every stamp issued by the United States (plus many from foreign countries, honoring the people and events that made our country grow). There are illustrated stories of many of the people and events shown on the stamps. It's a great way to learn American history. It's more than a storybook; it's a catalog of the stamps of the U.S., supplying issue dates and values for every stamp shown. The book is a stamp collector's manual; it contains page after page of collecting advice and suggestions, ranging from how to start collecting to how to identify stamps and how to start your own stamp club. $2.00 at your post office.

More Exciting Stamps at Your Post Office

You can buy all the new commemorative stamp issues as they come out. Also, the post office also has Souvenir Mint Sets—special portfolios of the commemorative stamps issued in 1972, 1973, and 1974—and the 1975 portfolio will be ready as soon as all the 1975 commemoratives have been issued. Commemorative stamp issues and sets are available at your post office. 1972 and 1973 sets $3.00, 1974 set $3.50. If you prefer, Stamp-Collecting Kits, *Stamps & Stories,* and Souvenir Mint Sets can be ordered by mail. SEND check or money order payable to Philatelic Sales Division, United States Postal Service, Washington, D.C. 20265. Add 50 cents for postage and handling.

Other Things You Can Get from the Post Office

You can get the beautiful American commemorative series, a collection of panels on which blocks of four commemorative stamps are displayed in clear plastic philatelic mounts and surrounded with related illustrations and the story of the person or event commemorated by the stamp.

You can get souvenir pages (there's one produced for each new commemorative stamp). A souvenir page is an 8½ x 11 inch sheet to which the new commemorative is affixed and canceled at the originating post office on the first day of issue. It's a wonderful page to put into an album, or to frame and hang on your wall.

Almost FREE Stamp Collecting

Here is a fact- and picture-filled *Boys' Life* reprint booklet, *Stamp Collecting:* 24 pages including everything a beginner could want to know—tips on collecting; the story back of the stamp; stamps related to many topics; quotations on stamps; philately—a many-faceted hobby; books about stamps; and so on. For your copy SEND 50 cents to Boys' Life, North Brunswick, N.J. 08902.

Model Railroading

Here is a *Boys' Life* reprint with 24 pages jam-packed with photographs and diagrams that show you an HO model railroad you can build; wiring your HO railroad; HO freight yards; building an engine terminal; how to make rock formations and mountains; how to build a farm supply store; a concrete loading platform; a timber trestle; and so on. This is a great booklet for model railroaders. To get your copy ask for *Model Railroading* (No. 26–030), by Glenn Wagner, and SEND 50 cents to Boys' Life, North Brunswick, N.J. 08902.

Model Railroading Books

Here are some books on model railroading that will intrigue you.

Model Railroader's popular "Dollar Car Projects" series is now a book, *Easy-to-Build Model Railroad Freight Cars.* Here are 24 simple-to-detailed scratchbuilding projects all made easy with plans, diagrams, model and prototype photos, and material lists. $3.00.

N Scale Model Railroad Track Plans shows exactly how many track pieces you need for each plan, and how large a space the plan will occupy. Full details on scale and gauge, radius and degree are included. Edited by Russ Larson. $2.50.

Stations, enginehouses, platform sheds are all a part of the real railroad scene. Complete your pike with prototypical accuracy by using the how-to hints and drawings in *Easy-to-Build Model Railroad Structures.* Includes scale rules. $2.00.

Basic data for everyone's needs from table construction to rolling stock to simplified wiring is contained in *HO Primer,* the beginner's book that makes model railroading a lifetime hobby. $3.50.

Eight-stage construction plan book *HO Railroad That Grows* starts you with a simple 4 x 8 foot layout which you can develop into a two-level, two-train, scenicked pike. List of materials included. $3.00.

Plans for eight small pikes are featured in *Small Railroads You Can Build,* a how-to-do-it account by leading hobby track planners. Construction is explained for railroads as small as 4 x 6 feet and up to 5 x 10 feet. $2.00.

Easy-to-read diagrams in *How to Wire Your Model Railroad* unravel wiring mysteries. This book instructs in the wiring of simple to complex model railroad track plans, including cab control, switches, power packs, block signals. $3.50.

Linn Westcott, editor of *Model Railroader,* tells how to choose your track plan, how to build, how to change plans to suit changing needs, and gives a fine selection of N, TT, HO, S, and O scale plans in *101 Track Plans for Model Railroaders.* $3.00.

Track Plans for Sectional Track has 144 track plans including lists of pieces for rug, table, and custom layouts, with design instructions for building better track plans. HO, O-27, S, and O gauges. $2.00.

In chapter 1 of *Scenery for Model Railroads,* Bill McClanahan asks, "Why scenery?" The answers are in nine more chapters on planning, railroad geology, materials and techniques, zip texturing, and hard-shell scenery. $4.00.

Flying switches and flying junctions, the coefficient of lurch, and the mischief of curves, lap sidings and gantlet tracks–there's a complete repertory of prototype railroading described for the modeler in *Track Planning for Realistic Operation.* $3.50.

Have you ever stopped to puzzle out some signaling problem on your pike, or tried to figure out some aspects on the prototype line in your area? Then *All About Signals* is for you. It explains fully everything from primitive manual block to today's complex CTC. $2.00.

In ordering a book be sure to list its exact title and include 50 cents for postage and handling. WRITE TO Kalmbach Books.

Collecting Stamps

A millionaire can go out, spend a quarter of a million dollars, and come home with one stamp. A young person, however, can go out —maybe even to the same stamp dealer—spend a quarter, and come home with over a hundred stamps. The beauty of stamp collecting is that it is only as expensive as you want it to be, and it is within reach of every youngster in the world. In *Collecting Stamps,* Paul Villiard discusses basic topics for the beginning collector: symbols used in collecting; ways to collect stamps; stamp identification; types of collections; rarities; gimmick stamps; and a glossary. Completely illustrated, this is a fine beginning to a rewarding hobby. Doubleday & Co. $5.95. Paperback from Signet (an imprint of New American Library). $1.95.

A Child's World of Stamps

A delightful book with full-color enlargements of postage stamps from around the world that illus-

trate a fascinating collection of tales, customs, recipes, and verse. One stamp showing youngsters building a snowman is captioned "Winter Joys"; a United Kingdom Christmas stamp is captioned "Boy with Toy Train"; and a stamp from Singapore that celebrates Hari Raya, a joyous religious holiday, is captioned "Children Flying Kites." Age 9 and up. The publisher is Parents' Magazine Press in cooperation with the U.S. Committee for UNICEF. For your copy of *A Child's World of Stamps: Poems, Fun and Facts from Many Lands,* by Mildred DePree, SEND $4.95 plus 25 cents for postage and handling to UNICEF, 331 E. 38th St., New York, N.Y. 10016.

Junior Philatelic Society of America

Membership dues are $2.00 per year. Applicants must be 21 or younger; if younger than 17, parents' permission is required. Members receive the *Philatelic Observer,* the Society's bimonthly publication designed to aid young stamp collectors, which features a pen-pal department that supplies members with names and addresses of junior stamp collectors in every country in the world. For full information and an application WRITE TO Junior Philatelic Society of America, Central Office, P.O. Box 9634, Midtown Plaza Sta., Rochester, N.Y. 14604.

Dairy Farm Panorama Kit

Want to visit a dairy farm without leaving home? Write for this fabulous kit. You get a large full-color wall panel (40 x 27 inches) that shows a dairy farm in action, a phonograph record that has all the sounds of a farm, and a story about the children who live there. There are 15 large (15½ x 9½ inch) black-and-white photographs that help you "see inside" the farm buildings. Pictures include "In a Milking Barn"; "Feeding Calf"; "Putting Hay into the Barn"; and "Cows Drinking Water." (A mother-teacher guide comes with the kit too.) The full-color dairy farm wall panel will make a beautiful decoration for your room. For your Dairy Farm Panorama Kit SEND $3.50 to National Dairy Council, 111 N. Canal St., Chicago, Ill. 60606.

Urban Panorama Kit

When you open the kit the first thing you'll see is a very large full-color poster of a city, almost 3½ x 3½ feet. (Besides learning about all of the things in a city, you can use this colorful poster as a background for your model railroad or model car set.) And it will all fit in, for trains and cars are part of the city too.) There's a record for you to play: the story of a city boy named Johnny and the different sounds he hears and people he meets. More pictures come with the kit: 16 large black-and-white sketches (14½ x 10¼ inches) of city buildings and scenes like an ice-cream vendor at the zoo, the interior of a fire station, the post office, and taking out books from the public library. A mother-teacher guide offers help to give youngsters who are too young to read by themselves. To get the Urban Panorama Kit SEND $3.50 to National Dairy Council, 111 N. Canal St., Chicago, Ill. 60606.

Short-Wave Listening

This *Boys' Life* reprint can literally open up the whole world to you through short-wave listening and ham radio. Says Ken Boord in the opening article:

> From the results of SWLs [short-wave listeners] whom I know, more interest is needed than investment! A friend of mine in Brooklyn, N.Y., has brought in real DX from all corners of the globe on a 6-tube, 15-year-old Philco table set, even though he's in a noisy reception area. He has verified such countries as Mozambique, Thailand, Indochina, Hong-Kong and the Fiji Islands.

The reprint includes sample log sheets, SWL cards, U.S. and continent call area checklist, Boys' Life Club certificate, and the requirements for seven Radio Club awards. For a copy of *Short-Wave Listening* SEND 50 cents to Boys' Life, North Brunswick, N.J. 08902, and be sure to state that you are ordering reprint No. 26–091.

There's a Mint of Fun in Coin and Currency Collecting

According to the U.S. Bureau of the Mint there are 10 million numismatists in the U.S. today. What makes coin and currency collecting such a popular hobby? First there's the fun and excitement you get from organizing and expanding your collection. Then there's the thrill that comes with suddenly discovering and acquiring that one special coin you've been searching for, or learning that one of the coins in your collection is rare and valuable.

There are many kinds of collecting: some people collect only ancient coins (the world's first coin was struck in the 8th century B.C. and there are a lot of old coins around; note that the first postage stamp wasn't issued until 1840). Others collect coins only of a certain century, from a certain country—or of a specific denomination, such as dimes, pennies, shillings, pound notes, or dollars.

What do you do with currency or coins as you collect them? Some people mount them in coin albums; others tuck them into paper or clear plastic envelopes made especially for the purpose. There are "you-do-it" coin holders which slip into pockets on plastic pages that fit into regular three-ring binders. You'll find them all at your local coin shop.

How do you get into coin collecting? Go to your local coin shop. Some dealers sell coins by the pound at surprisingly low prices. A pound of coins will often contain a lot of duplicates, but it'll give you a good start, and set you up with coins you can trade with other collectors for coins you want. Later on you can branch out and start buying specific coins to fill in a certain series. There are all kinds of places to find coins: coin stores, coin expositions, antique stores, flea markets—even in the ground. The pages of coin publications, such as *Coin World,* the coin hobbyist's weekly newspaper, are filled with ads from dealers offering coins by mail order.

Coins come in three classifications: circulated, uncirculated, and proof. Circulated coins have been used and show some degree of wear. The value of even a rare used coin decreases in proportion to the amount of wear it has experienced. Uncirculated coins are in mint condition; they've never been used as money. Proof sets are coins, usually commemorative coins, that are specially struck for collectors.

Catalogs are important to the currency and coin collector. You use a catalog to identify coins in your collection and to determine their current value. Among the best available:

Standard Catalog of World Coins, by Chester L. Krause and Clifford Mishler, lists coins from virtually every country, providing all numismatic data plus current values for a range of coin conditions. It is revised and updated annually to incorporate new issues and new valuations.

Scott Catalogue & Encyclopedia of U.S. Coins, by world-famous coin expert Don Taxay, lists every U.S. coin minted from colonial times to the present. This book is also revised annually.

Scott Standard 1976 U.S. Coin Catalogue, a brand-new presentation of up-to-date coin prices together with Scott catalogue numbers. The latest buy and sell prices and mint records of all U.S. coins, 1973 to date. Included are gold coins and silver commemoratives. Fully illustrated.

Scott Standard 1976 Paper Money Catalogue. This fully illustrated book provides you with an accurate record of all U.S. paper money buy and sell pricing from 1861 to date.

You can find these four books at any coin shop or, if you want to order them by mail, write to: Scott Publishing Company, P.O. Box 208, Floral Park, New York 11002. For the *Standard Catalogue of World Coins,* SEND a check or money order for $12.50 plus 75 cents for postage and handling; for *Scott's Catalog and Encyclopedia of U.S. Coins,* SEND a check or money order for $15.00 plus 75 cents for postage and handling; for *Scott's Standard 1976 U.S. Coin Catalogue,* SEND $1.25 plus 25 cents for postage and handling; for *Scott's Standard 1976 U.S. Paper Money Catalogue,* SEND $1.25 plus 25 cents for postage and handling.

FREE Electronics Catalog

Radio Shack's 164-page catalog describes the company's complete line of products for home entertainment, for hobbyists, and for experimenters, featuring everything from calculators to stereo sound systems. For the young experimenter there are all kinds of Science Fair kits: the AM Radio Kit, Shortwave Radio Kit, FM Radio Kit, and Electronic Organ Kit. You can also get the Electronic Digital Computer Kit, Optical Lab Kit, and AM Radio Broadcasting Kit. For the future spaceman there are over 100 experiments in the Aeronautical Lab Kit. With this catalog you can start on many learning projects or develop a new hobby. You can get your catalog free from any Radio Shack store; or WRITE TO Radio Shack, Dept. R-20, 2617 W. 7th St., Fort Worth, Tex. 76107.

FREE Foreign Coins

To get you started in coin collecting, Jolie Coins will send you five coins now being used in Austria, Finland, Japan, Peru, and Yugoslavia. You'll also get a free eight-page catalog. You'll learn about scarce U.S. coins that are worth many times their face value (maybe you have one now and don't know it). For your free foreign coins WRITE TO Jolie Coins, P.O. Box 50 FP, Brooklyn, N.Y. 11224. Be sure to enclose 25 cents for mailing and handling.

FREE Paper Money

That's right—free money! The paper money of China, Hong Kong, and Japan ranks high in beauty and artistry of engraving. To start you on an exciting and educational hobby, Jolie Coins will send you five paper money bills from China, Japan, Hong Kong, Korea, and Yugoslavia. The artwork is beautiful: junks in full sail, exotic long-tailed birds of longevity, Chinese and Japanese temples. You'll also receive a free eight-page catalog. To get started on this hobby WRITE TO Jolie Coins, P.O. Box 50 FP, Brooklyn, N.Y. 11224, for your free paper money. Be sure to enclose 25 cents for mailing and handling.

American Numismatic Association

Junior members (ages 11 to 17) receive *The Young Numismatist* and many other benefits. Annual dues are $9.00. For further information and an application WRITE TO American Numismatic Association, P.O. Box 2366, Colorado Springs, Colo. 80901.

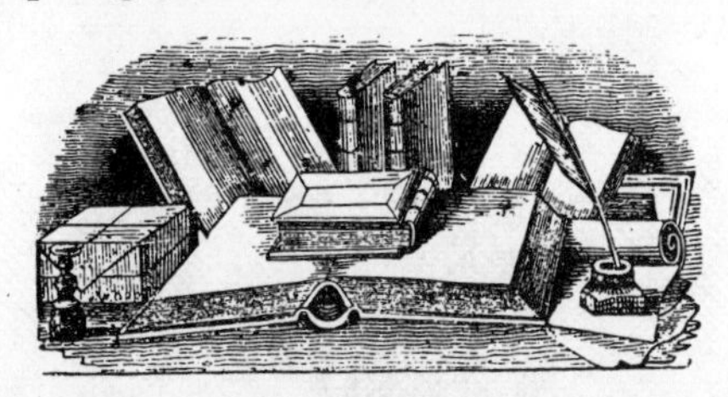

Make-a-Book

Here's a great idea for the young writer and illustrator: the new Make-a-Book series that lets you be creative—in some books as the writer; in some, as the artist who illustrates the book.

In *Giants Everywhere,* by Frances G. Scott, the pictures are already drawn and *you* make up the story—whatever you think goes with the pictures. There isn't any "right" way to do it. It's yours to do in your own way. You may find you want to make some of the writing very little, and some very big. You could make one word take up almost a whole page. Like "OH!" or "WHO?"

The Day It Snowed Colored Snow, by Ruth Cavin, is the other part of a book. When you put your pictures in the spaces left for them it will be a complete book. Read the story first. Then make any kind of pictures you like—with pencil or pen, markers or crayons, or paint. Or you can cut pictures out of magazines and paste them in.

These are fun books for the very young writer and the budding artist. Age 6 and up. Other titles of the Make-a-Book series include *Building a Clubhouse* (Carol Kropnick, illus.); *Picnic Pickles* (Ruth Cavin); *Litter, Rubbish, Trash* (Sally Brodie); *House Keep the Weather Out* (Jennie Soble); *Maybe I'll Be . . .* (Frances Scott); and *Scat Cat Finds a Friend* (June Ciancio). Two Continents Publishing Group/Sun River Press. $1.95 each.

Books for You to Make

Have you ever wanted to make your own book, to illustrate and bind a story you wrote yourself? Anyone of any age can enjoy the craft of bookmaking. It's not as difficult as you think, and you can use everyday materials. In *Books for You to Make,* Susan Purdy makes it easier with her illustrated instructions. Why not do *My Book by Me?* Discover the fun of writing, editing, and illustrating your own work; then fasten it together in one of several suggested bindings, from quick-to-make stapled pages with a paper cover to professional-looking sewn signatures in a full-bound case. Then complete your book by making and illustrating the dust jacket.

The last chapters of this guide discuss the professionally published and printed book, explaining its history, its various parts, and methods of technical production. Having created your own book, you will recognize many of the stages of this process as a manuscript and its artwork are prepared for the printer, then mechanically printed and bound, and presented to the public on publication day—the book's official birthday. J. B. Lippincott Co. $5.95.

A Printing Press?

Would you like the fun of operating your own printing press? It could be a hobby: making your own cards, personal stationery, bookplates, and so on; or you could start your own printing business; or maybe you want to set the type and print your own book. The Kelsey Company has been in business for over 100 years. It sells printing presses (the smallest one is a 3 x 5 inch press, price $59.75), type, supplies, cards, paper, and so on. For a free copy of their catalog, *Do Your Own Printing,* WRITE TO Kelsey Co., Meriden, Conn. 06450.

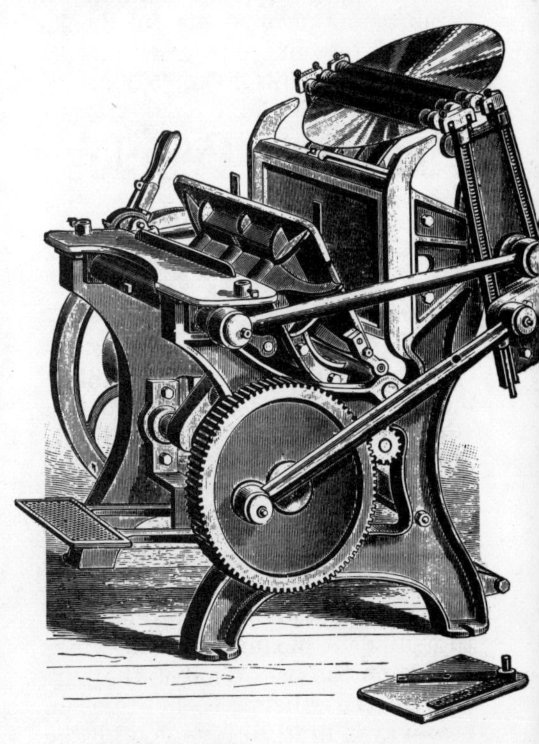

How to Make Your Own Books

If you are going on a trip or plan to write the story of your life, if you like to sketch or if you have stamps, newspaper clippings, or photographs that you are especially fond of, this is the book for you. The author tells you how to make your own books—not the printed books found in libraries, but personal, one-of-a-kind books anyone can put together.

Here are instructions for making travel journals, diaries, photo albums, stamp albums, scrapbooks, scrolls, and sketchbooks. If you can draw stick figures, you can create zany comic books or nonsense books with such titles as "How I Built the George Washington Bridge." If you have some thin paper and a wax crayon, you can make a book of rubbings from manhole covers, details on tombstones—even frozen fish.

How to Make Your Own Books gives advice on choosing paper, cutting and folding, binding, glueing, making covers and marbleized papers; also advice on styles of writing to use in the finished book. With authority and infectious enthusiasm, Harvey Weiss tells you everything you need to know to produce handsome, useful books tailored to suit your exact needs. Thomas Y. Crowell Co. $5.50.

The Young People's Thesaurus-Dictionary

Mark Twain said: "The difference between the right word and the almost right word is the difference between lightning and the lightning bug." In the pages of this excellent book by Harriet Wittels and Joan Greisman you will find hundreds of lists of words. Each word is related in meaning to or the same in meaning as the other words on the list. What is the purpose of these lists? To help you in two ways: first, to give you all the words that are similar in meaning, so that you can choose the word among them which seems just right for the thought you wish to convey; second, to introduce you to words you do not already know. *The Young People's Thesaurus-Dictionary* will help you express yourself. Grosset & Dunlap. $5.95.

The Perfect Speller

If you want to make your own book, you'd better be *The Perfect Speller*. There has never been a complete spelling reference book until this one by Harriet Wittels and Joan Greisman. When you couldn't spell a word, the advice was always "Look it up in the dictionary." But how do you find the word in the dictionary if you can't spell it? Here is a reference book that is easy to use. It lists thousands of words in alphabetical order. Simply look up a word any way you think it is spelled. If your error is a common or phonetic error (spelled the way the word sounds), you will find it entered in the left-hand column in black. In the right-hand column you will find the correct spelling in red. If you look up the word under its correct spelling, you will find it entered on the left, in alphabetical order, in red. Remember: correct spellings are always printed in red.

acored	**accord**
acorn	
acount	**account**
acownt	**account**

In the example above, the words in bold face (correct spellings) are printed in red in the book. **The** *Purfict* **Speller will be** usefull too peeple **of all** agez; their shood **be one on** every **desk!** Grosset & Dunlap. $5.95.

FREE Film Catalog

Just as we were gathering material for *The Whole Kids Catalog* we received the following letter:

> Many young people today are interested in nostalgic films. It's a great hobby!
>
> Jason, our 14-year-old son, is very much into Charlie Chaplin. He orders much of it from the enclosed catalog.
>
> His full-length feature of *The Gold Rush* is going to be used as a scholarship fund-raising event for a local church. Jason is most excited about the publicity he is getting as the Chaplin "expert"!

Collecting nostalgic films is a great hobby. For the hobbyist, Blackhawk Films puts out an illustrated 72-page bulletin every month. In each issue some of the films that Blackhawk offers are described and often pictured. Over a period of 3 to 4 months, the bulletins describe all Blackhawk Films releases. The index shows such names as Lon Chaney, Charlie Chaplin, Walt Disney, Doug Fairbanks, Laurel and Hardy, W. C. Fields, Buster Keaton, the Little Rascals, Harold Lloyd, Mary Pickford, Mack Sennett, Ben Turpin, and Rudolph Valentino.

To receive your free monthly bulletins WRITE TO Blackhawk Films, Eastin-Phelan Corp., 1235 W. 5th St., Davenport, Iowa 52808.

Godzilla Versus the Smog Monster *Brenton Gregory, 6*

If you like to write stories or poems or plays, do book reviews, draw pictures, then you should know about . . .

Stone Soup

What Is It?

Stone Soup is the only literary magazine written by children. Each issue contains stories, poems, plays, book reviews, and drawings by children ages 4 to 12. *Stone Soup* introduces children to creative work by their peers and reinforces children's confidence in the value of their natural expression.

This magazine provides a forum where children can express themselves freely, in their own idiom and through their own fantasies. Here they can read and enjoy the literary and graphic work of other children from all parts of the United States and Canada. *Stone Soup* prints stories and poems ranging in content from caribou hunts to adventures in inner city gangs to dreams of wild horses. Each issue contains some writing by non-English-speaking children. Translations are included.

In *Stone Soup* children review the books that are written for them. The editors believe that creative artists ought also to be critical readers and that children should have the opportunity to express their opinions on books meant for them.

A children's editorial board assists the adult editors in choosing work for the magazine. The editors are also advised by a board of consultants made up of professional educators and classroom teachers. *Stone Soup* is 80 pages printed on high quality paper and paperbound. It is produced by a nonprofit organization.

An Invitation from the Editors

Here is an invitation to you from the editors of *Stone Soup:*

> Everything you write is okay for *Stone Soup*. If you have any poems, or pictures of race cars or platypuses, or stories about witches or about what your father does at work, or anything, we would like to see your work. *Stone Soup* is entirely supported by you, our readers. Every issue is made up of material we are given. It is up to you to make *Stone Soup* the magazine you want it to be. If you want to see more poems or more articles on pets or science projects, or if you want games and puzzles, send them to us. If you are interested in reviewing books, write us a letter. We will forward an appropriate book from our library, if you are accepted as a reviewer.

Show This to Your Teacher . . .

The *Editors Notebook* is an informal guide for teachers published to accompany each issue of *Stone Soup*. It is designed to help teachers who subscribe to *Stone Soup* to integrate the journal and its philosophy into their language arts program.

The *Notebook* suggests concrete and meaningful ways teachers can use *Stone Soup* to stimulate children's creativity. Each issue includes an index to the most recent issue of *Stone Soup* and discusses significant topics relating to the writing and artwork of children. It is available by subscription only, $2.00 a year for three issues.

There are no deadlines for contributing writing and artwork to *Stone Soup*. All stories, poems, and drawings must be labeled with the name and age of the author/artist. The editors will only return contributions which are accompanied by a self-addressed stamped envelope.

Stone Soup is published in November, February, and May. A single copy costs $2.00; a one-year subscription $5.00; a two-year subscription $8.50.

To submit contributions or to subscribe to *Stone Soup* or the *Editors Notebook,* WRITE TO Stone Soup, P.O. Box 83, Santa Cruz, Calif. 95063.

CARPENTRY

Fun Projects for Dad and the Kids

Do you like building things? *Fun Projects for Dad and the Kids,* by David R. Stiles, contains some great ideas. It's a "what-to-make" as well as a "how-to-make" book with delightful drawings of tree houses, a spook house, jungle huts, wagons, rafts, a whaleboat, and a sailboat. It gives you hints on the best way to construct your project but leaves the size and materials up to you. Age 8 and up. If you're interested in building things (with your dad) buy the book. Arco Publishing Co. $2.95.

Make a table, a chair, a toolbox, a scooter... even a Soap Box Derby downhill racer. Plus tree houses, spook houses, a jungle hut, and a raft. A few boards and a saw, some nails and a hammer, and you're on your way to woodworking wonders!

How a House Happens

As you read *How a House Happens*, by author-artist Jan Adkins, and look at the pictures, you live with a house from the moment it is conceived in the mind of an architect until it takes root and grows in a hole in the ground. Then, with the help of a lot of people and a lot of material, a house happens.

The publisher says on the flap of the dust jacket:

> Jan Adkins doesn't let us do much. He thinks up books he wants to make, he writes them and he draws all the pictures. He designs the pages and sometimes he even letters them, as he has lettered this book. He let us print the book and he gives us this flap to write on. Big deal. We put up with stuff like this because we like him and we like his books. We think you will like this book especially.

For a copy of this book (which we think you'll like, too) SEND your check or money order for $5.95 plus 25 cents postage and handling to Walker & Co.

Learn to Be a Carpenter

If you like making things, you can't do better than to use one of nature's oldest and best materials: wood. It has natural strength and beauty of grain, and can be worked with simple tools. *Carpentry Is Easy—When You Know How*, by John Simmons, gives you detailed plans and directions for building over 20 exciting projects selected for ease and simplicity of construction. There are full-color photographs and easy-to-understand instructions on every page.

You'll read how to build a folding workbench, a cheese board, a stool for the garden, a puppet theater, a shoe-cleaning box, a bookrack, and many other things that will be worthy additions to your room. Age 8 and up. Arco Publishing Co. $4.95.

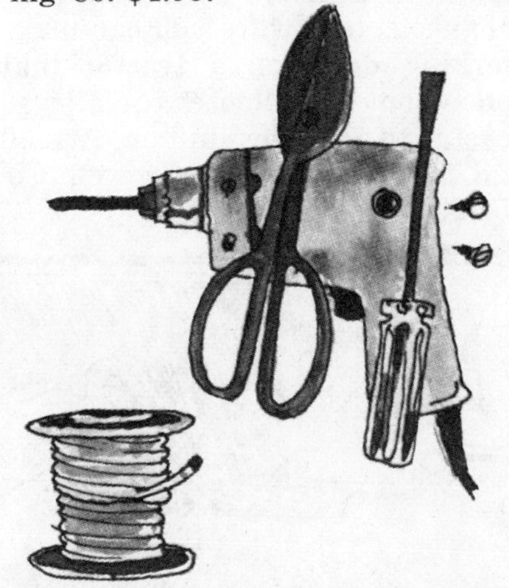

A "TIN LIZZIE"

Model Cars and Trucks and How to Build Them

This book tells how to make seven basic model cars and trucks that you can change in any number of ways to get exactly the vehicle you want. In *Model Cars and Trucks and How to Build Them,* by Harvey Weiss, there are instructions, with detailed drawings and photographs, for putting together elegant racing cars, an open-back truck to cart around your odds and ends, and an old-fashioned trailer truck you can turn into a fire engine. With a few simple tools, some wood, and miscellaneous hardware you can make working derricks, a tractor that can become a bulldozer, or a large coaster to ride downhill on. Age 10 and up. Thomas Y. Crowell Co. $5.50.

If I Had a Hammer

Want to begin woodworking? The first thing you have to know is how to work with tools. This book begins by showing you how to correctly use seven basic tools: hammer, saw, try square, tape measure, C-clamp, Surform plane, and drill.

Once you've acquired a little skill, you'll find six projects you can build: a potholder hanger, a desk-top bookrack, a hanging planter, a handy tool carrier, a reversible shelf, and a pet bed.

When you feel you've really mastered the tools is when you can begin to create projects of your own. Excellent step-by-step photographs make Robert Lasson's *If I Had a Hammer* a good basic book for the young woodworker. E. P. Dutton & Co. $7.95.

Like to Build Things?

Here are four little books with instructive and amusing illustrations by Ray Brock that tell you exactly how to make scooters, a table, a chair, a toolbox, even a car. If you would like to make a super toolbox, follow the simple instructions in *Now You Need a Toolbox.* If you would like to make furniture, *You Can Build a Table and Chair Too* tells you how. If you have an old roller skate, get *Scooters Are Groovy and You Can Build Your Own.* And *If You're Ready, Here's the Car* will give you the great experience of constructing one. Age 10 and older. (Younger, with help.) Dial Press. $1.50 each.

Toy Patterns

With Stanley Pattern Series you can learn how to construct toys. Directions are easy to follow because they are not blueprints and do not require arithmetic. Get Six Toy Patterns (No. P1: insert toy, paddle-wheel boat, clacker, tumbling clown, fighting roosters, cradle) or Six Toy Patterns (No. P2: spinning clown, marble game, wren birdhouse, propeller toy, periscope, dump wagon). Each set of toy patterns costs 25 cents. To order WRITE TO Stanley Tools, Advertising Services Dept., P.O. Box 1800, New Britain, Conn. 06050.

Instructions for Constructions

If you like to build things and enjoy carpentry, here are some patterns and directions that will help you make storage units, a clubhouse, trucks (which are really storage boxes on wheels), a table for a model train that can slide under your bed, a portable sandbox, and a rocking horse. Each set of directions is 50 cents. To get one or all of them WRITE TO American Plywood Association, 1119 A St., Tacoma, Wash. 98401.

The Buffy-Porson

If you would like to build your own downhill racing car, *The Buffy-Porson . . . a Car You Can Build and Drive* is made to order for you. The Buffy-Porson (see picture) is a handsome downhill coaster created by the father-and-son team of Peter and Mike Stevenson. Their clear and helpful layouts, diagrams, photographs, and instructions (together with their reassurance that everything will turn out all right in the end) make this a project you can take on with confidence. The pictures in the book will show you that you can get as much sport coming down a hill in a well-built coaster as you can in an Olympic bobsled, or giant slalom race.

You don't need a whole workshop full of tools to build a Buffy-Porson. Ordinary, household tools will do the job. One building aid that you *will* need once in a while is an Older Person to lend a hand when a special hole needs drilling or a tricky curve has to be cut. Don't forget that grown-ups often have very short attention spans (especially if there's a big game on TV).

It's hard to go wrong when you build your Buffy-Porson; there's a complete materials list, including lumber, fasteners, and hardware. But as the Stevensons say: "Don't let anybody kid you; building a coaster is a big project. It takes the kind of person who can stick with a job to build a Buffy-Porson (but then, maybe that's what separates the Buffy-Porson drivers from those who run along behind)." Charles Scribner's Sons. $5.95.

Make It and Ride It

Would you like to enter a soapbox derby? Every racer must have a custom-built car made according to his size and weight. Did you know that your racer should be long, low, as narrow as possible, and safe for you to ride in? Since many races are won by inches or by a fraction of a second, every little detail is important when you construct your racer. And here you'll find not one, but three plans for soapbox racers. When you're finished with your racing car you can learn how to build a bike trailer, a wagon, a jeep, and a scooter. What's nice is that you use only inexpensive materials (such as wood from fruit crates) and home or school workshop tools. Get on your mark, get set, go . . . for the soapbox derby with *Make It and Ride It,* by C. J. Maginley. Harcourt Brace Jovanovich. $5.50.

Catalog from the Workshop for Learning Things

This creative 40-page catalog not only gives you a variety of unusual things to send for, but shows in photographs and drawings what to do with them. There are books on cardboard carpentry. Heavy cardboard is a material which is fun to work with and is easy on the hands. The catalog tells you where to get the cardboard, what tools you'll need, and how to begin; also includes printing kits, soapstone carving and tools, microscopes, and so on. To get your catalog SEND 50 cents to Workshop for Learning Things, 5 Bridge St., Watertown, Mass. 02172.

Cardboard Carpentry

Here's a book from the Workshop for Learning Things, which designs things for learning. *Further Adventures of Cardboard Carpentry,* by George Cope and Phylis Morrison, tells about heavy cardboard and its many uses in classrooms, with many plans for making useful and exciting things. There are hints about the use of cardboard as a versatile structural material, details about the use of tools on cardboard, and over 300 drawings and photographs. A good book to use at home, too. To ORDER, make out your check or money order for $3.50 payable to Workshop for Learning Things, 5 Bridge St., Watertown, Mass. 02172.

More Cardboard Carpentry

Building with Cardboard is a good booklet on building with triple-corrugated cardboard. It shows various methods of half-slot and tab/slot construction, folding with and against flutes, tying together, taping, gluing, and finishing. Projects to make include a puppet stage, playhouses, a dollhouse, a table, a stool, a bookcase, a chair, a cradle, a sandbox, and an easel. Anthony Sharkey (ed.), Joan Green (illus.). For your copy SEND 60 cents plus 15 cents for postage and handling to EDC Distribution Center, 39 Chapel St., Newton, Mass. 02160. (Also ask for *Building with Tubes*, same price, same postage; and *Building with Tires*, $1.00 plus postage.)

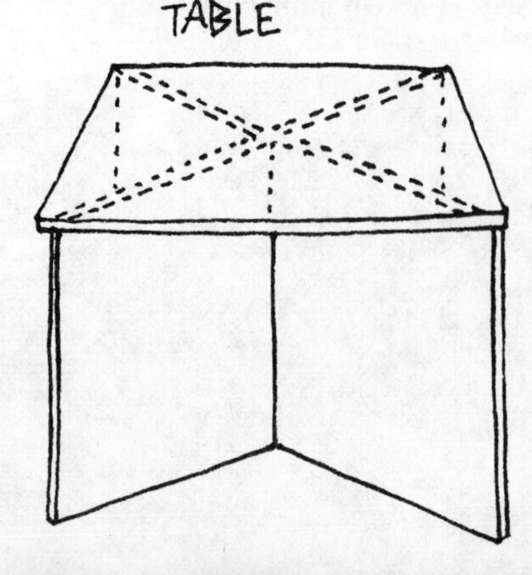

Making Children's Furniture and Play Structures

A bonanza of inexpensive play structures, spaces, and furniture that can be constructed without hammer and nails. Airships, school buses, play houses, domes, tents, igloos, and more. Areas to lace together with vinyl to define space for running, shouting, jumping, snuggling, sleeping, thinking, coloring, sulking, day-dreaming! Benches, tables, chairs, desks, stools, and beds from standard shipping tubes and packing cartons. A wondrous collection of objects that children help to build themselves. For your copy of Bruce Palmer's *Making Children's Furniture and Play Structures* SEND $3.95 (paper) or $8.95 (cloth) plus 50 cents for postage and handling to Workman Publishing Co.

Carving Animal Caricatures

If you enjoy whittling and wood carving you may be ready for caricature carving, the subject of this book. Animals become amusing with long ears for a donkey, a big mouth for a hippopotamus, funny eyes for a dog. There are 24 animals for you to have fun with, including Gulliver Goat, Clementine Cow, Esmerelda Elephant, and Dmitri Dinosaur.

The profile patterns presented in *Carving Animal Caricatures,* by Elma Waltner, are full size; they need only be transferred to the wood. Detailed sketches show the appearance of the figure from several angles and photographs show each step of the way. For older children. For your copy SEND $2.50 to Dover Publications.

F.A.O. SCHWARZ

Welcome to the greatest toy emporium on the face of the earth! See 10,000 wonders from all over the globe. Everything from a yo-yo for 65¢ to a $3000 electric car! The next best thing to Santa's workshop is a by-mail catalog from this palace of toys.

The World's Most Unique Toy Store

F. A. O. Schwarz is not only one of New York's most famous stores, but one of the city's great entertainment attractions. Shopping is pure fun in Schwarz's, where you really get the feeling that you are "somewhere over the rainbow."

The world's most unique toy store is 113 years young. Its founder, Frederick August Otto Schwarz, entered the toy business in 1862 six years after he arrived in America from Westphalia. The store traditions have changed little since the days of its founder. Between 10,000 and 12,000 different toys—the finest one can buy—some manufactured by Schwarz itself for its own stores, and others exclusively sold by Schwarz and imported from all parts of the world, are carried on its shelves.

You may find a puppet show going on to amuse small customers, browse through a children's book department as large as many a public library, cuddle a menagerie of stuffed animals from mini- to life-size, and admire antique toys from yesteryear. F. A. O. Schwarz is famed for its fantastic range of toys: you can buy anything from a yo-yo for 65 cents to a $3000 electric car; but half the items carried by the store retail for less than $10.00.

To get your full-color catalog SEND 50 cents to F. A. O. Schwarz, 5th Ave. at 58th St., New York, N.Y. 10022.

F. A. O. SCHWARZ SATELLITE STORES

Suburban Square
Ardmore, Pa. 19003
(215) MI 9–5048

Lenox Square
3393 Peachtree Road, NW
Atlanta, Ga. 30326
(404) 233–8241

Bal Harbour Shops
9700 Collins Ave., #143
Bal Harbour, Fla. 33154
(305) 865–2361

40 Newbury St.
Boston, Mass. 02116
(617) 266–5101

The Mall at Short Hills
Short Hills, N.J. 07078
(201) 376–8140

700 White Plains Road
Scarsdale, N.Y. 10583
(914) 725–3619

1370 Northern Blvd.
Manhasset, N.Y. 11030
(516) MA 7–7780

314 Royal Poinciana Plaza
Palm Beach, Fla. 33480
(305) 832–1698

The Fashion Center
Ridgewood Ave.
Paramus, N.J. 07652
(201) 652–3703

180 Post St.
San Francisco, Calif. 94108
(415) 391–0100

2755 Somerset Mall
Troy, Mich. 48084
(313) 643–6445 and 643–6447

Needlecrafts

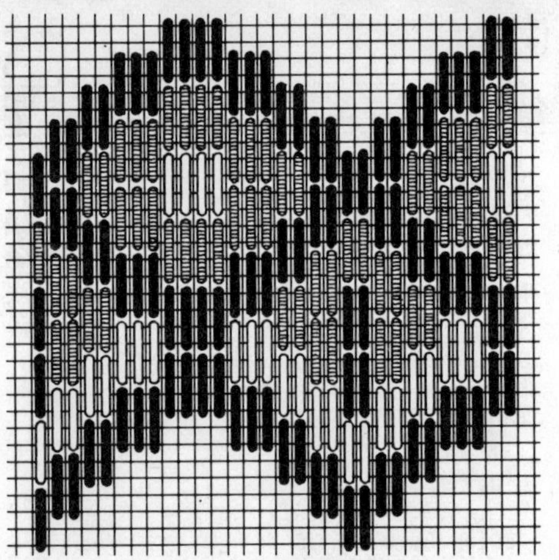

Step by Step Bargello

A needlework craze is sweeping the country. In keeping with the mounting interest, *Step by Step Bargello,* by Geraldine Cosentino, has been published: a complete, easy-to-follow instruction guide for beginner and experienced needleworker alike, which proves that bargello is fast and easy, and what's more, fun to do.

The only equipment needed is canvas, needles, and yarn. The book covers all the essentials from basic stitches and traditional patterns to finishing and sewing techniques. The projects you can do in this book range from pillows, belts, and handbags to a necktie, a trinket box, and a caftan with attractive bargello trim. Golden Press. $4.95 cloth, $2.95 paper.

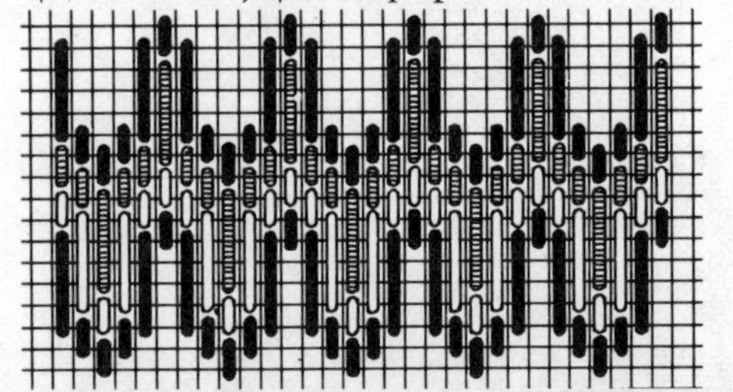

Step by Step Spinning and Dyeing

This is a fascinating new book for everyone who enjoys working with texture, fibers, and colors—in knitting, weaving, macrame, crochet, or other crafts. Included is a section on the equipment needed (plus instructions for making a simple hand spindle); directions for spinning yarns from wool, cotton, flax, and other fibers; and a section on how to obtain a glorious range of subtle, unique colors from natural dyestuffs. Written by Eunice Svinicki, whose work has appeared in numerous leading national magazines, *Step by Step Spinning and Dyeing* is packed full of detailed illustrations and full-color photographs—and contains a complete list of sources for all supplies. Golden Press. $2.95.

FREE Learn to Knit

It's easy to learn to knit with these instructions. More than 30 closeup diagrams show you exactly what to do. A first project for you to start on is a head hugger, a cozy, becoming little number that's easy to make and can be finished in a few hours. If you don't know how to knit, be sure to get a free copy of *The ABC of Knitting:* SEND a large self-addressed stamped envelope to Coats & Clark, P.O. Box 1010, Toccoa, Ga. 30577.

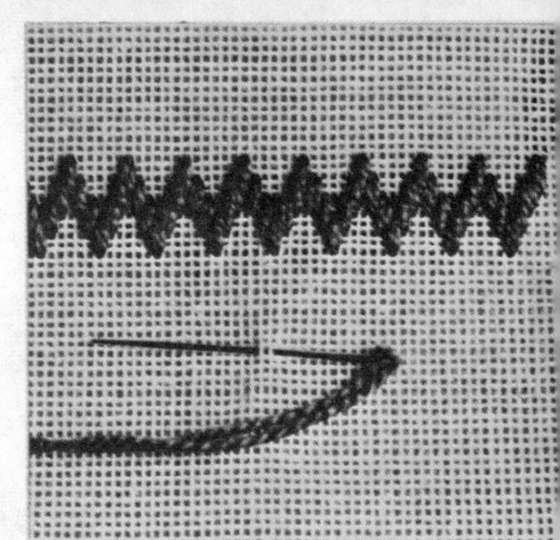

Have fun with yarn: make a checkered vest, a pom-pom beret. Learn bargello, how to knit, weave, spin, or crochet. Make your own gladrags, perk up a hand-me-down, embroider your jeans.

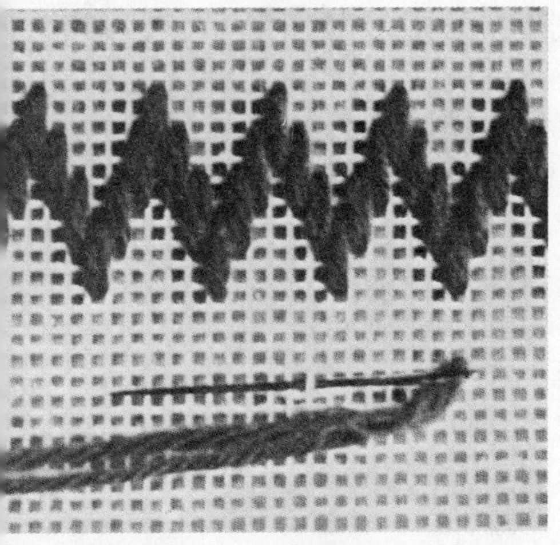

Step by Step Knitting

A new and exciting approach to the craft of knitting is introduced in this handsome, fully illustrated book. Mary Walker Phillips, whose work has been exhibited internationally, leads you with step by step instructions to the art of knitting, with directions for left- and right-handed learners. Besides stoles, scarfs, hats, and handbags there are beginners' projects in making rugs, blankets, pillow covers, and place mats. *Step by Step Knitting* is an exciting book to help you start knitting and to help you understand knitting directions. Golden Press. $2.95.

Knitting for Beginners

If you would like to learn how to knit but there is no one around to teach you, this is the book for you. It shows in photographs (by Edward Stevenson) each step of how to start, how to cast onto the needle, how to make that first stitch, and how to go from there to actually knitting. It will teach you the language of knitting and its abbreviations. Then when you have some confidence, it gives you projects to knit: a headband, belt, pincushion, pillow, and afghan. *Knitting for Beginners* is by Jessie Rubenstone. J. B. Lippincott Co. $5.95 cloth, $2.95 paper. *Crochet for Beginners,* same author and publisher. $5.95 cloth, $2.25 paper.

A Stitch in Time

Stitch by Stitch teaches the art of fine needlework, a skill that can bring pleasure for a lifetime. The author, Carolyn Meyer, begins with the simple running stitch and progresses through cross-stitching, the lazy daisy, the blanket stitch, French knots, and finally needlepoint. At each step there is an attractive practice project using the new skill—potholders, towels, pillowcases, and doll blankets are among the interesting things you make—and you are encouraged to use your own ideas and create your own designs. Harcourt Brace Jovanovich. $5.50.

FREE Dye-Craft Projects

Dye-craft is the creative application of color to fabric. It's of special interest because dye-craft designs relate to the symbols and patterns of the American Indian. Here's a fun one-page, full-color montage of designs. All projects can be completed in less than an hour. Full instructions for six designs are included. For your copy SEND 10 cents for mailing and handling to Dye-Craft for Groups, P.O. Box 307, Coventry, Conn. 06238.

This is Miss Patch. She is learning to sew.
You will see that if Miss Patch can sew, anyone can.
(Anyone except, of course, her dog Charlie.)
If you have two pieces of cloth,
you can sew them together to make something.

Miss Patch's Learn-to-Sew Book

A funny little lady named Miss Patch, with frizzy hair and glasses that slip down her nose, is learning how to sew. And as author Carolyn Meyer says, "If Miss Patch can sew, anyone can." Beginning with how to thread a needle and make a knot, she makes a wonderful pillow and goes on to learn how to cut newspaper patterns, sew a fine seam, and eventually how to read more complicated patterns. Miss Patch will teach you how to make scarfs, patchwork quilts, and simple clothes for yourself, as well as a lovely wardrobe for your doll, a gingerbread boy, an "I Love You" pincushion, and a hand puppet. Mary Suzuki's delightful drawings help too. *Miss Patch's Learn-to-Sew Book* is for ages 5 to 10. Harcourt Brace Jovanovich. $4.75.

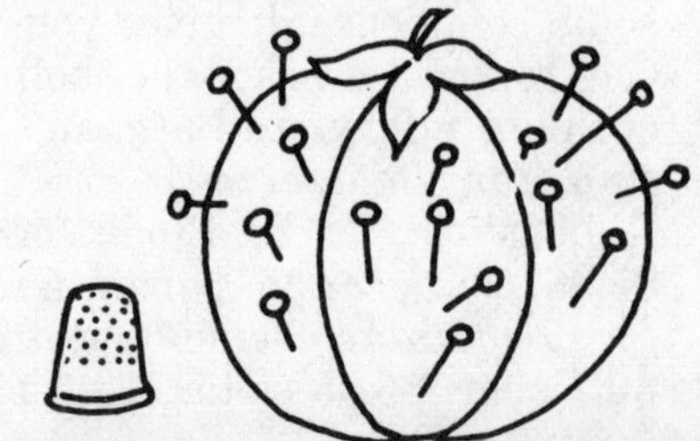

Sew Easy

For the young beginner who is fortunate enough to have this book as a guide, it is indeed *Sew Easy.* The author, Peggy Hoffmann, has well-organized instructions that read almost like a story. Every question that might occur to you seems to have been considered and answered. There are 12 projects, each introducing some new sewing skill; none of them are so complicated as to become boring.

You will learn to make a luncheon cloth and napkins, a change purse, bean bags, and a scarf, and you'll learn how to sew on buttons, snaps, and hooks and eyes.

This is primarily a book about hand sewing, but you will also learn how to operate a sewing machine. Diagrams and sketches guide you throughout and excellent photographs show the finished projects. E. P. Dutton & Co. $5.95.

The Lucky Sew-It-Yourself Book

If you don't know how to sew, you have to start somewhere—and this is the simplest of how-to-sew books for the very young. Some of the questions it answers are: How do you make a running stitch? How do you make a tack? When do you make a tack? How do you measure your cloth? But it's not all work; you'll learn how to make a scarf, a pocketbook, an apron, a pincushion in *The Lucky Sew-It-Yourself Book,* by Camille and Bill Sokol. Scholastic Book Services. 75 cents.

I Love to Sew

If you are a beginner at sewing, this is a book you will want to own. With *I Love to Sew,* by Barbara Corrigan, you can make fringed place mats and napkins, a drawstring bag, a ruffled hat, curtains, pillows, stuffed toys, hand puppets, skirts, pants, tops, an apron, and doll clothes.

As you sew you learn by doing. There are pictures and drawings throughout the book to help you every step of the way. Doubleday & Co. $4.95.

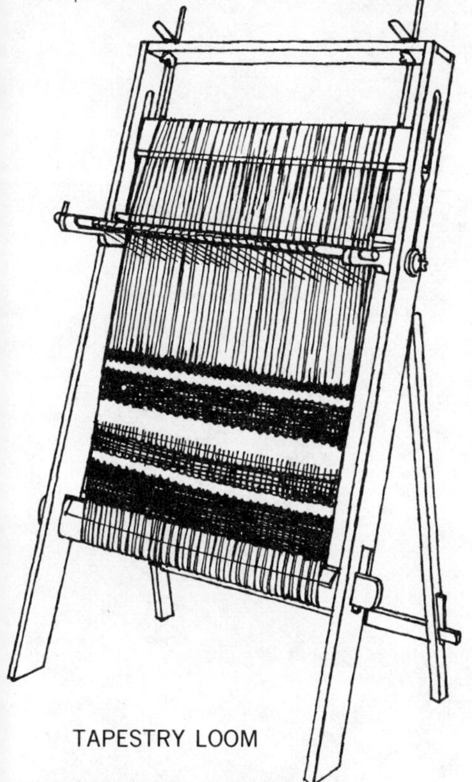

TAPESTRY LOOM

Step by Step Weaving

In this exciting, fully illustrated book, the world of weaving is opened for the beginner. Nell Znamierowski, free-lance designer and teacher, brings a freshness to her designs which show the influence of studies in Finland and in the Greek mountains.

Step by Step Weaving introduces you to the frame and four harness looms. The dyeing of yarns, using both commercial and natural dyes, is explored. With the help of this book you will be able to create a Rya rug, a poncho, place mats, bags, a tapestry, pillows, and stoles. For further instruction, there is a school directory; the glossary will help you to learn weaving terms; and there is a list of suppliers of materials and of books to read about weaving. Golden Press. $2.95.

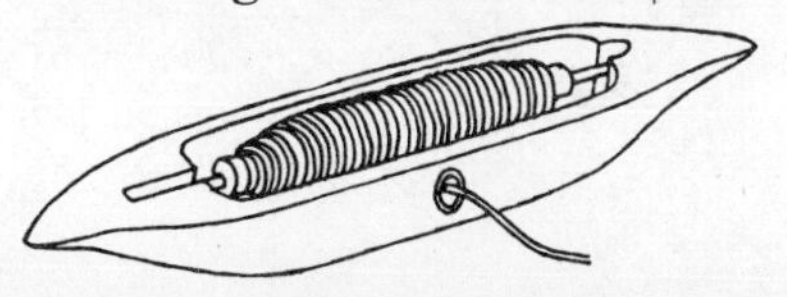

FREE The ABC of Crochet

You can learn how to crochet a lot of pretty things by following the instructions. Two free, illustrated leaflets—one for the left-handed, one for the right-handed—will teach you about lacy stoles, gloves, slippers, jaunty hats, sporty handbags, and bedspreads. An excellent introduction to the art of needlecraft. To get a free copy of *The ABC of Crochet* SEND a large self-addressed stamped envelope to Coats & Clark, P.O. Box 1010, Toccoa, Ga. 30577.

Learn How Book

Do you want to learn how to crochet, knit, do tatting or embroidery work? Then here's a book you'll enjoy. 50 pages crammed with simplified instructions, dozens of easy-to-understand diagrams, and photographs of finished pieces. Here are a few of the things you'll learn to make: afghan, tray cloth, crocheted luncheon set, mittens, and sweaters. The price of the *Learn How Book* is just 55 cents. WRITE TO Coats & Clark, Educational Mail Dept., P.O. Box 1010, Toccoa, Ga. 30577.

Learn Embroidery

If you have shied away from embroidery because you thought it was too complicated, here's a leaflet which will change your mind. These simple step-by-step diagrams of 13 stitches and their variations are all you need to learn a satisfying new pastime: embroidery. The leaflet has instructions for transferring designs and for beginning and finishing; it includes directions for a gingham pincushion. For the free leaflet *ABC of Embroidery* WRITE TO Coats & Clark, P.O. Box 1010, Toccoa, Ga. 30577.

Upright rug loom

Step by Step Rugmaking

If you have ever wanted to make a rug but didn't know how to get started or what technique to use, *Step by Step Rugmaking* can answer your needs. It has ideas and projects for making rugs using many different techniques. The author, Nell Znamierowski, guides you with clear diagrams and instructions in making rugs by embroidery, latch hook, crocheting, knitting, braiding, hooking, and weaving. Ideas for designing, color planning, and sources of inspiration are included. The color photographs of examples of rugmaking are inspiring. Golden Press. $2.95.

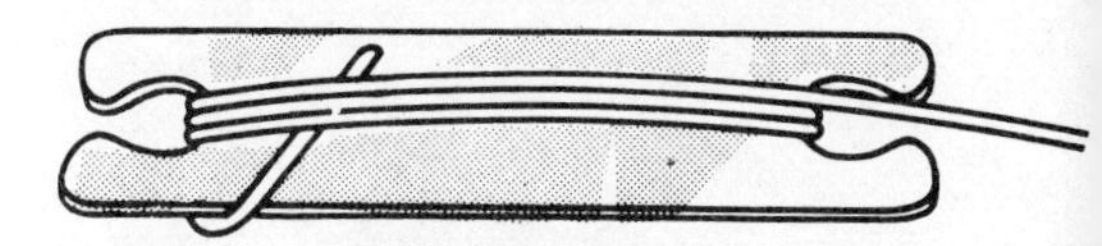

Fun with Wool

This book has lots of color drawings and photographs that show you how to have fun with wool. There are simple directions for making a checkered vest, a pompom beret, pillows, lots of animals, and a puppet. When you have made all the things in this book you will have learned many different kinds of needlecraft: knitting, crocheting, embroidery, sewing. *Fun with Wool,* by Kate Pountney will help you produce a lot of colorful things with wool. Grosset & Dunlap. $3.95.

Native Funk & Flash

Many of us have hungered for a cultural identity strong enough to produce our own "native" costumes, masks, or ritual objects comparable to, let's say, those of Afghanistan or Guatemala. Alexandra Jacopetti has recorded, in *Native Funk & Flash: An Emerging Folk Art,* what people are doing in this country to fulfill that hunger.

200 lavish color photographs by Jerry Wainwright and a lively text document the wild personal statements of a generation that has emerged from the psychedelic sixties. Embroidering, appliquéing, batiking, and adding patches to the old and reusable is the craft we show from our newfound culture.

This is not a how-to craft manual but a book about what has followed the last decade of cultural and political upheaval. A staggering number of ideas show us where that creative spirit is today and why. There are no patterns here, for the authors say that patterns are within us. We're sure you'll find yourself picking up needle and thread and the nearest paintpot too. Those most private dreams and even the wildest demons are longing for expression. For your copy SEND $7.50 to Scrimshaw Press, 149 9th St., San Francisco, Calif. 94103.

Sew It and Wear It

Would you like to make attractive skirts, tops, caps, capes, and other articles? They are all in this informative book, *Sew It and Wear It,* by Duane Bradley, and can be cut without a pattern (or with a simple newspaper pattern) and completed in a short time. Anyone who can run a sewing machine will be able to make ponchos, kimonos made from a bed sheet, or a personal beach tent.

Most important, this book will stimulate you to experiment, to develop sewing shortcuts of your own, and to add your individual touch to a garment by imaginative choice of fabric and trim. The techniques are basic; after reading this book the conventional printed pattern will be much less mystifying. Illustrated by Ava Morgan. Thomas Y. Crowell Co. $4.50.

Superjeans

Want to own the most stunning denim wardrobe in school without spending a lot of money? Here's how. Use easy-to-follow instructions, diagrams, and drawings (by Jeanne Kovalak) in *Superjeans* to recycle and decorate your jeans. Chapters include "Painting Jeans"; "Embroidering Jeans"; "Appliquéing Jeans"; "Patchwork on Jeans"; "Studs on Jeans"; "Glitter on Jeans"; "Sew-Ons, Iron-Ons, Fuse-Ons"; and "The Living End." There's a page from history: do you know where the word "dungarees" came from—and the word "jeans"? A super book, by Donna Lawson. Scholastic Book Services. 95 cents.

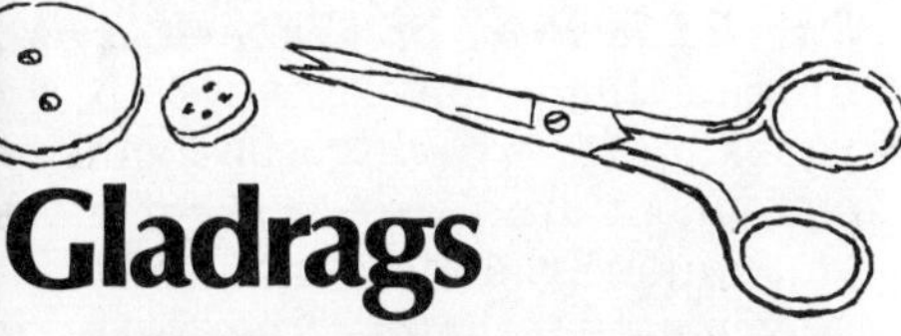

Gladrags

Redesign, remake, refit all your clothes to become the clothes you want: old clothes into "new," plain clothes into fancy, styleless clothes into the latest look, and tight or loose clothes into a perfect fit. *Gladrags,* by Delia Brock and Lorraine Bodger, has a spectacular improvement plan for every item in your wardrobe: sweaters, blouses, pants, skirts, jerseys, and accessories.

Based on a revolutionary approach to creative home sewing, suited to beginners as well as advanced sewers, this exciting new book shows you how to remake, decorate, take a hand-me-down and perk it up, use your imagination and—presto!—turn out a new, made-to-order, guaranteed-to-please garment. The sewing and craft techniques for every transformation are explained in detail, and charming, vividly clear line drawings illustrate every step of the way. Simon and Schuster. $8.95.

FREE

Dye Up a Wardrobe

Like tie-dyeing and batik? Then these two *Glamour* and *Redbook* magazine reprints are for you. Full-color pictures, fully detailed instructions—and the philosophy for T-shirts is yours: "The idea is to do what you want, on whatever you want: a scribble on a T-shirt, a snappy geometric on a sweater or even a city skyline brightened up with rhinestones." You'll learn how to make six simplified batik outfits; the shortcut is called "brush dyeing," and you can do it! For this pair of "how-tos" SEND 20 cents for mailing and handling to *Dye Up a Wardrobe* and *Batik Outfits,* P.O. Box 307, Coventry, Conn. 06238.

Leather!

Soft Suede, Supple Leather is an exciting book which gives a whole new approach to working with leather. Drawing from his wide experiences as a teacher and craftsman, R. K. Furst describes and illustrates practical and imaginative ways of leatherwork. He shows basic patterns that never go out of style, and explains methods such as hand stitching, studding, gluing, and hole punching, which require no more than a few easily obtainable tools and the kitchen table as a workshop. Complete instructions tell how to make beautiful wallets, belts, pants, jackets, pillows, flight bags, and many other attractive objects. Simon and Schuster. $7.95.

A Beginner's Book of Patchwork, Appliqué & Quilting

Have you ever seen a beautiful quilt and thought how great it would be to make something like that yourself? If so, you are probably enjoying the current revival of the crafts of patchwork, appliqué, and quilting. *A Beginner's Book of Patchwork, Appliqué, & Quilting,* with numerous diagrams, photographs, and patterns, is an easy-to-understand introduction to these age-old crafts, especially designed for the beginner.

Learning how to turn scraps of leftover material into a piece of patchwork clothing, how to make blocks for a quilt top, and how to create an appliqué wall hanging is only part of this book. Constance Bogen also covers the various stitches, cutting the basic pieces, materials needed, planning a design, and much more. There is a section of projects you may want to undertake, from a simple quilted potholder to a more complex log cabin quilt. Dodd, Mead & Co. $5.95.

Creative Soft Toy Making

Can you imagine being able to make these special creatures: a small furry dormouse, a splendid lizard over 4 feet long, an elephant, a frog, an owl and a pussycat, a penguin, and others from woodland, wild, and seashore? They can all be made with the help of *Creative Soft Toy Making,* by Pamela Peake. Over 70 animal friends are beautifully photographed in this book, with patterns to help you make them. Your imagination can take over after becoming familiar with the fundamentals in this book. There are chapters to help you with fabrics and designs, toymaking techniques, and character-making eyes, whiskers, and painted features. All you need is the desire and you can make any animal you want. Bobbs-Merrill Co. $7.95.

Tie Dyeing and Batik

Tie dyeing and batik are fine related crafts that are quickly mastered. You can make simple batiks with crayons and tempera, and use food coloring for tie-dye design. You can make beautiful tie-dye and batik by following a few basic directions. The craft experience helps you experiment with basic design: color, form, and pattern. *Tie Dyeing and Batik,* by Astrith Deyrup, is an excellent book. You can begin experimenting with fabric craft with tie dyeing or batik. The suggested projects are carefully explained, but you can always substitute your own ideas. Doubleday & Co. $4.95.

Do-It-Yourself Dinosaurs

Make your own dinosaurs. It isn't difficult, even if you are a beginning sewer. You will need a few materials like thread, pipe cleaners, and felt to produce make-believe monsters complete with scales and horns. Brenda Morton offers thoroughly illustrated instructions for using a combination of sewing techniques and ingenuity to create sinister teeth out of white thread, a toy dinosaur stuffed with cotton, hooves made with felt, or wiry fingers derived from a pipe cleaner. You can choose from patterns for all your favorite dinosaurs: Brontosaurus, Diplodocus, Tyrannosaurus, Triceratops, Iguanodon, and many others. Create a monster menagerie with *Do-It-Yourself Dinosaurs: Imaginative Toycraft for Beginners.* Taplinger Publishing Co. $6.95.

The Basic Book of Fingerweaving

If you are into macrame and like it and want to go farther, try fingerweaving. Instantly appealing, this unusual collection of weaving techniques eliminates the frustrations and disappointments a beginner can experience. Not only does *The Basic Book of Fingerweaving* provide the simplest ways to set yarns up for weaving (no looms or equipment are necessary); it gives you uniquely satisfying direct involvement of hands and yarn or string. The author, Esther Warner Dendel, gives you excellent step-by-step line drawings and photographs and an easy-to-understand text to explain the basic concepts of fingerweaving all over the world —American Indian, African tribal weaves, Chinese and Mexican braids, and round braids from Egypt. Simon and Schuster. $7.95.

POLK'S HOBBY CATALOG

Introducing to one and all the world-famous house of hobbies . . . *plus* a colossal order-by-mail catalog of wonders to amaze and amuse you!

This year, Polk's, the world-famous four-story hobby shop on Fifth Avenue in New York City, is celebrating its 40th anniversary. Hobbies is one of the fastest-growing businesses in America. It is difficult to turn a corner, walk down an avenue, or stroll through a mall without finding a hobby shop. But Polk's hobby shop is the granddaddy of them all. And whether or not there is a hobby shop near you, Polk's catalog is must reading for all hobbyists: here, for you to pore over and read and reread, is the whole hobbycraft spectrum in a single buyer's guide of almost 150 pages. Most major manufacturers and many smaller specialized suppliers are included. The items in Polk's catalog represent the very best in hobbies that their buyers can find worldwide. Airplanes and boats, trains and cars, scenic supplies and buildings, slot racing, military miniatures, science kits, collector vehicles, crafts, tools and supplies—all these and more are pictured and described to whet your hobby appetite. To order your copy of Polk's *Bluebook of Hobbies*© SEND $3.95 to Polk's Hobby Department Store, 314 5th Ave., New York, N.Y. 10001.

Here for your "window shopping" are a few of the thousands of hobby items Polk's can offer you. Please write for their catalog before you order, so you can be certain about postage and handling charges.

COOKING

Betty Crocker's Cookbook for Boys and Girls

Fun in the kitchen is what *Betty Crocker's Cookbook for Boys and Girls* is about. Step-by-step recipes make it simple, along with quick tips and color photographs (by Len Weiss) to help first-time cooks learn their way around the pots and pans. You can learn how to prepare everything from a snack to a whole meal, and every recipe is a winner because each one has been tested by boys and girls around the country. Golden Press. $3.95.

Merry Metric Cookbook

The *Merry Metric Cookbook* is designed to help you learn the meaning of a few metric units. It also helps you make some very delicious things to eat. The recipes are easy and by the time you finish cooking all the dishes in the book you will know what is meant by a kilogram and a milliliter. The book is related to nursery rhymes; Alice in Wonderland, the White Rabbit, and the Mad Hatter lead you all the way, so take a tasty trip through metric land with Mary Miller and Toni Richardson. For younger cooks. For your copy SEND $3.00 plus 50 cents postage to Activity Resources Co., P.O. Box 4875, Hayward, Calif. 94540.

Junior Cookbook

The Saturday Evening Post Junior Cookbook (for grades 1 to 10) is a collection of over 100 of the best recipes from *Children's Playmate, Child Life, Jack and Jill,* and *Young World* magazines. You'll find favorites that children have cooked and loved for years, plus many new delicious recipes. Want to make sloppy Joes? Coleslaw? Pineapple upside-down cake? Peanut oatmeal cookies? You can learn to make all these and more if you send for this cookbook. It's $1.00 plus 25 cents for postage and handling. WRITE TO Mary Alice Simpson, Saturday Evening Post, Youth Division, 1100 Waterway Blvd., Indianapolis, Ind. 46202.

Attention all hungry kids and junior gourmets! Enjoy the adventure of cooking! Scores of mouth-watering recipes for everything from juicy hamburgers to spaghetti dripping with sauce. Want a chocolate cake for dessert? Be a master chef!

The Bread Book

The fascinating history of one of our most basic requirements, bread, is long and colorful. Kingdoms have been won and lost for bread; religious ceremonies are focused on bread. Bread has been used for money, for tableware, for sculpture, for magic. The book has many humorous drawings of an international variety of people eating their versions of bread. In *The Bread Book: All About Bread and How to Make It*, by Carolyn Meyer, you learn that the Mexican eats tortillas, the Indian (from India) eats chapati, the Frenchman has his baguette, Swedes and Finns eat knäckebröd, Italians nibble on crunchy grissini, Jewish people like bagels, and the English are fond of crumpets. There are amusing tales about bread, and, of course, recipes. You will learn how to make fun buns, Cuban bread, and a lovely wreath bread for Christmas. Harcourt Brace Jovanovich. $5.95.

The Kid's Cookbook

Hi, Cooks and Cooks-to-be! We hope you'll like this cookbook, and will enjoy both the adventure of cooking and the fun of sharing what you make.

The recipes are easy, and anyone can do them—as long as they can read and follow directions. If you are just learning to read, get someone to help you with the reading part. There are drawings to help you figure out cooking tools and directions.

That's the way the amusingly illustrated *The Kid's Cookbook* starts off. Patricia Petrich and Rosemary Dalton give you a few cooking rules, and then the fun begins. You learn how to make Jingle Bell Salad, Pigs in Blankets, Zoo Sandwiches, Merry-Go-Round Cakes, and Magical No-Cook Candy. There are over 100 recipes for you to try for breakfast, lunch, dinner—snacks too! Nitty Gritty Productions. $3.95.

COOKING

I like to cook
From a book
I like to bake
A beautiful cake
I get my kicks
When I mix
A gushey pudgey
Chocolate fudgy!

Monica Sabty, 11
Coolidge Community School
Flint, Michigan
From *Stone Soup* (Vol. 2, No. 2), February 1974

Candies, Cookies, Cakes

With the same easy-to-follow format as Aileen Paul's other cookbooks, *Candies, Cookies, Cakes,* by Aileen Paul and Arthur Hawkins, includes recipes for brownies without baking, orange chocolate chips, butterscotch thins, lollipop cookies, peanut brittle, old-fashioned peppermints, pulling taffy, easy chocolate cake, spice cake, and many more. If you're a sweet-tooth cook, this book is for you. Doubleday & Co. $4.95.

The Natural Cook's First Book

Here is a natural foods cookbook for beginners; vegetable fried rice, whole wheat spaghetti with fresh clam sauce, simple salad, apple brown betty, and granola are some of the mouth-watering things you can cook on your own. The best part of it is that they're all good for you. Says Carole Getzoff in her introduction to *The Natural Cook's First Book:*

> When I was young I didn't like to eat. I especially didn't like to eat vegetables or rice or anything that was good for me. But that was because everything that was good for me didn't taste very good. When I grew up I found out how to make food that was good for me taste good, too. And now I could practically eat a vegetable sundae, I love vegetables so much. And I love rice and cereals, too. I bet you don't believe it. Well, try some of the recipes in this book and see if the same thing doesn't happen to you.

Good idea! The illustrator is Jill Pinkwater. Dodd, Mead & Co. $4.50.

Kids Are Natural Cooks

This is a great book for beginner cooks to use at home. *Kids Are Natural Cooks* demonstrates what a group of energetic and enthusiastic parents and teachers can do when they believe in natural foods —and in kids. First published privately in 1972 as a fund-raising project for the Parents' Nursery School in Cambridge, Massachusetts, the book has since been expanded, and illustrated charmingly by Lady McCrady.

It is a treasure trove of information. After a convincing discussion of the advantages of natural foods over the glop too many kids eat, it goes on to practical pointers on how to cook with kids for maximum fun and involvement.

Here are dozens of ingenious recipes, arranged by seasons, that let kids really get into the act. No special equipment is necessary, just small and busy hands following simple, well-illustrated instructions. Three cheers for Liz Uraneck, who had the ideas and worked with the kids, and Roz Ault, the mother who wrote the text. Houghton Mifflin Co. $5.95 cloth, $3.95 paper. (The proceeds of the book continue to go to the Parents' Nursery School's education and scholarship program.)

SESAME HONEY CANDY

When the ancient Roman soldiers set out for battle, their main food during the long marches was a mixture of sesame seeds and honey. This candy may not make you as strong as a Roman soldier, but it does taste good.

In a big frying pan, over low heat, melt

¼ cup BUTTER

Stir in

½ cup SESAME SEEDS
1 cup grated COCONUT

Stir the mixture around over low heat for about 5 minutes.

Take the pan off the stove. Add:

½ teaspoon VANILLA
¼ cup HONEY

When the honey is all mixed in, put the candy in a cold place till it gets stiff enough to shape into balls. This will take about an hour in the refrigerator or half an hour in the freezer. Or set it in a pan of ice water.

Roll the candy into little balls. You will have about 3 dozen. Keep the ones you don't eat right away in the refrigerator.

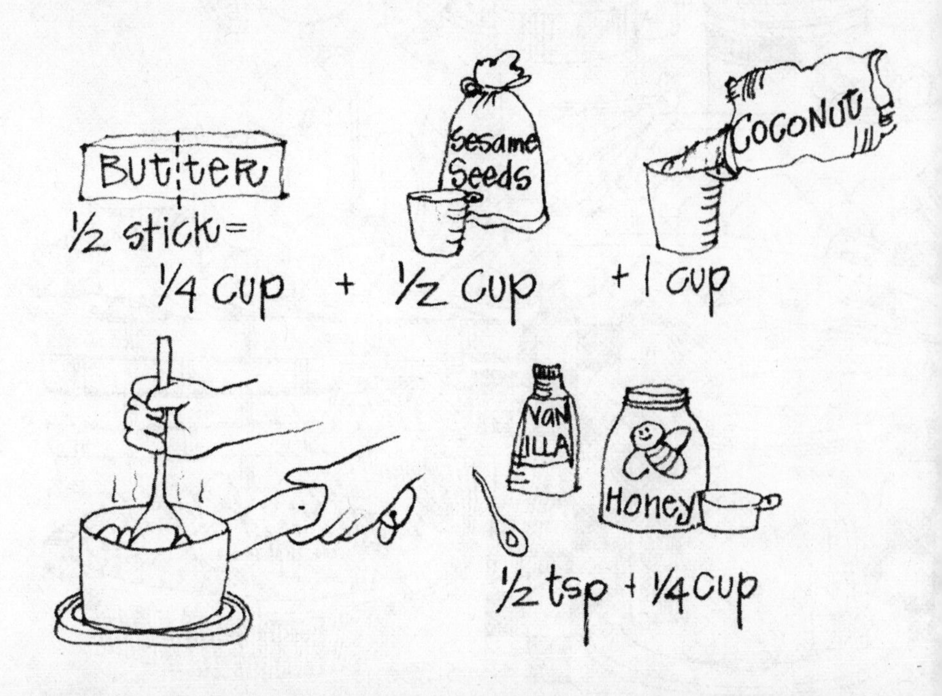

Many Hands Cooking

This is an international cookbook for boys and girls, with delicious, easy-to-make recipes from 40 countries, illustrated in full color by Tony Chen. The spiral binding and board covers allow the book to stand up for consultation while cooking. Here are some of the dishes authors Terry Touff Cooper and Marilyn Ratner have cooked up for you: guacamole (Mexico), groundnut soup (Nigeria), croque monsieur (France), meatballs (Sweden), and so round the world.

Each recipe has a code: (one hand) recipes are the easiest to make; (two hands) recipes are easy to make; (three hands) recipes are not hard, but do take a little extra time and work. If you've never cooked before, start with a one-hand recipe. You'll soon find you can easily make even those rated three hands. If you can follow instructions, you can learn to cook anything in *Many Hands Cooking*. (Published by Thomas Y. Crowell Co. in cooperation with the U.S. Committee for UNICEF.) For your copy SEND $4.95 plus 25 cents for mailing and handling to UNICEF, 331 E. 38th St., New York, N.Y. 10016.

From China
Egg Flower Soup
Serves 4

Watch an egg turn into a flower. Chinese cooks say that the cooked shreds of egg afloat in this soup look like flower petals. With a few secrets of Chinese cooking you can make this happen.

One is to have all the ingredients out, measured, and ready to use *before* you start to cook. Another is to serve food right after cooking. Some foods look and taste better this way. Egg Flower Soup is one—so serve it promptly.

NOTE: You may know this soup by another name—Egg Drop Soup.

Ingredients
1 tablespoon cornstarch
2 tablespoons cold water
1 egg
3 cups clear canned chicken broth
1 teaspoon salt
1 teaspoon chopped scallion or parsley (optional)

Equipment
measuring spoons
2 small bowls
fork
medium saucepan
mixing spoon
soup bowls

How to Make:

1. Put the cornstarch into one of the small bowls and gradually add the water, stirring it with the fork until you no longer see any lumps.
2. Break the egg into the other small bowl and beat it with the fork.
3. Pour the broth into the saucepan. Bring it to a boil over high heat.
4. Add the salt.
5. Give the cornstarch-and-water mixture a quick stir with the fork. Add it to the soup.
6. Stir the soup with the spoon until it thickens and becomes clear—about one minute.
7. Slowly pour the beaten egg into the soup. The egg will cook in the hot soup and form shreds.
8. When all the egg has been added, stir once. Turn off the heat.
9. Pour the soup into four soup bowls. Top the soup with the chopped scallion or parsley for decoration.
10. Pick up the bowl with both hands and sip the soup or eat it with a spoon.

Cooking Is Easy—When You Know How

Here's another one of those great easy-when-you-know-how books. This time the subject is cooking: scores of mouth-watering recipes are given so you can prepare everything from main courses to snacks, soups, and desserts, and they are simple enough for young people age 7 and up to follow. It's an attractive introduction to the art of cooking for every youngster who wants to plan and cook a meal for the whole family. The color drawings on every page will tempt you to try your hand at pancakes, baked Alaska, sausage rolls, chocolate sauce, and homemade bread. A useful and instructive book that will be used again and again, *Cooking Is Easy—When You Know How* was written by Isabelle Barrett and illustrated by Joy Simpson. Arco Publishing Co. $4.95.

Hippo Cook Book

Here's a trip around the world for the hamburger lover with many kinds of hamburgers you can make: spanishburger, stroganoffburger, jacquesburger, kyotoburger, bombayburger, tahitianburger, svenska kottbullar (Swedish meatballs), piroshki (Russian small pie) filled with you-know what, Chinese-style hamburger hash. And there are all the mixes, sauces, and side dishes that make this a hamburger heaven at home: western barbecue sauce, Spanish sauce, pizza sauce, Old English game sauce, teri sauce, hippo bleu cheese dressing, Iranian salad dressing, Sardi's meat sauce —and toasted garlic French bread to add a continental touch to your delightful meal! To add to the fun of making all these hamburgers there are pictures of that flippant hippo "Wolo" dillydallying around the kitchen with you! Jack Falvay was inspired enough to open a restaurant in California and nice enough to pass on his racy recipes in the *Hippo Cook Book*. Nitty Gritty Productions. $3.95.

It's More Than a Cookbook

Here is a whole kit of stuff to help kids learn to cook—even if they haven't learned to read. It has 36 recipe cards (7 x 10 inches), containing words and pictures which show each step. Any kid can learn how to make stuff like banana split salad (really!), jack-o'-lantern bread, all-American bar-b-que beans, and any-kind-of-dog sandwiches. The kit also includes a set of eight kitchen posters, a small easel, a book of worksheets with 40 pages of activities, and a book for teachers or parents. *It's More Than a Cookbook*, by Em Riggs and Barbara Darpinian, was written for schools, and all the recipes were tested with kids (even 5- and 6-year-olds), but it works just as well for older kids who do their cooking at home. To get the kit SEND $29.95 plus $2.00 postage and handling and 6 percent sales tax (if ordering from California), to Learning Stuff, P.O. Box 4123, Modesto, Calif. 95352.

Kids Cooking

The subtitle is *A First Cookbook for Children*. This is a practical how-to book giving young cooks a sound background in the kitchen. All the dishes are well-established favorites. Dinner dishes include chili con carne, hamburgers, meat loaf, roast beef, and spaghetti and meat. Regional cooking includes California avocado salad; Colorado-Denver sandwiches; New England baked beans; Pennsylvania Dutch funnel cake; and southern hominy grits. The ingredients and equipment for each dish are listed on a left-hand page, and the cooking instructions on the right-hand page. You will become acquainted with the pleasure of cooking, and the pleasure of sharing and entertaining. *Kids Cooking* is written by Aileen Paul and Arthur Hawkins. Doubleday & Co. $4.95.

Aileen Paul has also written *Kids Cooking Complete Meals* (illustrated by John Delulio), complete with menus, recipes, and instructions. Same publisher, same price.

SKILLET SPAGHETTI

1 tablespoon vegetable oil
1/2 cup chopped onion
1 pound ground beef
3 cans (8 ounces each) tomato sauce
1 3/4 cups water
1 1/2 teaspoons salt
1/2 teaspoon dried parsley
1/2 teaspoon dried basil
1/8 teaspoon pepper
8 ounces uncooked spaghetti

1. Heat vegetable oil in a large heavy skillet.
2. Cook chopped onion and ground beef in the oil until brown, stirring and mashing the meat with a fork to break up lumps.
3. Stir in tomato sauce, water, salt, parsley, basil and pepper. Heat to boiling.
4. Break the spaghetti in half and stir it little by little into the sauce. (Stir to keep it separated.)
5. Cover tightly. Lower heat and simmer 25 minutes.

4 servings.

What do you raise on a farm that's two miles long and a half inch wide? Spaghetti.

The guy who invented spaghetti sure used his noodle.

SPAGHETTI

The Lip-Smackin' Joke-Crackin' Cookbook for Kids

This is no ordinary cookbook. The graphics are wildly colorful. Jokes and riddles are everywhere. "When is a cook bad? When she whips the cream and beats the eggs!" There are songs to wash dishes by, instructions on how to grow a mini-garden, helpful safety hints, and interesting food facts. Did you know that the Indians surprised the Pilgrims with popcorn at the first Thanksgiving feast? *The Lip-Smackin' Joke-Crackin' Cookbook for Kids,* by Wicke Chambers and Spring Asher, is a zany book, just right for you if you are beginning to experiment with home cooking. Golden Books. $3.95.

Ask a bookstore to show you *The A to Z No-Cook Cookbook,* by Felipe Rojas-Lombardi. It's one of those no-stove, no-oven, but lots-of-fun cookbooks! Same publisher, same price.

"ON TOP OF SPAGHETTI"

On top of spaghetti
All covered with cheese,
I lost my poor meatball
When somebody sneezed.
It rolled off the table,
And onto the floor,
And my poor little meatball,
It rolled out the door.

(sung to the tune of
"On Top of Old Smokey")

GARDENING

Mother Earth's Hassle-Free Indoor Plant Book

This book has amusing illustrations and is a complete how-to for the beginner or experienced indoor plant grower. The authors, Lynn and Joel Rapp, unabashedly love their plants (every single one, even the temperamental fussy ones) and have given them names like Dexter, Baldwin, and Reginald. You will quickly learn to share their enthusiasm.

The authors own the Mother Earth Plant Boutique in Los Angeles and write out of their own plant-growing experiences. Simple to read, with easy-to-follow instructions, *Mother Earth's Hassle-Free Indoor Plant Book* will teach anyone and everyone what kind of plants grow best indoors, and how to properly care for them. J. P. Tarcher. $6.95 cloth. Bantam Books. $1.75 paper.

The Kid's Garden Book

Did you ever stop to think that you can make plants grow from things you throw away: pineapple tops, avocado seeds, lemon, orange, or grapefruit seeds, or the tops of beets and carrots? With *The Kid's Garden Book,* by Patricia Petrich and Rosemary Dalton, you'll learn how and get special instructions on herbs, terrariums, growing plants in water, making a park for your turtle, and growing midget fruits and vegetables in containers on your windowsill. You'll learn to plant and watch marigolds, pansies, tulips, and daffodils come out in wondrous colors in your garden. Nitty Gritty Productions. $3.95.

Garden Flowers Coloring Book

With your crayons, colored pencils, or felt-tip pens you can create all the colorful sights you see in a flower garden. 40 of the most important garden flowers, including morning glory, hyacinth, delphinium, iris, tulip, peony, and marigold, are handsomely drawn by author Stefen Bernath—and you provide the colors. The flowers are reproduced in their natural colors on the covers. To get your copy of *Garden Flowers Coloring Book* SEND $1.35 plus 35 cents for mailing and handling to Dover Publications.

GAILLARDIA. (*Gaillardia aristata*)

Growing up green is easy. Save a pit and grow a tree . . . lemon, lime, orange, or avocado. Garden from seed to harvest, a pumpkin or a petunia, a window box or a bottle garden. A jelly-glass farm or an underwater garden. Raise a root top!

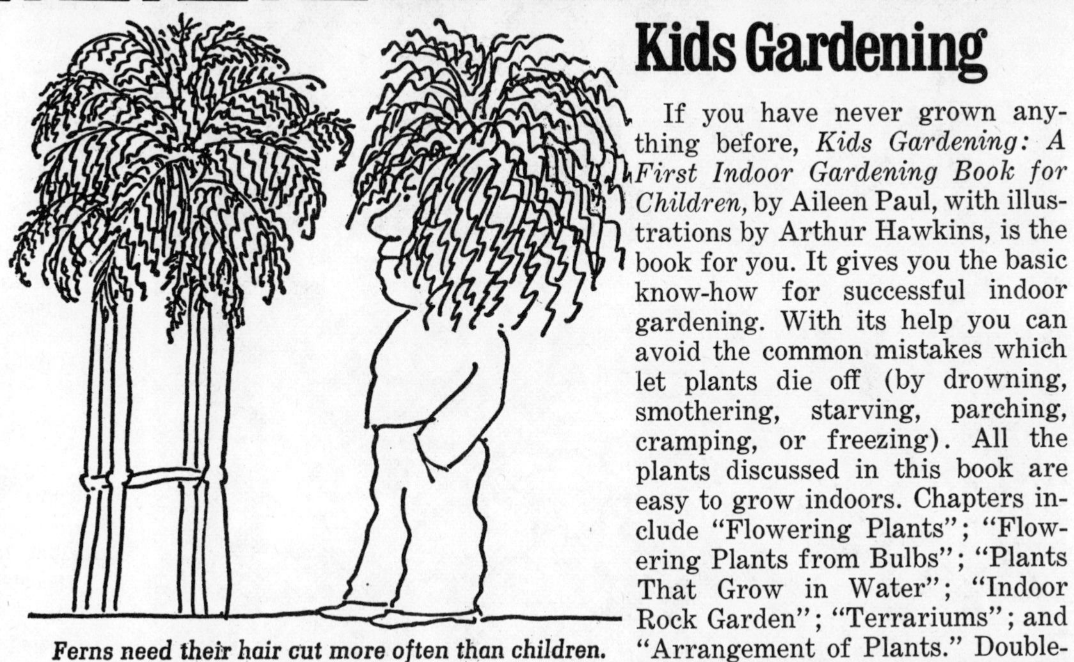

Ferns need their hair cut more often than children.

Kids Gardening

If you have never grown anything before, *Kids Gardening: A First Indoor Gardening Book for Children,* by Aileen Paul, with illustrations by Arthur Hawkins, is the book for you. It gives you the basic know-how for successful indoor gardening. With its help you can avoid the common mistakes which let plants die off (by drowning, smothering, starving, parching, cramping, or freezing). All the plants discussed in this book are easy to grow indoors. Chapters include "Flowering Plants"; "Flowering Plants from Bulbs"; "Plants That Grow in Water"; "Indoor Rock Garden"; "Terrariums"; and "Arrangement of Plants." Doubleday & Co. $4.50. In paperback, the book is titled *Kids' Indoor Gardening*. Archway. 75 cents.

Knowledge Through Color

Here are books on gardening from the Knowledge Through Color series. Each book is magnificently illustrated in full color and written by a noted authority.

Flowers of the World classifies into families, genera, and species flowers from around the world. Methods used by gardeners to increase their stock of plants are described along with the suitable type of soil for each plant. By Sandra Holmes.

House Plants has descriptions of commercially available plants throughout the world, with advice on the care and growth of hundreds of species suitable for a wide range of light and temperatures. Joan Compton also provides useful tips on propagation, pruning, cleaning, and protection. Both Bantam Books. $1.95.

We bathe our plants every Saturday night whether they need it or not.

How Does Your Garden Grow?

Gardening Is Easy—When You Know How, by Diana Sommons, illustrated by Marilyn Day, is one of those you-just-have-to-see-it books; you have to see it to realize it has an enthusiasm in words and pictures that makes it special for beginner gardeners. The dust jacket says:

> From starting a cacti collection and making a bottle garden to pressing flowers and making a window box, here is everything a boy or girl needs to know about growing plants, vegetables and flowers for fun and for the table, superbly illustrated with four-color photographs on every page. Included are scores of unique projects every youngster will enjoy: plate, basket and underwater gardens; growing and drying herbs; preserving leaves; planting in pebbles; drying and keeping flowers; raising roottops; growing tropical trees; raising avocado plants from pits; and much more. The perfect introduction to one of life's most useful as well as enjoyable activities.

It is! It is! Arco Publishing Co. $4.95.

Growing Up Green

Want to learn to grow pumpkins and petunias . . . to arrange flowers in the Japanese way . . . and all about wind and weather, trees and birds? Alice Skelsey and Gloria Huckaby tell you about famous naturalists, how to garden from seed to harvest, forcing bulbs, how to make lots of plants from one. And there's more: learn about food-plant studies, spore and seed prints, spiderwebs, stump rubbings, leaf patterns, pressing flowers, drying herbs, growing fruits and vegetables indoors, collecting ferns and cacti, bonsai-ing trees, making terrariums, and growing gourds. The book will show you how to treat cut flowers kindly, take a tree census, and celebrate a beautiful day!

Growing Up Green: Parents & Children Gardening Together is quite a book; it grows on you—and with you. Age 5 and up. To have a copy in your house SEND $4.95 (paper) or $8.95 (cloth) plus 50 cents for postage and handling to Workman Publishing Co.

"Now get growing."

How to Grow a Jelly Glass Farm

You don't need tractors or plows or acres of land to have a farm. All you need is some potting soil from the supermarket and a few things you can find in the kitchen and other corners of the house—things like coffee cans, paper cups, and pits and seeds and other leftover parts of the foods you eat.

Jelly glass farms may not grow a carrot or an orange, but even the youngest kids will find that with enough sunlight and water and plenty of love they can easily grow 15 plants, including pot of sunshine, alligator tree, green feet, prickly mountain, and other leaves, vines, and sprouts. *How to Grow a Jelly Glass Farm* is by Kathy Mandry and Joe Toto. Pantheon Books. $4.50.

FROM *GARDEN FLOWERS COLORING BOOK*

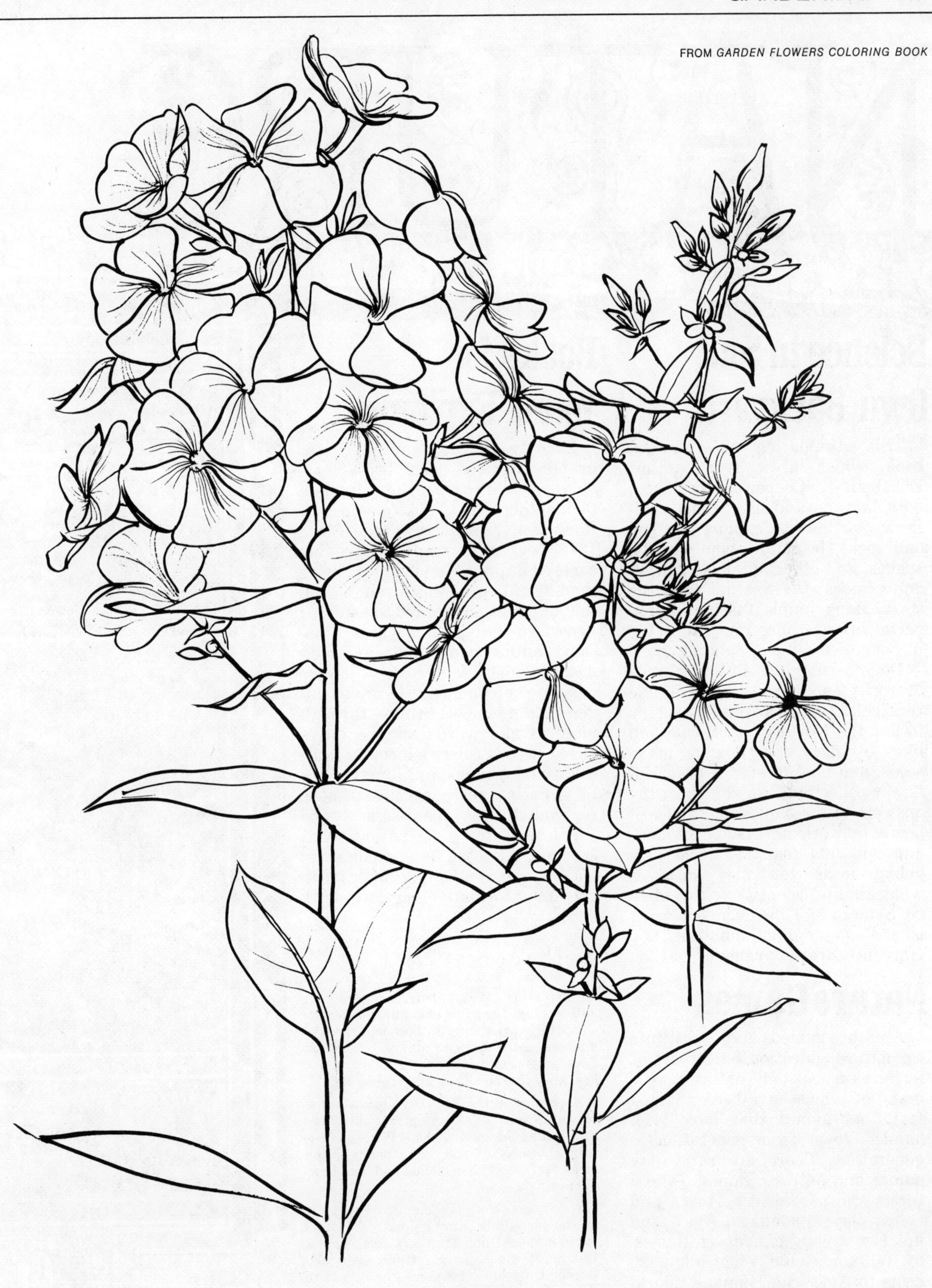

PHLOX. (*Phlox paniculata*). White, pink, red, lilac, or purple. Perennial. Mid-summer.

Science in Your Own Backyard

This stimulating and unusual book, which grew from author Elizabeth K. Cooper's experiences with her son and his friends, will show you how to explore the soil and rocks, learning about insects, snakes, and other animals; flowers and grasses; the weather; and the stars. Many simple, fascinating experiments are here for you to try in your own outdoor laboratory.

Do you know the different kinds of rocks and how to start a fine rock collection? Do you know how to find the North Star? Would you like to know how flowers make seeds, and how to keep a beautiful flower collection? As you learn the answers to these questions you will begin to work as a scientist does and will find many exciting new hobby ideas—you may discover what kind of scientist you want to be. *Science in Your Own Backyard* is a Voyager Book published by Harcourt Brace Jovanovich. $1.35.

Nature Games

A nature game is not a substitute for nature education but a part of it. Here's a book of 86 nature games, many of which have been adapted from old games that have been handed down from generation to generation. There are rainy day games and outdoor games. Take a forest census; identify birds and trees; play camouflage. Age 7 and up. For a copy of *Nature Games,* by William Gould Vinal, SEND 25 cents to American Humane Education Society, 180 Longwood Ave., Boston, Mass. 02115.

Books for Young Explorers

Each year the National Geographic Society asks 4-to-8-year-olds across the country to help them pick subjects for their new set of *Books for Young Explorers.* Since 1972 beginning readers have become the proud owners of more than 3 million of these volumes on the subjects they wanted to explore: American Indians, honeybees, dinosaurs, animal camouflage, and pandas, to name a few.

Now you can have the joy of understanding new things through pictures and words and be introduced to the Society's incomparable color photographs and illustrations by ordering these excellent books for your library. There's a special supplement for your parents that accompanies each four-volume set —a 24-page adult guide with background information on each book subject.

Set I: No. 00125

Lion Cubs. Watch tiny, playful cubs with spotted fur grow to regal adolescence —carefully practicing to become ferocious hunters on the African plains. Learn how lions protect and care for their young and how they live together in groups.

Treasures in the Sea. Dive into the underwater world of sunken pirate ships and long-lost gold doubloons. Witness the fragile beauty of the ocean floor: living coral reefs, rare seashells, and rainbow-colored fish.

Dinosaurs. Here is the engrossing story of the incredible giants that roamed the earth many millions of years ago. Some of the finest dinosaur paintings ever produced bring these creatures vividly to life. Come face to face with 50-ton vegetarian brachiosaur, or with fearsome *Tyrannosaurus rex,* which had a head 4 feet long.

Discover the world around you. Animals, rocks, clouds, and bugs. Why leaves fall and flowers bloom. Which bird is that? Learn about shells and honeybees, how animals hide, the reason for seasons, and why whales migrate. Be a curious naturalist!

Dogs Working for People. See the many different canine breeds at work. Rewarded with praise and loving care, man's best friend does so many jobs well, including guarding the home, finding lost children, herding sheep, guiding the blind, performing circus stunts, and pulling sleds.

Set II: No. 00145

How Animals Hide. Using camouflage in its many forms, seals, insects, birds, spotted fawns, chameleons, and octopuses hide for protection or to catch their prey. Enjoy the challenge of trying to find the camouflaged animals in the remarkable photographs in this book.

Namu: Making Friends with a Killer Whale. Relive the exciting true story of Namu, a dangerous predator who becomes a friendly companion to his owner. Watch Namu feed from his trainer's hand and let his new pal ride on his back.

Honeybees. What happens to a honeybee's nectar and pollen when she returns to the hive? Trace her journey and watch other members of the highly efficient community—the queen, the workers, and the drones—work together hatching eggs and building honeycombs for the common good.

Pandas. See Hsing-Hsing and Ling-Ling, the only two giant pandas in the western hemisphere, perform roly-poly antics, munch bamboo shoots, and splash in a tub of water at the National Zoo in Washington, D.C. Compared with their natural habitat in China's misty forests, the zoo is a very different way of life.

Set III: No. 00160

Three Little Indians. Live in a tepee with a Cheyenne boy named Little Knife; plant corn with She-Likes-Somebody, a girl from a Creek farm village; and go salmon fishing with Center-of-the-Sky, a Nootka boy. Enjoy riding a pony, hunting whales, and attending a New Fire ceremony.

Creepy Crawly Things. Learn about reptiles and amphibians, cold-blooded creatures that come in an incredible variety of sizes and shapes. Take a closeup look at the coral snake, alligator, sea turtle, flying lizard, strawberry frog, African chameleon, and Komodo dragon.

Cats: Little Tigers in Your House. Read the charming story of two playful and endlessly curious kittens during the first two months of their lives. Follow their frisky adventures; watch them making friends with the family dog.

Spiders. Meet a wide assortment of these fascinating eight-legged spinners of silk: the web-spinning orchard spider, sharp-eyed wolf spider, jumping spider, trap-door spider, water spider, and poisonous black widow. Contrast an inch-long garden spider with the huge tarantula.

Each hardbound 8¾ x 11¼ inch book contains 32 or more pages of full-color illustrations, plus clear, easy-to-follow text; the books are printed on heavy paper, and are available only from the National Geographic Society.

To order WRITE TO National Geographic Society, 17th and M Sts., NW, Washington, D.C. 20036. Specify set I, set II, or set III. The Society will bill you $6.95 for each four-volume series at time of shipment, plus postage and handling. If you are not satisfied, you may return the complete set without payment.

Outdoor Things to Do

This book, by William Hillcourt, is chock-full of *Outdoor Things to Do* that will provide 9-to-14-year-olds with year-round nature fun. Projects and activities include making a vivarium (which means, with living creatures); leaf printing; casting animal tracks; setting up a weather station; identifying trees, birds, insects, and water life; stargazing; nature photography; collecting shells, rocks and minerals, butterflies, and wild flowers; and building an observation ant colony. Golden Press. $4.95.

Peterson Field Guides

Don't forget the Peterson Field Guides when you go on a field trip (even one in your backyard). They are a great help in identifying the plants, animals, minerals, shells, and so on, around you. Explorers and discoverers of all ages have considered these guides essential equipment for decades. Edited by internationally famous wildlife authority Roger Tory Peterson and sponsored by the National Audubon Society and National Wildlife Federation, they are the most authoritative identification guides you can find. You can find the facts you want in seconds, whether you're a beginner or an expert naturalist. Fully illustrated and jam-packed with information.

Some guides in paperback are: *A Field Guide to the Birds* (Eastern, $4.95); *to Western Birds* ($4.95); *to Shells* (Atlantic Coast, etc., $4.95); *to Animal Tracks* ($4.95); *to the Ferns* . . . ($3.95); *to Trees and Shrubs* ($4.95); *to Reptiles and Amphibians* (Eastern and Central North America, $6.95); *to Western Reptiles and Amphibians* ($4.95); *to Wildflowers* . . . ($4.95); and *to the Insects* . . . ($5.95).

If you can't find the guides you want at your bookstore SEND the correct amount plus 50 cents postage and handling for each book to Houghton Mifflin Co., Mail Order Dept.

The Reasons for Seasons

And—can you imagine—the subtitle is *The Great Cosmic Megagalactic Trip Without Moving from Your Chair*. It's a fun book and here are some of the editor's notes:

The Reasons for Seasons invites you to open your eyes to the great, big wondrous magic show that our planet offers up all year 'round. Whether you live in the country or city or in between (as long as you're an earthling) this book is for you. If you like to inspect, collect, watch living things grow, get to the bottom of mysterious holiday customs and sudden star showers in the night sky, if you like the idea of turning some fruit skins tnto marmalade and others into sculptures, of learning age-old natural methods to divine your future, if you like games and puzzles and much much more, you have 365 days worth of treats in store.

The Reasons for Seasons, written and illustrated by Linda Allison, is a Brown Paper School book. Little, Brown and Co. $5.95 cloth, $2.95 paper.

Golden Introduction to Nature Books

The Life of Birds describes how birds mate and build a nest, why eggs differ in shape and color, how birds fly and how they use stars to navigate in far-flung migrations; much other bird lore is explained by Maurice Burton. You will learn about plumage and colors, birdcalls and their meanings, and how birds eat. One of the projects is to make a bird table feeder.

The Life of Fishes explores the shallow and the deep seas and traces the evolution of underwater life. It shows the amazing variety of color, shape, and size of fishes, explains how they feed, and describes their often brilliant camouflage; it looks at their scales, spines, and armor, their defenses, color changes, and the long migrations of fishes. One of the projects tells you how to keep goldfish. By Dr. Maurice Burton.

The Life of Insects tells you about insects that build cities and maintain armies, that fly backward, and that disguise themselves as twigs and sticks. Here are bees and ants, butterflies and beetles. And Maurice Burton will teach you about anatomy, wings and flight, stings, poisons, and the enemies of insects. Keeping ants is one project you can carry out.

The Life of Meat Eaters ranges from dogs and cats to lions, tigers, bears, wolves, and pandas. It reveals many intriguing facts: some meat eaters "talk" to each other in body language; foxes sometimes play by themselves to "charm" their prey. Life spans, speeds, tracking, and protection of endangered species are discussed. One of the projects discussed is kitten and puppy care. By Maurice and Robert Burton.

Filled with full-color pictures and written in clear and simple language. For younger children. Golden Press. $2.95 each.

Look What I Found

This is the young conservationist's guide to the care and feeding of small wildlife. You are, no doubt, becoming more aware of the importance of conservation and the vital role it will have in your future. In this book Marshal Case, executive director of the Audubon Society of Connecticut, explains why we should understand the animal life in our woods, fields, and waters and shows you, the young naturalist, how to care for and study certain wild creatures, then release them to their native habitats.

The text, photographs, and drawings describe many kinds of insects, marine animals, amphibians, reptiles, birds, and mammals which may be brought home or to school as short-term pets. *Look What I Found* gives you the basis for an enjoyable, constructive experience learning about the fascinating animal world around us. Chatham Press. $4.95.

Nature Books for the Young Reader

The Child's Golden Science Books are a series designed as a beginner's introduction to the nature of living things. Simple but meaningful answers are given to the question posed by each title: *What Is a Bird?; What Is a Flower?; What Is a Tree?; What Is a Mammal?; What Is a Fruit?; What Is an Insect?* Age 5 and up. Golden Press. 95 cents each.

Journey of the Gray Whales

The migration of the California gray whale along the Pacific Coast is called the Moby Dick parade. Thousands of people gather on the headlands to watch the whales. The California coast is the only place in the world where so many whales can be seen so close to land. They roll in the surf, "tread water" on their tails, and float in the water taking naps.

In San Diego and Los Angeles harbors, excursion boats have regular schedules for taking visitors out for a 2-hour trip in the midst of the whales. The water excursion is the prize field trip of the year for many schoolchildren.

Each year a dramatic journey takes place when the huge gray whales go northward. They are returning to their summer feeding grounds in the Bering Sea. Gladys Conklin brings that journey to life.

A whale is born in a quiet Mexican lagoon, where his education in breathing and swimming takes place. Then the long trip begins. We see the danger of death from the hungry killer whales, the strange eating methods of the grays, how they clean their bodies of clinging barnacles, their arrival in the Bering Sea, and the weaning of a young male as he begins to reach adulthood. He, in turn, will travel back to the Mexican coast to father young gray whales.

Journey of the Gray Whales is written by Gladys Conklin and illustrated by Leonard Everett Fisher. Holiday House. $5.50.

Science on the Shores and Banks

Whether you live near the Atlantic or the Pacific Ocean, or near a lake, river, stream, pond, swamp, or irrigation ditch, you will find information in *Science on the Shores and Banks* that will help you discover the fascinating plant and animal life that abounds in, on, and near bodies of water. Author Elizabeth K. Cooper leads you directly to projects you can do alone: how and what you can collect, what you can keep in aquariums or underwater gardens, what you can dry and mount, and what experiments you can perform with the simplest equipment. The contents include "Tide Pool Exploring"; "How to Be a Fish Watcher"; "Shells and Shell Collecting"; "Exploring the Water's Edge"; and "Crusty Crustaceans." A Voyager Book published by Harcourt Brace Jovanovich. 60 cents.

FREE

Birds, Flowers and Trees of the United States

You probably know your birthstone, but do you know your state flower, bird, and tree? *Birds, Flowers and Trees of the United States* lists them all in alphabetical order by state. For your free copy WRITE TO National Wildlife Federation, 1412 16th St., NW, Washington, D.C. 20036.

The Doubleday Nature Encyclopedia

A world full of plants and animals that begins with discovering nature, then takes you through the wonderful world of microscopic life, insects, life on the seashore, life in the sea, fish, amphibians, reptiles, birds, and mammals. *The Doubleday Nature Encyclopedia,* by Angela Sheehan, tells *where* animals live: in the polar regions, in the northern forests, on the African plains, in the deserts, in the rain forests, and in the mountains. It shows *how* the animals live: how they move, breathe, feed, court and reproduce, hibernate, migrate, and protect themselves. You learn about the plants that live in the sea and on land, and about the biggest plant of all—the tree. The last part of the book is about man and nature: domestic animals, studying nature, the balance of nature, and conserving nature. The book has more than 1000 illustrations in wonderful, vivid color and is a most comprehensive and informative guide to the enthralling world of plants and animals. Doubleday & Co. $6.95.

Animal Posters

Why not decorate your room with some posters of animals? Each one is a big colored picture. The titles include "Androcles and the Lion"; "Your Good Friend"; "Kindness Is For Real"; "I'm Theirs and They're Mine"; "Help Us to Help Them"; "Be Kind to Animals"; and "Horses Need Love, Too." The cost? 20 cents each. WRITE TO Education Department of the ASPCA, 441 E. 92d St., New York, N.Y. 10028.

The Dinosaur Coloring Book

Do you know how old dinosaurs really are? Some 190 million years ago, in the Mesozoic era, the dinosaurs began their reign. For 130 million years these giant reptiles dominated the earth, eating, fighting, reproducing, and changing into many curious and spectacular forms; then they became extinct. Man has ruled the globe for only 1 million years. It is fascinating to study dinosaurs and find out what life was like during their existence. Meet—and color—Diplodocus (which means "double beam"), a 20-ton plant-eating reptile; Stegosaurus ("covered lizard"), which wore a row of heavy bony plates and two pairs of 2 foot long tail spikes to protect itself; Pteranodon ("toothless wing"), with a wing spread of over 20 feet; and many many others of this strange group of animals that were here so many years before us. You'll find these fascinating creatures in *The Dinosaur Coloring Book,* illustrated by Winston Tong. Troubador Press. $2.00.

National Audubon Society

If you are a student you are invited to join the National Audubon Society and receive a year's subscription to *Audubon,* the most beautiful nature magazine in the world. Each issue has over 100 pages devoted to our precious wilderness—its beautiful vistas, natural wonders, and varied wildlife—provocative articles by foremost natural history writers and scientists; magnificent full-color photography by the most creative artists photographing the natural world today; drawings and paintings by world-renowned nature artists; up-to-the-minute bulletins on the conservation front, featuring investigative reporting on threats to the environment.

If you are a believer—if you have not given up hope that much of our wilderness can be preserved, that more wildlife sanctuaries can be established, that man can conserve instead of waste—the Society is for you. To join and receive a year's subscription to *Audubon* SEND the student membership fee of $7.00 to National Audubon Society, 950 3d Ave., New York, N.Y. 10022.

Nature Gadgets

Audubon Bird Call

This is a must for bird lovers, and simple enough for a child to use. A turn of the handle and you imitate the calls of the birds—attract them to your own backyard. Small enough to fit in your pocket or to take with you on hikes in the woods. $2.50 plus 25 cents for postage.

Plastic Magnifier

Here's a magnifying glass you can slip into your pocket and take everywhere with you. Use it to look at rocks, designs on shells, insects, leaves—whatever you want. Comes with its own case. 35 cents plus 25 cents for postage.

SEND check or money order to Field Museum of Natural History Bookshop, Roosevelt Road at Lake Shore Drive, Chicago, Ill. 60605.

Audubon's Birds of America Coloring Book

This is more than a coloring book, it is art, reproduced so that you can be the coloring artist. 46 species of birds from all parts of the United States are included. The pictures have been faithfully redrawn by Paul E. Kennedy from originals by John James Audubon, the most famous American painter-naturalist. Audubon's original plates, reduced in size, have been reproduced in color on the covers of the book. They are numbered to correspond to the pages inside; if you follow them, you will not only have a great deal of coloring pleasure, but you will learn how to identify many birds. The pictures are fine enough to frame; or you may want to transfer them to a canvas and try them in oils. Also, they would make beautiful needlepoint. Let the artist in you express itself through *Audubon's Birds of America Coloring Book.* To get your copy SEND $1.35 plus 35 cents for mailing and handling to Dover Publications.

Nature Guides

The Golden Nature Guides are an introduction to the world of nature, the most common, most easily seen, and most interesting aspects of the world around us. The 160-page books (paperback, 4 x 6 inches) are profusely illustrated in full color and loaded with concise information that makes identification and understanding easy and enjoyable.

Here is a list of available titles from Golden Press at $1.95 each.

- Birds
- Butterflies and Moths
- Cacti
- Cats
- Exotic Plants
- Fishes
- Flowers
- Fossils
- Game Birds
- Insect Pests
- Insects
- Mammals
- Non-Flowering Plants
- Orchids
- Pond Life
- Reptiles and Amphibians
- Rocks and Minerals
- The Rocky Mountains
- Seashells of the World
- Seashores
- Spiders and Their Kin
- Stars
- Trees
- Weeds
- Zoo Animals

The Bug Club Book

Did you know that there are probably a thousand different kinds of insects (every one of them fascinating) in your backyard? This book tells you how to collect, observe, and raise your own bugs, and how to organize a bug club as a center for nature activities. The author, Gladys Conklin, organized her long-successful bug club at the Hayward, California, Public Library. Her down-to-earth suggestions cover not only collecting but bug tools, methods of mounting, identifying, and raising insects from the egg and larval stages, cocoons, galls, insect anatomy, winter activities (including bug games), and public bug exhibits. *The Bug Club Book* is a highly readable handbook for young bug collectors and other naturalists. Holiday House. $4.50.

North American Wildlife Coloring Albums

It's a new way to learn; a colorful one, too! There are four large (10 x 12 inch) natural science albums with beautiful drawings for you to color, and detailed descriptive legends.

The four titles are *North American Birdlife Coloring Album, North American Sea Life Coloring Album, North American Wildlife Coloring Album,* and *North American Wildflowers Coloring Album.* (The first three are edited by Mal Whyte; the fourth, by John L. Kipping.) There are over 40 wildflowers, from blue violets to Venus's-flytrap; 15 wild animals, from the cougar to the bison; 18 forms of sea life, from the octopus to the sea horse; and 16 wild birds, from the osprey to the horned puffin.

These are not outline coloring figures, but richly detailed drawings (by Gompers Saijo) of the animal and its environment. The result, when the pictures are all colored? Your own illustrated natural science encyclopedia—and you can use crayon, coloring pencils, colors, or whatever you choose. Troubador Press. $2.00 each.

Knowledge Through Color

Here are books about nature from the Knowledge Through Color series. Each book is magnificently illustrated in full color and written by a noted authority. Collect your favorite subjects from this series and build your own encyclopedia. Knowledge Through Color is published by Bantam Books.

Rocks and Minerals contains comprehensive studies of minerals, their properties, rock formations, crystals, gems and gem minerals, with tables of chemical elements and mineral characteristics. By Joel Arem. $1.95.

Sea Shells has full-color illustrations which provide explanations and descriptions of rare shells, their inhabitants, and their human value in ornamental and practical use. By S. Peter Dance. $1.45.

Fishes of the World is an overview of the main categories of fish, with emphases on feeding, sense perceptions, reproduction, and mass migrations. By Allan Cooper. $1.45.

Snakes of the World. Popular misconceptions about snakes are corrected in this study by John Stidworthy of basic traits and of the religious-medical-economic ties between snakes and man. $1.95.

Prehistoric Animals carefully describes the history of the ancestors of modern fishes, amphibians, reptiles, mammals, outlined against a 500-million-year span of changing climates. By Barry Cox. $1.95.

The Animal Kingdom covers the whole range of animals from the smallest protozoan to man, studying diverse habitats, evolutionary trends, worldwide variation, and basic zoological principles. By Sali Money. $1.95.

Monkeys and Apes is an authoritative survey of primates, ranging from tiny tree shrews to gorillas, with focuses on their characteristics, locomotion, and family and group behavior. By Prue Napier. $1.95.

American Birds has photographs from the collection of the National Audubon Society which enhance the value of this history of loons, grebes, pelicans, waterfowl, raptors, and herons. Includes guidelines to bird-watching, bird groupings, names, and general life zones of North America. By Roland C. Clement. $1.95.

Sea Birds is illustrated with more than 200 full-color drawings. The text discusses the appearance, habits, and distribution of every species within major bird families. Also explored is the threat pollution poses to the sea bird's existence. By David Saunders. $1.95.

Birds of Prey. This authoritative book by Glenys and Derek Lloyd discusses all the world's living birds of prey, their life cycles, hunting and nesting behavior, geographic ranges; it describes how to train birds of prey for the sport of falconry. $1.45.

The Plant Kingdom describes the immense varieties of plant life with analyses of how plants feed and reproduce, compete, and cooperate for survival. By Ian Tribe. $1.45.

Trees of the World gives a description of trees from around the world grouped by type and botanical group: the way trees grow, how forests develop, and where different kinds of trees are found. Each tree is illustrated by a full-color photograph and described in detail, giving height, diameter, and shape of leaves. By Sandra Holmes. $1.95.

Salt-Water Aquariums

This book, by Barbara and John Waters, tells you how to make your own "ocean" indoors, how to buy or make equipment, how to collect or order marine invertebrates and fish, and how to keep them healthy and active in artificial seawater. *Salt-Water Aquariums* also gives advice on handling small marine animals and performing harmless, instructive experiments with them. Most important is the comprehensive appendix, which offers a list of equipment, supply sources, animal groupings, suggested reading, and many other helpful details. Age 9 and up. Illustrated by Robert Candy. Holiday House. $4.95.

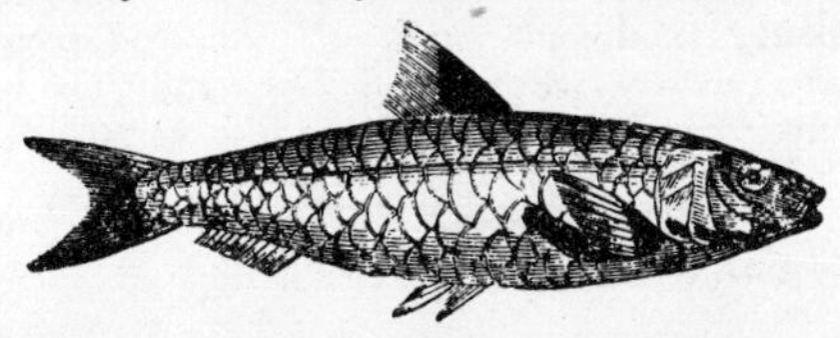

Be a Discoverer

The Discovering Nature Series contains some of the most popular Science Workshop articles from *Nature and Science,* the young people's magazine published for the American Museum of Natural History. Written by experienced people in their field, the articles are illustrated with many color diagrams and photographs; they give you an exciting invitation to investigate the life which shares the earth with man.

Discovering the Outdoors. Where do you begin to look for the nest of a blue jay? How do you go about catching and studying the small mammals of the forest? What can you find out about the life in a pond? The editor of this book, Laurence P. Pringle takes you on explorations to fields, forests, and ponds, telling you of the plants and animals that live there and suggesting scientific investigations you can make. $4.95.

Discovering Nature Indoors. Sooner or later nearly everyone has a goldfish, turtle, or other small animal pet. But too often the lives of these fascinating animals remain a mystery. Now editor Laurence P. Pringle invites you to go beyond the mere keeping of pets. The book tells you how to investigate the lives of these creatures—how to study fish and other water animals in aquariums, how to build and use a simple microscope, how to test the intelligence of mice, and much more. $4.95.

Discovering Plants: A Nature and Science Book of Experiments. Do you know why leaves fall in fall, or why they change color? Why do some flowers bloom in spring, others in autumn? Why does an oak tree growing in a forest look completely different from one growing in an open field? How do plants eat and breathe? Do you know that one stalk of corn may drink 20 gallons of water a day? How does water get from the roots of a 300 foot tree to the top of its branches? Richard M. and Deana T. Klein will help you observe and ask questions about plants and will help you find the answers. $4.95.

Discovering Rocks and Minerals. A rock you hold in your hand could contain evidence of a land once underwater or filled with dinosaurs. It is from the study of rocks, and the fossils which they sometimes contain, that we have been able to learn what we know about the earth's past—such things as how long it takes a river to form a valley or when the Rocky Mountains were formed. Ray A. Gallant and Christopher J. Schuberth have written a handbook for those who want to learn how to collect and identify rocks and minerals, discussing what tools and maps are needed, where to collect, and how to categorize and file specimens. $4.50.

All four books are published by Natural History Press.

The Curious Naturalist

What is it? This magazine, for beginning nature lovers of all ages, will help you find out. Each issue (nine per year) will give you facts and projects to help you discover more about the exciting world of nature. Why do some frogs live in trees? How do living things get water in the desert? Why can some plants live in cities while others can't? Lots of pictures and easy-to-read text make finding the answers fun: you add to your knowledge and skill while having a good time. Age 6 and up. To subscribe to *The Curious Naturalist,* cosponsored by the National Audubon Society, SEND $3.50 to the publisher, Massachusetts Audubon Society, Lincoln, Mass. 01773.

Full-Color Animal Pictures

Here's a portfolio of 12 full-color animal photographs to decorate your room. They are large (11 x 14 inches) and mounted on heavy board. Elephants, black bears, camels, giraffes, leopards, calves, and even some animals like the banteng and the sitatunga that you may only see in zoos, are photographed. On the back of each photo there is a full story about the animal or animals shown: habitat, size, gestation, and dozens of other colorful facts for your animal information library. For this truly handsome portfolio, *We All Like Milk,* SEND $2.00 to National Dairy Council, 111 N. Canal St., Chicago, Ill. 60606.

Roger Caras Nature Quiz Books

Is the jaguar the largest American wild cat? Do you know the shape of a falcon's wing? Roger Caras is a widely respected writer and filmmaker on natural history subjects who believes that nature —the world around you, animals, plants, clouds, all of it—is more than beauty and wonder: it's fun. The fun in nature is in learning about it. And after you have learned, what could be better than showing off?

That is what these books are about: showing off what you know, and finding out what you don't know. Each book has 101 quizzes on all aspects of nature. Some tests of your knowledge you will find simple, others more difficult; some you will not be able to answer. But the important thing is to relax and enjoy yourself. That is what the natural world is—the best form of recreation there ever was. Bantam Books. No. 1, 95 cents; No. 2, $1.25.

The Animals Next Door

Here is a pioneering, invaluable guide to the zoos and aquariums of the Americas, written by Harry Gersh and published in cooperation with the National Recreation and Park Association (Washington, D.C.). *The Animals Next Door* has three sections: (1) an up-to-date, complete introduction to the real meaning of zoos; (2) a directory of zoos: their location, species population, visiting hours, entrance fees, and so on; (3) a unique listing of worldwide endangered animals —birds, fish, amphibians, and reptiles—with their native countries and regions. Fleet Academic Editions. $6.95 cloth, $3.50 paper.

ECOLOGY

Save The Earth!

There are lots of books on ecology, and conservation, and saving the earth. They are all good; it's difficult to write a bad book on this subject.

One of the best of these books is *Save the Earth!: An Ecology Handbook for Kids.*

And this is what author Betty Miles says in her introduction:

> Every person can help to save the earth. You can.
>
> This book shows some of the earth's problems, and tells about ways that people have started to solve them.
>
> Learning about the problems, and inventing solutions, is one step toward saving the earth.
>
> The next step is working hard to make the solutions happen. This book gives you suggestions about things you can do that will really make a difference.
>
> It is terrible to know that land on the earth is stripped and torn apart, air is filled with poisons, and water is fouled. It is exciting to know that many people care enough about the earth to fight for it. Right now, people across this country are working for ecology. You can work with them. You and your friends can become part of the most important movement on earth: the movement to save it.

The book has four sections: "Land"; "Air"; "Water"; and "How to Do It." In each section you will learn the facts, and then there are projects for you to do. For example, in the second section:

Water

People and the Water
Everywhere Is Somewhere
Every Person's Water
The Dirty Water Blues
Taking Action
The Story of the *Clearwater* and the Hudson River
Saving the Water

Projects About Water

1. Water Count
2. Catching Rain
3. Tracking Water Pollution
4. Catching a Leak
5. Bath or Shower?

In the "How to Do It" section you'll read about specific things that you can do to work for ecology. If enough people read *Save the Earth!* and take action, we may just do it: save the earth. Illustrations are by Claire A. Nivola. Alfred A. Knopf. $2.50.

Kids! Earth is a crowded spaceship in serious trouble. Air full of poison. Water that's dirty. Animals in danger. Not enough fuel or food. Somebody do something! Join the Sierra Club. Fight pollution. Help save the only earth we've got!

The Air We Breathe and The Water We Drink

Water fulfills many needs in our daily lives: we use it for drinking, washing, cooking—and for many sports, such as swimming, fishing, boating, and waterskiing. Without water, life on earth could not survive. We must learn to conserve what fresh water we still have. *The Water We Drink!* by Enid P. Bloome, tells how you can do your part in this vital effort. In *The Air We Breathe!* Mrs. Bloome makes us aware of the ever-increasing pollution of our vital air, and how we can contribute to the fight for cleaner air. Both are excellent books for young readers. Doubleday & Co. $4.95 each.

PHOTO: DAN LEVIN

Classroom Ecology

A good place to learn about ecology is in school. Ask your teacher to include ecology as a class project. Have everyone bring magazine or newspaper articles that discuss ecology and share them. Some of the projects in this book would be fun to do in class. You can use the ecologist badge as a classroom award.

It's Your World

A basic ecology guide for young children with projects, activities, and pictures to color. Charmingly written and illustrated by Claudia Chargin. *It's Your World* is a Fat Cat Fun Book. Troubador Press. $1.50.

Earth: Our Crowded Spaceship

Enjoy the opening of the preface, by C. Lloyd Bailey, to Isaac Asimov's exciting book *Earth: Our Crowded Spaceship:*

> Just for a moment, imagine that you are a first-class passenger on a huge spaceship traveling at a speed of 100,000 kilometers per hour. You discover that the ship's environmental system is faulty. Some passengers are dying due to poisonous gases in their oxygen supply. Also, there is a serious shortage of provisions—food supplies are being used up and the water supply is rapidly becoming polluted due to breakdowns in the waste and propulsion systems.
>
> In the economy sections passengers are crowded together. Conditions are bad, especially for children. Many are seriously ill. The ship's medical officers are able to help few of the sick and medicines are in short supply.
>
> Mutinies and fighting have been reported in some sections. Hopefully this conflict can be contained, but there is fear that the violence may spread into the other compartments.
>
> . . . You are on such a spaceship right now—Spaceship Earth!
>
> The idea of our planet Earth as a spaceship may seem like science fiction, but what we have learned in this recent "space age" makes the idea more important as science fact. Do you know that everyone born since 1957—when the Russians launched the first satellite—is a member of the first generation of the space age? If you are seventeen years or younger, you are a part of that new generation, and this book is written especially for you.
>
> We believe in young people. We believe that you want to know more about population and resource problems aboard our Spaceship Earth. We believe that you will think carefully about these important ideas as you plan your own families and plan the wise use of resources. The future depends on you—the new space age generation. Therefore, we are proud to join with the John Day Company in bringing you this book by Isaac Asimov, one of the world's leading writers of science fiction and science fact.

This book discusses people, industry, food, energy, growth rate, and limits—especially, the limits of the population explosion—but most of all it discusses the changes that must be made if we are to live on this our Spaceship Earth.

Age 10 and up. For your copy SEND $2.50 plus 25 cents for postage and handling to UNICEF, 331 E. 38th St., New York, N.Y. 10016.

A Story of Pollution

This important story, which everyone should read, is called *The World You Inherit:* the story of our polluted planet, the world we have all inherited. Our air is so unclean that at times we can scarcely breathe it. Our freshwater lakes, streams, and ponds are rapidly dying from the pollutants we pump into them. Our vast oceans are becoming clogged with waste oil and other chemicals. Our landscapes are polluted with junked autos, neon signs, and other types of human litter.

The author, Dr. John G. Navarra, gives us a stern warning: we must do something about these conditions before we are drowned in our own garbage. This book is a plea for common sense: it's must reading for anyone who cares about his environment. For older children. Natural History Press. $5.50.

Science Fights Pollution

Two exciting books by Seymour Simon tell you how to set up simple experiments you can do at home or in the classroom to help you join the fight against pollution. *Science Projects in Ecology* (illustrated by Charles Jakubowski) includes 20 projects to help you study living things in their environment. You can set up a miniature living community in a jar to observe at home. A pencil, a notebook, and your curiosity are all you need to get started.

Knowledge of ecology is helpful in understanding the facts about pollution, since there are no longer unlimited supplies of pure air, pure water, and good land. Through simple experiments, *Science Projects in Pollution* explains the causes of pollution. For example, with a pack of index cards, a jar of vaseline, and a magnifying lens you can check how the exhaust from cars gives off carbon monoxide, lead, hydrocarbons, and nitrogen oxides. Strong and simple suggestions are given about what you can do about air, water, and earth pollution. Holiday House. $4.95 each.

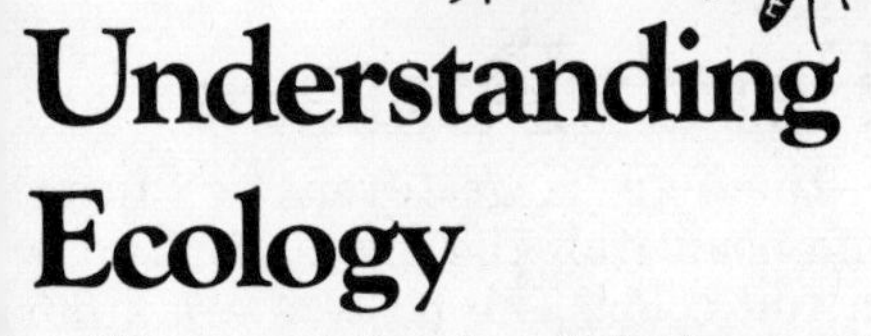

Understanding Ecology

The word "ecology" is in everyday use now, not only in classrooms but in newspapers, magazines, television, and radio. This book leads to an understanding of the science of ecology: the study of the relationship of living things to each other and to their environment. The author of *Understanding Ecology*, Elizabeth Billington, has done an unusually good job in presenting the material to the intermediate reader. As the *Instructor* said: "The book deserves to be widely distributed and read, if we are to retain many of the parts of the environment we take too much for granted." Frederick Warne & Co. $4.50.

Join the Sierra Club

John Muir founded the Sierra Club in 1892 to enable more people to explore, enjoy, and cherish the wild lands that are their heritage. He felt that man should come as a visitor to these places—the mountains, rivers, canyons, coasts, deserts, and swamps—to learn, not to leave his mark.

If you feel the need to know more about nature, and know that this need is basic to man, you are invited to join the Sierra Club. You can be as active as you want. There are 45 chapters and 200 local groups all over the U.S. You will receive a subscription to the illustrated, timely *Sierra Club Bulletin;* a discount on the more than 60 Sierra Club books in print; and the opportunity to participate in hundreds of outings that are organized each year. Juniors (up to age 13) and students (age 14 to 23) can join for the reduced fee of $8.00. Membership information is free if you WRITE TO Sierra Club, 1050 Mills Tower, San Francisco, Calif. 94104.

Be a Friend of Animals Arm Bears!

Whatever the arguments about the right to bear arms, one thing is certain: there is nothing in the Constitution that gives any person the right to murder animals.

Be patriotic. Display this red, white, and blue decal, "Support the right to ARM BEARS," on your car, window, shopping bag, book bag, anywhere and everywhere. Also use the black and orange bumper sticker: "Warning—I brake for animals." It's printed on permaplastic nite-glo for visibility in the dark. Each decal costs $1.00.

Save the Animals

Join the fight against cruel treatment of animals, birds, and other creatures by supporting the activities of Friends of Animals. You can become a junior member for $5.00 per year. Write for an application.

Also write for the new scriptographic booklet, *Save the Animals.* Its philosophy is that now is the time to take the time to stop time from running out for animals. You can help stop animal exploitation. Four copies cost $1.00. Keep one and give three to friends.

Friends of Animals Buttons

Do you want people to know you are a friend of animals? Wear a Friends of Animals button. There are three kinds: "Save the Porpoise" (shows two porpoises leaping from the water); "Boycott Fur-Bearing People" (shows a baby seal); and "Friends of Animals" (shows a polar bear). Four buttons for $1.00.

Animals R/4 Loving Patch

Now you can wear your heart on your sleeve with a heart-shaped patch that shows two hearts and a seal, and says: "Animals R/4 Loving . . . Not Wearing." $1.00 each.

SEND your orders to Friends of Animals, 11 W. 60th St., New York, N.Y. 10023.

Endangered Species Poster

We all hear about endangered species these days, but do you really know what they are and why they are endangered? Now this beautiful 28 x 43 inch full-color poster will teach you about endangered species. You will see for yourself 27 animal species currently on the endangered list. Suitable for hanging in a den or bedroom, the poster should help you become aware of which animals especially need your attention and protection.

SEND $1.00 to Endangered Species Poster, Crosman Arms Co., P.O. Box 355, Newark, N.Y. 14513.

Friends of the Earth Posters

These are large (over 3 x 2 feet) and beautiful; the photographs are in full color. The posters are from the series "The Earth's Wild Places." Put them up on your wall and, because the proceeds of these posters are allocated to conservation lobbying in Washington, D.C., you will be helping preserve the beauty of our earth. The posters available are "Haleakala Crater"; "Kukui Trees Maui"; "Midnight Rainbow"; "Dall Sheep Alaska"; "Caribou and Tundra"; "Waterfall Fluella Pass"; "Point Lobos Abalone"; "Big Sur Kelp"; and "Big Sur Surf." Each Friends of the Earth poster costs $2.50 plus 25¢ for postage and handling. SEND your order to Friends of the Earth, 529 Commercial St., San Francisco, Calif. 94111.

He prayeth best who loveth best
All things both great and small;
For the dear God, who loveth us,
He made and loveth all.

—S. T. COLERIDGE
Rime of the Ancient Mariner

Sparrows Don't Drop Candy Wrappers

Sparrows don't drop candy wrappers. Beavers build no billboards. Trees don't shed tin cans. Dolphins don't dump chemicals into the water. Bears don't belch carbon monoxide. And butterflies need no pesticides.

Here are simple dos and don'ts telling how each of us can help in small but vital ways to make this a better world to live in. Written by Margaret Gabel and delightfully illustrated by Susan Perl, *Sparrows Don't Drop Candy Wrappers* is dedicated to "an earthful of young people, with the hope they'll grow up in a cleaner, happier world." Dodd, Mead & Co. $3.95.

Somebody Do Something!

This is one of the Sierra Club notebooks for young people published six times a year. There are fact-stories about birds (the golden eagle) and animals (an elephant seal pup), reports on conservation efforts by children in school, things that you can actually do, field trips you can take, drawings, games, and so on.

Here's a letter from some sixth graders that was published in *Somebody Do Something!*:

How would you like to go to an outdoor school for three days? With your whole class? I tell you, I was there, and it was great. We went to Magee Ranger Station's Trail Creek Camp for two nights. We did studies of the water, rocks, trees, and animals. In the oxygen test we tested the water to see if fish could survive in it. We looked for tracks of wild animals then we poured plaster in the tracks and we had a cast of the tracks . . . I think that more kids should be able to go to outdoor education school across the country. The point of telling you about this school is to give *your* teachers ideas so you might have an opportunity to do something terrific like this.

To subscribe to this notebook for young people SEND $1.00 for six issues to Information Services, Sierra Club, 220 Bush St., San Francisco, Calif. 94104.

FREE A Wildlife Library

If you want to learn more about ecology and conservation, here are enough free bokets for you to start your own wildlife library. The conservation policy of the National Wildlife Federation has these objectives:

> To create and encourage an awareness of the need for wise use and proper management of those resources of the earth upon which the lives and welfare of men depend: the soils, the waters, the forests, the minerals, the plantlife, and the wildlife.

FREE No Pelican's Bellican Take DDT

If you like to work in the earth; if you like to grow flowers and vegetables; if you are concerned about the state of the environment and want to know more about pesticides, two free booklets you should write for and read are *Pesticides and Your Environment* and *Pesticides Are Perilous.*

FREE Ranger Rick Reprints

Ranger Rick's Nature Magazine free reprints will make ecology fun for you. All the who, what, why, where, when, and how behind the vital issues you want to know about, such as air and water pollution, pesticides, and litter. Colorful illustrations and lively writing explain the facts and explore their meaning in a simple, clear manner. You'll be motivated toward meaningful action at home and in the community. You can even write a school paper on ecology. If you want to learn and to help improve your environment write for and read *The Mess We're In; Recycling; Air Pollution; What Would We Do Without It;* and *The Best Present of All.*

FREE Special Report on Energy

Energy is going to be the most-talked-about topic during the next 30 years. Do you know that the energy in the fuel we use now is equivalent to each American having 80 servants working for him 24 hours a day? If you want to learn about our problems, and our big hope, write for a free copy of *Special Report on Energy.*

FREE Help Save Our Estuaries

Estuaries, which form where river meets ocean, are among the most productive environments on earth. This new 20-page booklet introduces many of the wild creatures that inhabit these priceless coastal areas to you. The answer to the question "How do we save our estuaries?" is: "With difficulty." To find out more write for your free copy of *Estuary.*

FREE Your Wildlife Heritage

Just about everyone likes to be outdoors doing things. If you do, why not learn the names and habits of the wild animals you see there and find out how you can help preserve and protcet your valuable wildlife heritage for future generations to enjoy. You probably have seen many of the animals described in these booklets in Walt Disney movies and TV nature shows; here's a chance to learn some details about them. Write for a copy of three excellent booklets: *Wildlife of Forests and Rangelands; Wildlife of Farm and Field;* and *Wildlife of Lakes, Streams, and Marshes.*

FREE Invite Wildlife to Your Backyard

Whether you have a small or large backyard, this new free 12-page color reprint from *National Wildlife* magazine will show you in detail how to develop your area into an inviting habitat for wildlife. You'll learn what kinds of large and small trees and shrubs you can plant to help provide three of the basic requirements for wildlife: food, cover, and safe areas for rearing young. You'll also find out how you can receive a backyard wildlife registration certificate and how you can become eligible to receive a bronze medallion award for backyard wildlife improvement. Write for a free copy of *Invite Wildlife to Your Backyard.*

FREE More Wildlife Booklets

Here's a wonderful chance to build your wildlife library for absolutely no money—and everything you read will help you in your schoolwork. These three sets of *Wildlife Notes* each contain interesting facts about an animal or related wildlife topic, brief enough to appear on both sides of a single sheet. There are three sets of nine notes each.

Set I includes "The Whales"; "The Polar Bear"; "The Bald Eagle"; "The American Alligator"; "Things to Know About Bird-watching"; "The California Condor"; "Birds of the City"; "The Tree Squirrel"; and "Creep Up on Nature." Set II includes "The Whooping Crane"; "The Black Footed Ferret"; "The Cougar"; "The Woodpeckers"; "Wild Horses"; "The Masked Bobwhite"; "The Pronghorn"; "The Whitetail Deer"; and "Setting the Table for Wildlife." Set III includes "Wildlife Is Amazing"; "The American Peregrine Falcon"; "The Prairie Chicken"; "The Devil's Hole Pupfish"; "The Wolf"; "The Brown Pelican"; "The Kit Fox"; "The Florida Everglade Kite"; and "Animal Communication."

One copy of each set is free.

All National Wildlife booklets are free. List the titles of the booklets you want and write to Educational Servicing Section, National Wildlife Federation, 1412 16th St., NW, Washington, D.C. 20036.

FREE

Once There Lived a Wicked Dragon

This delightful environmental coloring book tells a story; you color the drawings. It begins: "Once upon a time in a not-so-far-away kingdom there lived a wicked dragon. For many years this dragon dwelled quietly in his cave at the bottom of a hill, dozing and snacking a lot, a growing a little—ever so little—each year."

Do you know who the wicked dragon is? He's pollution! But fear not: the story has a happy ending, and will teach everyone who reads (and colors) it some of the things we have to do if we want to "smell the smell of watercress and see the sparkle of the moonlight on the water."

For a free copy of *Once There Lived a Wicked Dragon* be sure to list the title and its number, SW-105.1, and WRITE TO Solid Waste Information Materials Control Section, U.S. Environmental Protection agency, Cincinnati, Ohio 45268.

For Pollution Fighters Only

Do you want to be a pollution fighter? If so, this is the book for you. Pollution is everywhere: in the atmosphere, the water supply, and the land. The author, Margaret O. Hyde, describes the ecological systems which are affected, how our environment is polluted, and what parts of our environment are most important. People cannot continue to toss things into rivers and oceans or to spew smoke into the air; there are more intelligent ways to handle wastes, and they are discussed in this book.

You can be a pollution fighter. *For Pollution Fighters Only* shows you how to organize pollution fighting groups, offers projects and ideas for helping toward a cleaner world, and lists organizations you can join or contact for information. McGraw-Hill Book Co. $4.95.

Litter-the Ugly Enemy

Litter—the Ugly Enemy is an ecology story you should read. Author Dorothy Shuttlesworth says: "Man may be well on his way toward being the only species in the history of life on earth to become extinct by suffocating in his own garbage." The book presents the problem vividly, offers some solutions, and will make you want to join the fight against this national menace. *Litter—the Ugly Enemy* was awarded a special citation by Keep America Beautiful. Illustrated by Thomas Cervasio. Doubleday & Co. $4.95.

The Earth Book

This is a frightening book: the author, Gary Jennings, wants you to get frightened enough to do something to help save our earth from the danger mankind has caused. He has some good suggestions of specific things you can do.

Get a little frightened by this excerpt from *The Earth Book:*

> The pollution and uglification of our once-beautiful earth can be blamed on many things—man's selfishness, greed, laziness, and ignorance. But these factors boil down to just one basic cause: too many people, and those people demanding more from the earth than it can provide. There is a limit to how much food the earth can grow to feed our bodies and how much oil the earth can provide to fuel our machines. But, long before we run out of food and fuel, we will have run out of other things that make life worth living—elbow room, privacy, independence, quiet, and peace of mind.

The book is only 32 pages long. You'll want to read it over and over again, to yourself, your parents, your neighbors, almost anyone you can get to listen. J. B. Lippincott Co. $4.95.

Our Noisy World

We have the distinction of living in the noisiest age in history. Hums, hisses, rumbles, pops, clicks, and sounds beyond description pollute the atmosphere. These noises are so prevalent that we seem almost unaware of their presence. But we only think we tune them out: the mounting racket may be momentarily excluded from conscious thought, but it makes its impact all the same on our mental, emotional, and physical well-being.

In *Our Noisy World,* Dr. John Gabriel Navarra tells of our newest pollution problem: noise; and indicates how we can solve the problem. The book was included in the Child Study Association list of Children's Books of the Year, (1969). Doubleday & Co. $4.95.

Antipollution Lab

Everybody talks about pollution, and now you can help solve this problem in your community. Newspapers, television, and radio repeatedly tell us how badly polluted our world is becoming. *Antipollution Lab: Elementary Research in Air, Water and Solid Pollution,* by Elliott H. Blaustein, provides you with the know-how to begin approaching the problem of pollution scientifically. You will learn how to locate the pollution sources in your community, to measure how much pollution exists there, to examine the nature of the pollutants, and finally, to gather information that will be useful to the people in your community who are fighting pollution in the social and political arena.

Let's stop being tolerant of pollution! A good first step is this book. Sentinel Books. $2.25.

Johnny Horizon Children's Kit

You will receive a "Dear Partner" letter from Johnny Horizon that begins: "Thank you for joining me in the campaign to clean up America for our 200th birthday. By pledging to work for a clean America and becoming a Johnny Horizon '76 Partner, you can help our nation improve and protect the environment."

This is what you'll get in your Johnny Horizon Children's Kit:

your "Dear Partner" letter
a colorful decal for your bicycle
a Johnny Horizon plastic litterbag
your Pledge Card
"Small Steps in the Right Direction"
"Energy and Kids"
a picture of Cicely Tyson

There are many ways you, your family, and your friends can help. You can start by cleaning up your own neighborhood, school grounds, or a nearby park; collect bottles, paper, and other material that can be recycled; learn about your environment; tell your friends about this campaign; and WRITE FOR your free kit to Consumer Information, Pueblo, Colo. 81009.

Let's Clean Up America For Our 200th Birthday

Pamphlets on Forestry and Conservation

Our Forest and You is a special teacher's packet containing the following pamphlets on forestry and conservation. A free set is available to teachers, or you may want to select a pamphlet or pamphlets and order them.

Logs into Lumber. A poster that is a pictorial description of the lumber manufacturing process. 25 cents.

Eleven Great Trees of the West. A 6 x 11½ inch folder that opens out into a two-sided 23 x 35 inch wall poster showing botanical information, growth range, inventory volumes, properties, and product uses of the 11 western commercial tree species most commonly used in the production of lumber and wood products. The folder has photographic and line sketches of trees, cones, barks, and needles. 25 cents.

A Regrowth Record of Western Forests. A 24-page, 31-year photographic history of clearcutting performance in the Clearwater National Forests and the Willow Creek area of the St. Joe National Forest, both in Idaho. 5½ x 7¼ inches. 10 cents.

The Wilderness Resource. The book contrasts what you may and may not do in modern wilderness areas. Provides a definition of wilderness as given in the 1964 Wilderness Act, outlines the present wilderness acreage, and asks, Can we afford it? 5 cents.

The Wilderness: Just How Wild Should It Be? A six-page reprint of a *Los Angeles Times* article by a southern California trial lawyer, an avid user of wilderness areas, who believes that enlightened conservationists—not purists—can increase access to our wild areas of America. 5 cents.

A Vital Aspect of Forest Ecology—Clearcutting. An eight-page explanation of clearcutting as a valuable technique of harvesting for regrowth, a discussion of erosion, fertility, wildlife, and harvesting practices. There is a series of photos showing the 31-year sequence in the life cycle of a forest. 5 cents.

The Forester in the Plaid Shirt. Outlines the educational qualifications, philosophy, and work of the industrial forester. 5 cents.

Our Forest Bank Account. The book presents the concept of sustained yield and concludes that sustained yield adds interest to our forest bank account. 5 cents.

If you want a copy of any of these pamphlets WRITE TO Western Wood Products Association, 1500 Yeon Building, Portland, Ore. 97204.

Boys' Life REPRINTS

Build a Boys' Life Reprint Library

Here's an easy, inexpensive way for boys and girls to build an "instant" library on dozens of subjects you want to learn about. Use the famous *Boys' Life* magazine reprints. Each one is 24 pages, completely illustrated with drawings and photographs; the articles are usually written by America's top authorities. Reprints are only 50 cents each; for $1.80 you can order a handsome reprint binder that holds all the booklets, and a special index sheet.

Just look at the different titles you can have in this "things to do and make" library:

26–023 *Webelos Scout Helps*
26–024 *Stamp Collecting*
26–026 *Stunts & Skits*
26–030 *Model Railroading*
26–033 *Be a Second Class Scout*
26–034 *Be First Class*
26–036 *First Aid Skills*
26–038 *Boats and Canoes*
26–041 *Cooking Skills and Menus*
26–042 *Hiking & Camping Equipment*
26–043 *Handicraft*
26–044 *Pioneering*
26–045 *Fishing*
26–046 *Toughen Up*
26–047 *Showman Activity Badge Helps*
26–048 *Outdoorsman Activity Badge Helps*
26–053 *Forester Activity Badge Helps*
26–054 *Naturalist Activity Badge Helps*
26–057 *Craftsman Activity Badge Helps*
26–058 *Scoutcraft Skills*
26–074 *Litepac Camping Equipment*
26–076 *Hiking Skills*
26–077 *Camping Skills*
26–079 *Traveller & Engineer*
26–081 *Slides of the Month*
26–082 *Geologist & Scientist*
26–085 *Bike Fun*
26–091 *Short-Wave Listening*
26–092 *Winter Activities*
26–094 *Nature Hobbies & Activities*
26–095 *Bill of Rights*
26–097 *Our Heritage of Freedom*
26–099 *Law and Justice*

For your *Boys' Life* reprints be sure to list the number and name of the reprint. WRITE TO Boys' Life, North Brunswick, N.J. 08902.

BICENTENNIAL

FREE Bicentennial Booklets

To help you understand more about the Bicentennial celebration, here are 10 booklets you should have. *The Declaration of Independence* deals with the events which led up to the Declaration and includes this great statement. You can have your own copy, in booklet form, of *The Constitution of the United States*. Do you know all the presidents? In *The Presidents of the United States* there is a capsule history of each and the main events of his administration.

Other booklets are *The Flag of the United States, The Story of the Pilgrims,* and *Patriotic Songs of America;* plus the lives of *George Washington, John Hancock, Abraham Lincoln,* and *Robert E. Lee*.

These informative booklets are yours free. WRITE TO John Hancock Mutual Life Insurance Co., Community Relations, 200 Berkeley St., Boston, Mass. 02117.

FREE Patriotic Songs of America

Now you can sing out for the Bicentennial with the favorite songs that are part of our heritage—the songs of patriotism that have been sung by Americans for the past 200 years. *The Patriotic Songs of America* contains the words and music to some of our country's oldest and best loved songs: "Yankee Doodle," "Battle Hymn of the Republic," "Dixie," "America the Beautiful," "Tenting on the Old Camp Ground," "Hail, Columbia," "America," and "The Star-Spangled Banner." There is also a brief history of each of the songs. For your free copy of the *Patriotic Songs of America* WRITE TO John Hancock Mutual Life Insurance Co., Community Relations, 200 Berkeley St., Boston, Mass. 02117.

Early American Life

Early American Life is the official magazine of the Early American Society, the national organization interested in advancing understanding of American social history and early arts, crafts, furnishings, and architecture. The magazine is published every other month. Membership in the Society (dues are $6.00 a year) entitles you to a subscription to *Early American Life* plus large discounts on "books, furnishings & sundries" (catalog will be sent to you when you join). Articles in one issue of the magazine included "The Art of Spinning"; "Kachina—Free Spirits of the Hopi Indians"; "Frontier America"; "Life in Newport"; "The Art and Mystery of Tanning"; "Rocky Mount in Ten-

Join the U.S. birthday fun . . . ride with Paul Revere, build your own Independence Hall, cut and color heroes of the Revolution. In the Spirit of '76 braid a rug, dip a candle, brew sarsaparilla, wave flags . . . celebrate 200 years of freedom.

nessee"; "Make Your Own History"; and "Mechanical Banks." The advertisements for early American reproductions are fun too. To subscribe to *Early American Life* and receive all the benefits of the Society SEND $6.00 to Early American Society, 330 Walnut St., Boulder, Colo. 80302.

Our Heritage of Freedom

This booklet is presented by the Boy Scouts of America as a reminder of our duty to preserve the wonderful heritage of freedom that our forefathers left to us—those who, in the words of Abraham Lincoln, "brought forth upon this continent a new nation, conceived in liberty, and dedicated to the proposition that all men are created equal . . . and that government of the people, by the people, and for the people, shall not perish from the earth."

The full-color, 15-page, comic-style booklet shows-and-tells about the Mayflower Compact, the Pennsylvania Charter of Privileges, the Colonists defending their rights, *Common Sense* and *The Crisis*, the eve of Independence, the Bill of Rights, and so on. To get your copy of this excellent *Boys' Life* reprint SEND 50 cents to Boys' Life, North Brunswick, N.J. 08902, and ask for *Our Heritage of Freedom* (No. 26–097).

Bill of Rights

This timely booklet (8½ x 11 inches) discusses the Bill of Rights —the first ten amendments to the Constitution. They are concerned with the individual—his rights and freedoms. They protect life, liberty, and the pursuit of happiness: in other words, the very heart of our democracy. Each amendment is stated and the reason for its existence is pictured in full-color drawings. For a copy of this just-right-for-the-Bicentennial *Boys' Life* reprint SEND 50 cents to Boys' Life, North Brunswick, N.J. 08902, and ask for *Bill of Rights* (No. 26–095).

FREE
Color Me Proud Coloring Book

Here's a fun and educational coloring book. The pictures accompany the words of famous American songs such as "The Star-Spangled Banner," "America," "You're a Grand Old Flag," "Yankee Doodle Boy," and "America the Beautiful." Here too is the Pledge of Allegiance and the opening of the Constitution. The *Color Me Proud* coloring book for the Bicentennial can be yours free. WRITE TO Color Me Proud, P.O. Box 7338, Chicago, Ill. 60680. (Available only until October 31, 1976!)

A Coloring Book of the American Revolution

The heroes of the Revolution are here—George Washington, Alexander Hamilton, Patrick Henry, Tom Paine, Paul Revere, and all the others. The drawings are filled with action and wit, and each of the 50 pages is annotated with historical notes and famous quotations from the men who were the voices of '76. A delightful history book. Alfred Frankenstein in the *San Francisco Chronicle* wrote about Bellerophon Books: "A serious effort—beautifully brought off. The underlying theory is to teach children something without telling them they are being taught." Bellerophon's coloring books are printed on good, sturdy paper and are bound in heavy, bright, and colorful covers. *A Coloring Book of the American Revolution* is available at your bookstore or SEND $1.95 plus 25 cents for postage and handling to Bellerophon Books.

Captain John Parker

The Golden Book of Colonial Crafts

Here's a great way to get into the spirit of the Bicentennial: get the comprehensive *Golden Book of Colonial Crafts* and start working on one of the projects that will take you back to early American days. Modern tools and materials can be used in braiding rugs, candlemaking, Indian beadwork, making cheese, canoeing, doing needlework, recaning old chairs, making toys, even brewing sarsaparilla! More than 400 photographs and illustrations accompany easy instructions. There is helpful information on sources and costs of materials, and on the time needed and degree of skill required for each activity. Golden Press. $8.95.

Build Your Own Early American Village

This book is made more important because of the Bicentennial and could be an interesting project for school. Author-architect Forrest Wilson does a show-and-tell on early New England houses: the different types, how they were built, furnishings. He discusses the makeup of a town, what life was like 200 years ago, and how you can plan and build your own "restoration" town with the colorful punch-out houses, barns, and sheds included in the book. If you like putting models together and want to learn some history, you'll enjoy *Build Your Own Early American Village*. Pantheon Books. $2.95.

LEXINGTON FIGHT, APRIL 19, 1775

O, what a glorious morning is this!
Samuel Adams

Major John Pitcairn

Paper Soldiers of the American Revolution

Each of these two unusual books contains 80 Revolutionary paper soldiers. The figures are standing, marching, kneeling, or on horseback, firing a musket, beating a drum, and so on. And *you* color them and cut them out. You can even set them up in ranks.

These paper soldiers are more than mere toys; they help recreate history and our battle for independence. The uniforms are authentic. Marko Zlatich, a member of the Company of Military Historians, has checked every figure for accuracy, and he has included historical notes and color instructions for each uniform. Alaen Averill, who is responsible for the ingenious figures and engineering, has included a short history of paper soldiers, which date back to the 15th century.

Be careful: Don't let your father get hold of sets I and II of *Paper Soldiers of the American Revolution* before you do, or *he'll* start coloring them. There are horses and cannons and soldiers from the Riflemen of Bucks County, Pennsylvania, and Colonel Samuel Webb's Continental Regiment . . . even Alexander Hamilton and General George Washington. What a handsome collection they will make—for your room, to show at school, or just to own. Set I contains American soldiers and their allies; set II contains British and Hessian soldiers. Each set (8½ x 11 inches) is sold separately. The books are available at your bookstore or SEND $2.50 per book plus 25 cents for postage and handling to Bellerophon Books.

About Early America

A great story-puzzle book for the Bicentennial: word finds, word grids, and crossword puzzles about early America. The information you need to solve the puzzles is in the book. You just have to find it somewhere in the stories. If you read the book in chronological order, you'll have all the facts you need. *About Early America* is by Charles Preston with illustrations by Kelly Oechsli. Golden Press. $1.50.

Happy Birthday to U.S.

This is a do-it-yourself Bicentennial book that makes sense for a do-it-yourself country like ours. Some of the activities invite you to write things in the book: stories about your family, facts about your school and town, songs, jokes, even dreams. There are also places to paste in photographs and drawings. Other activities will take you outside the book into the real world where you can get involved with people close to home or in distant parts of the country.

These activities are meant to help you discover the good things about America's past and present so you can cherish them. At the same time they may help you figure out what needs to be done now so that, like patriots of every age, you can work to make things better.

While some of the projects call for serious thinking and action, others are meant to be fun. After all, this is a birthday time for you and for all of U.S.

Happy Birthday to U.S., activities for the Bicentennial, is by Murray and Roberta Suid. Addison-Wesley Publishing Co. $3.96.

The Great Seal of the United States

front back

A Mini-Quiz About the Seal

You'll find the answers to this quiz printed upside-down on this page. After you find out how much you know, give the test to your friends.

1. What do the stars signify? (Hint: count them.)
2. What does *E Pluribus Unum* mean?
3. What is the eagle holding? What do these things mean?
4. What does the unfinished pyramid signify?
5. What does the eye above the pyramid stand for?
6. What does *Annuit coeptis* mean?
7. What does *Novus Ordo Seclorum* mean?

---

1. The thirteen stars stand for the thirteen original states.
2. *E Pluribus Unum* means "From Many, One."
3. The eagle is holding arrows and an olive branch. These symbolize the government's power to make war and peace.
4. The unfinished pyramid is an emblem of strength and suggests the idea that the work of building a great nation is never finished.
5. The eye of God above the pyramid stands for the hope that God will give favor to the nation.
6. *Annuit Coeptis* means "He has favored our undertaking."
7. *Novus Ordo Seclorum* means "A New Order of the Ages."

Now that you know how words and picture symbols are brought together to make a seal, try designing your own version of a seal for the United States.

My U.S. Seal

Foreign Stamps Commemorate the U.S. Bicentennial

America's Bicentennial has a lot of meaning to peoples of other countries. French, German, and Polish officers and men fought alongside our own in many of the famous battles of the Revolution. Because their histories are deeply intertwined with ours, few of the world's countries are allowing this exciting anniversary to pass unnoticed. Many countries have already begun printing stamps to honor our 200th year as a nation, among them Grenada, Upper Volta, Liberia, and Nicaragua. More will follow soon. The Bicentennial commemorative stamps may well become collector's items of the future. If you would like a selection of the foreign stamps already produced, SEND $1.00 plus 25 cents for postage and handling to Scott Publishing Co., P.O. Box 208, Floral Park, N.Y. 11002.

Stamps for the Bicentennial

Why not celebrate the Bicentennial with stamps showing historical greats such as Benjamin Franklin, George Washington, and Thomas Jefferson? The United States Stamp Collecting Kit is filled with exciting stamps that will take you back to the most colorful times in our country's past. You can get the United States Stamp Collecting Kit at your local post office. The cost is $2.00 and the kit comes with a stamp album to put your stamps in, hinges to attach the stamps, and a 32-page booklet that tells you how to start a stamp collection.

Flags of the American Revolution to Color

Here are 39 flags for you to color as you celebrate the Bicentennial. Each flag has color notes ("red field, yellow scrolls, fringe and outer edge of shield, blue inner edge of shield and ribbon above the shield. Grapes are purple, leaves green and ground brown"). There are also historical notes for each flag, for example: "The New York Beaver Flag was reported flown in 1775 on the armed ships of New York. The beaver had been on the seal of New Netherland, and is still seen on the seal of the city of New York." *Flags of the American Revolution to Color* is available at your bookstore or SEND $1.00 plus 25 cents for postage and handling to Bellerophon Books.

FLAG OF THE CONTINENTAL NAVY
(The Gadsden Flag)

Colors: Yellow field and fringe; black lettering; brown snake with red tongue; green grass.

The device below appeared during the French and Indian War and may have inspired the rattlesnake on several of our early flags.

Congress' first legislation on the subject of a federal navy was in Oct. 1775.

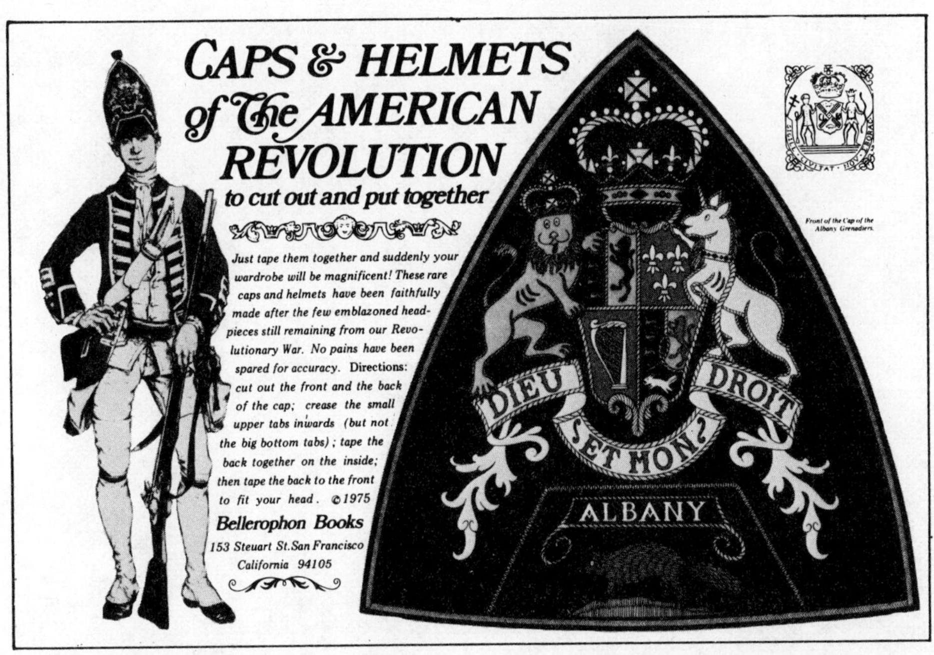

Caps & Helmets of the American Revolution

These tall, colorful helmets and mitered caps are emblazoned in gold and silver and bright red and blue, and no pains have been spared for authenticity. Once you've cut them out and taped them together, they'll fit your head. You will look magnificent as you wear these 13 inch high headpieces to celebrate the Bicentennial!

Detailed historical notes are included with each cap. And there are full-color pictures of soldiers of the Revolution in complete uniform, including the caps and helmets.

Set I contains the headgear of the Newport Light Infantry, the Rhode Island Train of Artillery, the Light Infantry Company, the Scottish Regiment, and the Providence Grenadiers. Set II contains the headgear of the Albany Grenadiers, the Connecticut Grenadiers, the Hessian Fusiliers (von Knyphausen), and the Newport Guards.

Collect! Wear! Celebrate! Great for all kids. You may want two of each: one for decoration in your room and one for helmets to wear. *Caps & Helmets of the American Revolution,* sets I and II, is available at your bookstore or SEND $2.95 plus 25 cents for postage and handling for each set to Bellerophon Books.

$50.00 Reward from Bellerophon Books!

Bellerophon Books is offering a $50.00 reward for information leading to the discovery of any other authentic extant American emblazoned, mitered caps, 1775–83, beyond those in sets I and II of *Caps & Helmets of the American Revolution.* Look in grandma's attic!

Give me liberty or give me death!

PATRICK HENRY

Historical Models

Now you can have a tabletop adventure in American history to celebrate the Bicentennial by putting together authentic colorful models of some of our most famous historic buildings. Old North Church, Faneuil Hall, and the George Wythe House are just a few of the landmarks you can recreate. Each kit contains a durable artist's illustration board printed in full color, and each section is die cut for easy assembly. The scale is 1 inch = 12 feet, and illustrated historical notes are included. All the Landmark Models™ carry the distinguished marks of Bicentennial commissions and preservation foundations across the country.

Old North Church. Completed in 1774, it became famous as the starting place for the midnight ride of Paul Revere. $2.50.

Faneuil Hall. In the annals of America, there is no place more distinguished for powerful eloquence than the Cradle of Liberty. $2.50.

Old State House. In 1776 the Declaration of Independence was first read to the citizens of Boston from its balcony. $2.50.

George Wythe House. This Williamsburg Model Kit contains replicas of the George Wythe House and ten outbuildings which form a plantation in miniature. $5.00.

Independence Hall. The building, in Philadelphia, in which the Second Continental Congress created and signed the Declaration of Independence. Here in 1787 the Constitution of the United States was drafted. $4.00.

Carpenters' Hall. Begun in Philadelphia in 1770, it was the site of the first Continental Congress and also the office of the first secretary of war, Henry Knox, from 1790 to 1800. $2.00.

Federal Hall. Built as New York's first City Hall in 1703. President George Washington took the oath of office on its balcony. $2.50.

The White House. The White House was built to house the second president of the United States, John Adams. Since then it has always been the familiar American landmark of the home of the president. $3.00.

To order any of the Landmark Models SEND a check for the correct amount plus 50 cents for postage and handling to Museum Shop, Museum of Fine Arts, Boston, Mass. 02115.

The Golden Book of the American Revolution

This colorful book was adapted by Fred Cook from *The American Heritage Book of the Revolution* for young readers. And it's filled with hundreds of beautiful full-color pictures. The whole man-by-man important story is here from "The War Begins" and "The First Campaigns" through "The Darkest Hours" and "The Making of an Army" to "The World Turned Upside Down." Bruce Catton ends his introduction to the book with:

> This book is an attempt to give flesh and blood to the war that gave us our independence, to get back to the reality beneath the legend—not only in words, but also as far as it is possible, through pictures drawn and painted by men who lived through these times. If it provides a clearer understanding of the Revolution—if it breathes a little life into the legend of the men who provided us with our freedom—it will have served its purpose.

The Golden Book of the American Revolution is a must book for these days of remembering history, our history. Golden Press (an imprint of Western Publishing Co.). $3.95.

DRAWING: PETER F. COPELAND

AMERICAN TROOPS

Drummer of the Sixth Connecticut Regiment, 1777

DRAWING: PETER F. COPELAND

BRITISH TROOPS

Pioneer of the 54th Regiment of Foot, the "West Norfolk Regiment," 1782

Uniforms of the American Revolution Coloring Book

How right for the Bicentennial! 30 lively drawings reproduce with complete authenticity uniforms from the American Revolution . . . British, French, German, and American. Each regiment is described briefly, with complete instructions for coloring the uniforms accurately. Peter F. Copeland, historical artist and consultant, has reconstructed the uniforms from early descriptions and other source material, illustrating many for the first time. There are 30 illustrations, 64 pages, 8¼ x 11 inches. For your copy of *Uniforms of the American Revolution Coloring Book* SEND $1.50 plus 35 cents for postage and handling to Dover Publications.

The Revolutionary War

This unique National Geographic book helps you relive the struggle of a "tea party" in Boston Harbor, of Lexington and Concord, and of the long, bitter conflict that led to Yorktown and gave birth to a nation. Magnificent illustrations help bring an era alive, as do text-and-picture visits to historic Revolutionary War sites as they are today.

To order your copy of *The Revolutionary War: America's Fight for Freedom,* by B. McDowell, WRITE TO National Geographic Society, 17th and M Sts., NW, Washington, D.C. 20036. The Society will bill you $4.25 at time of shipment, plus postage and handling.

Everyday Dress of the American Revolution

Another delightful coloring book by Peter F. Copeland, with 46 detailed drawings that reproduce the everyday dress of American colonists of the 1770s: broom seller, farmer and his wife, wagoner, cooper, gentleman, barber, doctor, fisherman, carpenter, rural mailman, wigmaker, and many others. Each authentic illustration is accompanied by a caption and a brief description of the clothing that was worn. 15 examples are pictured in full color on the covers. Color the drawings and use them for a Bicentennial project at school or to decorate your room. For your copy of *Everyday Dress of the American Revolution Coloring Book* SEND $1.35 plus 35 cents for postage and handling to Dover Publications.

American Revolution Posters

These are spectacular, giant posters; each one is 23 x 35 inches. There are five of them, and they recreate the turbulent days of the beginning of our country. These lithographs of famous paintings are "The Battle of Bunker Hill"; "General George Washington and His Horse"; "Drum Corps Leads Marching Colonists"; "Reading of the Declaration of Independence"; and "Heroes of the American Revolution" (Henry Knox, Nathaniel Greene, Horatio Gates, Israel Putnam, Marquis de Lafayette, Alexander Hamilton, Francis Marion, and Anthony Wayne).

For decoration for your own room or your game room or to take to school, these posters are just right to help you celebrate the Bicentennial. To get this beautiful set of American Revolution posters SEND $4.00 plus 25 cents for postage and handling to Bellerophon Books.

DRAWING: PETER F. COPELAND

The BROOM SELLER was a peddler who cried his wares through the streets of town. He wears a workman's jacket, a small round felt hat and breeches of mattress ticking.

CHILDREN'S MUSEUMS

The Children's Museum

It's hard to believe there are people who don't know there is such a thing as a children's museum, when in fact there are 70 of them in the U.S. today. Not the largest of these, but second oldest (it has already celebrated its 60th birthday) and surely one of the best known and best loved is the Children's Museum of Boston. A staggering total of 20 million children and adults have participated in stimulating educational experiences under its roof. It has served as a model for many of the children's museums that are springing up all over the country. The people who know the Boston Children's Museum the best use it the most, and love it the hardest are usually under 5 feet tall, fast on their feet, curious as gerbils, and as full of bounce as a can of brand-new tennis balls.

How do you describe a children's museum? It's a place designed to stimulate kids' powers of imagination; where playing and learning are paired and happen at the same time; where exhibits are experiences in which kids participate:

A simple pair of Eskimo snow goggles can tell us volumes about the harsh demands of the arctic, of relief from squinting at ice floes in the glare of a low spring sun, the craftsmanship of the Eskimo and even the shape of his face. But the goggles will not tell their story while locked inside a glass cage, even when "explained" by a neatly typed label. At the Boston Children's Museum snow goggles are not to look at . . . they are to look through!

Maybe the best description is also the briefest: "Do touch!" And grown-ups are invited to join children in exploration.

The Boston Children's Museum was founded in 1914 by a group of teachers who needed a museum that was relevant to the educational needs of children. Today it continues to explore new ways of offering intense, provocative, useful experiences to the child who must learn to cope with an increasingly tough and demanding world.

The Visitor Center at the Museum

The Visitor Center is a 7500 square foot, multilevel brick building that houses the Museum's principal exhibits, called "learning spaces." Here are computers; living things; grandmother's attic; Japanese home; Sitaround; zoetropes; Algonquin grownups and kids; wigwam; and a "what's new" exhibit theme (with changes every 2 to 3 months). Nathan Cobb wrote in the *Boston Globe:* "There is little doubt the word 'museum' is being stretched to include what is happening here. Although the facility still administrates a collection, it has evolved more and more into a home for self-expression and less and less into a cultural mausoleum." As a matter of fact, children in a sense serve as consultants in the process of exhibit development: what the Museum develops is what will work as a stimulant to their imaginations.

A favorite exhibit is: *How Movies Move.* Youngsters make their own animated movies. They draw pictures in a dozen small frames along a paper strip and then fit the strip like a hatband inside a zoetrope. With the aid of simple equipment and a free environment conducive to exploration, youngsters make a discovery . . . a series of still images rapidly viewed under controlled lighting produces the illusion of movement.

Youngsters who visit "Grandmother's Attic" are introduced to the world their grandparents knew. Here are old-fashioned dresses, an old-fashioned washing machine, a butter churn, and a 1927 Atwater-Kent radio —and all in working order.

Children learn about data processing by playing simple word and number games on the computor.

In an Indian exhibit, youngsters learn about Native Americans not by watching a movie or looking at artifacts, but by grinding maize in stone mortars, munching berries, and trying on skins, pouches, and moccasins.

In a related measurement exhibit, various types of scales and balances are used by the children to weigh anything from a coin to a rock, from a bar of candy to the children themselves . . .

Said one assistant: "We want to get away from traditional exhibits. Kids should be able to look at, touch and play with exhibits. This makes them solve problems and make discoveries of their own."

The Museum Does More!

The Visitor Center, the biggest attraction for the public, is only a third of today's Museum. The Resource Center houses a reference library; collections; circulating department (educational kits and units); teacher shop; and "recycle," an area filled with industrial scrap for sale, with regularly scheduled workshops on uses for

Explore a new kind of museum where the rule is: Please Touch! Look through some Eskimo snow goggles, not at them. Play with a raccoon. Find out about zoetropes. Program a computer. It's all adventure and discovery . . . and you're in the middle of it!

the materials. Community Services assists community centers, neighborhood houses, youth clubs, and community schools to develop materials and activities for learning, design and adapt space for educational and art uses, and help train leaders of youth programs and others in community-based organizations.

Visit a Children's Museum

The Boston Children's Museum is lively and colorful. Its annual audience approaches 200,000 and it is open year round (except for September, when it gets a thorough housecleaning and sprucing up after its rugged summertime use). Hours from October 1 to late June are 2:00 to 5:00 p.m. Tuesdays through Fridays and 10:00 a.m. to 5:00 p.m. Saturdays and Sundays; from late June through the end of August, 10:00 a.m. to 5:00 p.m. daily (except July 4). The Museum is closed Thanksgiving, Christmas, and New Year's Day, and on Mondays throughout the school year. Children's Museum, Jamaicaway, Boston, Mass. 02130.

Children's and Junior Museums

Academy of Natural Sciences of Philadelphia, Philadelphia, Pa. 19103

Akron Museum of Natural History and the Children's Zoo, Akron, Ohio 44307

Art Institute of Chicago, Chicago, Ill. 60603

Audrain County Historical Society and Museum, Mexico, Mo. 65265

Belmont County Museum, Barnesville, Ohio 43713

Brazosport Museum of Natural Science, Lake Jackson, Tex. 77566

Brome County Historical Museum, Knowlton, Quebec, Canada

Brookfield Historical Society, Brookfield, Wis. 53005

PHOTO: ED FITZGERALD

Brooklyn Children's Museum, Brooklyn, N.Y. 11216
Brueckner Museum of Starr Commonwealth for Boys, Albion, Mich. 49224

Cabin John Regional Park-Noah's Ark. Rockville, Md. 20854
Cherryvale Museum, Cherryvale, Kan. 67335
Children's Art Centre, Boston, Mass. 02118
Children's Museum, Boston, Mass. 02130
Children's Museum, Dartmouth, Mass. 02714
Children's Museum, Detroit, Mich. 48202
Children's Museum, Nashville, Tenn. 37210
Children's Museum & Planetarium at Sunrise, Charleston, W.Va. 25314
Children's Museum of Hartford, West Hartford, Conn. 06119
Children's Museum of Indianapolis, Indianapolis, Ind. 46208
Children's Nature Museum of York County, Rock Hill, S.C. 29730
Chisholm, A. M., Museum, Duluth, Minn. 55812
City of Palo Alto Junior Museum, Palo Alto, Calif. 94301
Clyburn Museum, Baltimore, Md. 21209
Columbus Museum of Arts and Crafts, Columbus, Ga. 31906

Dallas Health and Science Museum, Dallas, Tex. 75226
Danbury Scott-Fanton Museum and Historical Society, Danbury, Conn. 06810
Douglas County Historical Museum, Superior, Wis. 54870
Durham Children's Museum, Durham, N.C. 27704

Edmonton Art Gallery, Edmonton, Alberta, Canada
Elwood, Walter, Museum, Amsterdam, N.Y. 12010
Evanston Historical Society, Evanston, Ill. 60201

Fairbanks Museum of Natural Science, St. Johnsbury, Vt. 05819
Fenner, Carl G., Arboretum, Lansing, Mich. 48910
Fort Worth Museum of Science and History, Fort Worth, Tex. 76107
Franklin Institute Science Museum, Philadelphia, Pa. 19103

Glendora Historical Society Museum, Glendora, Calif. 91740
Goodhue County Historical Society, Red Wing, Minn. 55066

Harrison County Historical Museum, Marshall, Tex. 75670
High Museum of Art, Atlanta, Ga. 30309
Historic Hermann Museum, Hermann, Mo. 65041
Holyoke Museum-Wistariahurst, Holyoke, Mass. 01040
Houston, Sam, Memorial Museum, Huntsville, Tex. 77340

Imperial Calcasieu Historical Museum, Lake Charles, La. 70601
Jackson, Andrew, School Children's Museum, Kingsport, Tenn. 37660
Jacksonville Children's Museum, Jacksonville, Fla. 32207
Jacksonville Museum, Jacksonville, Ore. 97350
Junior Art Gallery, Louisville, Ky. 40203
Junior Arts Center, Los Angeles, Calif. 90027
Junior Museum of Oneida County, Utica, N.Y. 13503

Kalamazoo Nature Center, Kalamazoo, Mich. 49001

Lake Erie Junior Nature and Science Center, Bay Village, Ohio 44140
Lawrence Hall of Science, Berkeley, Calif. 94720
Libby Museum, Wolfebord, N.H. 03894
Liers' Otter Sanctuary, Homer, Minn. 55942
Lindsay, Alexander, Junior Museum, Walnut Creek, Calif. 94596
Litchfield Historical Society and Museum, Litchfield, Conn. 06759
Little, E. H., Gallery, Memphis, Tenn. 38111
Little Red Schoolhouse, Hazelwood, Mo. 63042
Louisiana Arts and Science Center, Baton Rouge, La. 70802
Louisiana State Exhibit Museum, Shreveport, La. 71109
Lutz Junior Museum, Manchester, Conn. 06040
Lyme Historical Society-Florence Griswold Association, Old Lyme, Conn. 06371
Lyon County Historical Society, Marshall, Minn. 56258

Massena Historical Center, Massena, N.Y. 13662
McAllen International Museum, McAllen, Tex. 78501
Memphis Pink Palace Museum, Memphis, Tenn. 38111
Mid Fairfield County Youth Museum, Westport, Conn. 06880
Mishawaka Children's Museum, Mishawaka, Ind. 45644
Monmouth County Historical Association, Freehold, N.J. 07728
Morris Museum of Arts and Sciences, Morristown, N.J. 07960
Muir House, Brownville, Nebr. 68321
Museum of American Architecture and Decorative Arts, Houston, Tex. 77036
Museum of the Philadelphia Civic Center, Philadelphia, Pa. 19104

Nankin Mills Nature Center, Westland, Mich. 48185
National Cowboy Hall of Fame and Western Heritage Center, Oklahoma City, Okla. 73111
National Society of Children of American Revolution Museum, Washington, D.C. 20006
Natural Science for Youth Foundation, New Canaan, Conn. 06840
New Britain Children's Museum, New Britain, Conn. 06051
North Shore Junior Science Museum, Roslyn, N.Y. 11576
Nutley Historical Society Museum, Nutley, N.J. 07110

Ocean City Historical Museum, Ocean City, N.J. 08226
Old Jail Museum and Felker House, Vienna, Mo. 65583
Old Princess Anne Days, Princess Anne, Md. 21853
Oregon State University Marine Science Center, Newport, Ore. 97365
Owensboro Area Museum, Owensboro, Ky. 42301

Peninsula Junior Nature Museum and Planetarium, Newport News, Va. 23601
Perelman Antique Toy Museum, Philadelphia, Pa. 19106
Peterborough Historical Society, Peterborough, N.H. 03458
Philadelphia Museum of Art, Philadelphia, Pa. 19101
Portland Junior Museum, Portland, Ore. 97201

Randall, Josephine D., Junior Museum, San Francisco, Calif. 94114
Rensselaer County Junior Museum, Troy, N.Y. 12180
Rock Creek Nature Center, Washington, D.C. 20011
Rocky Mount Children's Museum, Rocky Mount, N.C. 27801
Royal Ontario Museum, Toronto, Ontario, Canada
Rutherford Museum, Rutherford, N.J. 07070

San Mateo County Junior Museum, San Mateo, Calif. 94401
Santa Cruz Museum, Santa Cruz, Calif. 95060
Skenesborough Museum, Whitehall, N.Y. 12887
South Street Seaport Museum, New York, N.Y. 10038
Students' Museum, Knoxville, Tenn. 37918
Suffolk Museum and Carriage House, Stony Brook, N.Y. 11778
Sulphur Creek Park Nature Center, Hayward, Calif. 94541

Tallahassee Junior Museum, Tallahassee, Fla. 32311
Temple Mound Museum, Fort Walton Beach, Fla. 32548
Trailside Museum, Springfield, Mass. 01108
Travertine Nature Center, Sulphur, Okla. 73086

Utah Museum of Natural History, Salt Lake City, Utah 84112

Valentine Museum, Richmond, Va. 23219

Wainwright, Jonathan, Museum and Natural Resources Center, Fort Wainwright, Alas. 98731
Weinberg Nature Center, Scarsdale, N.Y. 10583
Worcester Science Center, Worcester, Mass. 01604

Youth Cultural Center, Waco, Tex. 76710
Youth Science Center of Monterey County, Salinas, Calif. 93901
Youth Science Institute, San Jose, Calif. 95127

Children's Museum Mini-Units

Becoming involved and doing and making and learning are the kind of things that the Boston Children's Museum is happily concerned with. You can participate whether you live in California or

PHOTOS: ED FITZGERALD

Texas or almost anywhere. How? Write for and use their four-page idea sheets, called Mini-Units.

The titles of the 12 Mini-Units are *A Balloon and Funnel Pump; Building Blocks from Milk Cartons; Exploration with Food Coloring; Making Large Bubbles No. 1; Making Large Bubbles No. 2; Making Simple Books; Organdy Screening; A Pie Plate Water Wheel; A Special Bubble Machine; A Spinning Top That Writes; A Tin Can Pump;* and *Stained-Glass Cookies.* The price for each Mini-Unit is 25 cents postpaid. If you want all 12, SEND $2.00 plus 50 cents for postage and handling, to Children's Museum, Jamaicaway, Boston, Mass. 02130.

Recycle

"Recycle" is the name of 40 art and educational project work sheets compiled by the Resource Center at the Boston Children's Museum: all the materials you need are things that you or somebody was going to throw out anyway. What can you make? Music makers like drums, tambourines, woodblocks, finger cymbals, and kazoos. The projects are endless: you can make a loom, an ant farm, a mask, a filmstrip movie, a terrarium, a hanging planter, a leather belt, mazes, a water microscope, and games and games and games. Robin Simons has prepared an imaginative push into hours of creative play. For your copy SEND $1.00 plus 30 cents for postage and handling to Children's Museum, Jamaicaway, Boston, Mass. 02130.

PETS

A Zoo in Your Room

If you have—or if you want to have—a zoo in your room, this book is must reading. Animal lovers will find a wealth of information in this entertaining guide to the care of a wide variety of small, often unconventional creatures. Roger Caras, famed naturalist, author, and television personality, has sound advice for you on how to house and feed over 30 species of mammals, birds, fish, reptiles, amphibians, and insects that can live comfortably in your room. Because the ideal home for an animal is usually one that resembles its natural habitat, the author focuses on the creation of attractive terrariums that can be populated with a variety of animals. His easy-to-follow directions will help you plant three distinct environments—woodland, desert, and bog—and make an appropriate selection of animals to live in each. *A Zoo in Your Room* also describes the equipment and procedures necessary to start and maintain a successful freshwater aquarium.

Whether you as a young zookeeper choose to buy your animals in a pet shop or capture them in the wild this handbook, illustrated with Pamela Johnson's lifelike drawings, will make your zoo a constant source of pride and pleasure. Harcourt Brace Jovanovich. $5.95.

Train Your Cat

If you've seen a kitten and wonder what it would be like to have a cat in your life, *The Common Sense Book of Kitten and Cat Care* will tell you. Harry Miller tells you how to care for the cat in your home: how to train it, what to feed it, how to change its diet as it grows older, how to prevent clawing of furniture, care and treatment of the sick cat, play and exercise. There are eight full-color illustrations of breeds of cat. Bantam Books. $1.25.

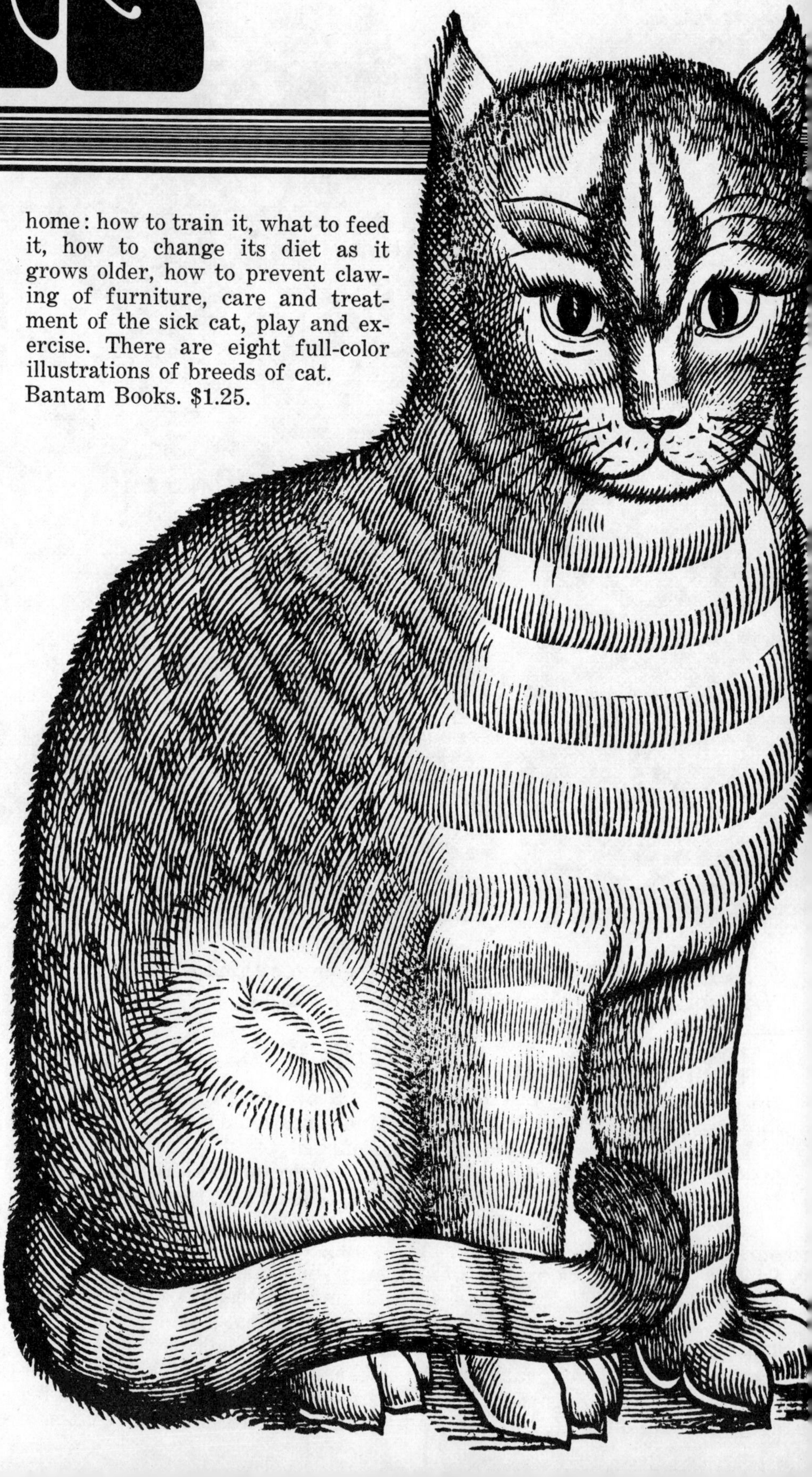

Have a zoo in your room. Catch a cricket and keep it happy. A clear plastic world for gerbils. How to buy and train a dog. How to select a kitten. A bird, a fish, a monkey, a skunk, and many more. Name that pet. It's here!

Pets and People

The subtitle is *How to Understand and Live with Animals.* What kind of pet do you think you want—a dog, cat, parakeet, gerbil? Perhaps a horse. If you live in the country, why not choose a pig, some ducklings or tame pigeons, a raccoon, or a flying squirrel? There are other pets to consider for city living: a salamander, turtle, or ant colony in a terrarium environment.

Before you can enjoy a pet and give it a good home, there are many responsibilities to consider. Dorothy Shuttlesworth has combined fascinating facts with amusing anecdotes about animals she has known. Most important, she cautions, a pet is "an animal kept as a companion and treated with affection." This well-known author of books on the animal world for young readers was the founder and editor of *Junior Natural History,* published by the American Museum of Natural History. If you have a pet (or are thinking about getting one) you should have a copy of *Pets and People.* E. P. Dutton & Co. $6.95.

The Pet Library

Do you have a pet—any pet? A bird? A dog, fish, monkey, or skunk? The more information and knowledge you have about your pet, the more you will enjoy it. The Know series of books will help you take better care of your pets and get more pleasure from them. Each Know book is 64 pages long, completely illustrated in full color. Feeding, grooming, breeding, and general care are extensively covered for particular pet, whatever it is:

Basset Hound
Beagle
Boston Terrier
Boxer
Bulldog
Chihuahua
Cocker Spaniel
Collie
Dachshund
Dalmatian
Doberman Pinscher
Fox Terrier
German Shepherd
Great Dane
Irish Setter
Maltese
Miniature Schnauzer
Pekingese
Pomeranian
Poodle
Pug
Retriever
Scottish Terrier
Shetland Sheepdog
Weimaraner
Yorkshire Terrier
Lovable Mutt
Shih Tzu
Lhasa Apso
Airedale
West Highland White Terrier
Cairn Terrier
Saint Bernard
Kerry Blue Terrier
Old English Sheepdog
Toy Fox Terrier
How to Train Your Dog
Welsh Corgi
Labrador Retriever
How to Choose Your Dog
How to Clip a Poodle
How to Groom Your Dog
First Aid For Dogs
How to Train Your Guard Dog
How to Raise & Train Your Puppy
Domestic and Exotic Cats
Persian Cat
Siamese Cat
Popular Cage Birds
Canary
Parakeets—Budgies
Parrot
Wild Birds
Aquarium
How to Breed Tropical Fish
How to Breed Egglayers
How to Breed Livebearers
Bettas
Goldfish
Guppies
Guinea Pigs
Hamster
Monkey
Gerbils

To get your Know books SEND the title of each book you want plus $1.50 to Pet Library, P.O. Box D, Harrison, N.J. 07029.

Handbooks on Cat Care

Are you aware that owning a cat was once a privilege reserved to royalty? If you have ever owned a cat, you know how special they are. If you would like to own a cat or are a new owner, here are some helpful pamphlets that will tell you what you need to know about feeding, handling, training, and caring for your cat.

Cat Care Handbook Offer. P.O. Box 9092, St. Paul, Minn. 55190. Enclose 25 cents for mailing and handling.

Kittens and Cats. Animal Welfare Institute, P.O. Box 3650, Washington, D.C. 20007. Enclose 10 cents for mailing.

Care of the Cat. American Humane Education Society, 180 Longwood Ave., Boston, Mass. 02115. Enclose 15 cents for mailing and handling.

The Cat. Education Department of the ASPCA, 441 E. 92d St., New York, N.Y. 10028. Enclose 10 cents for mailing.

Caring for Your Cat. Lowe's, North Edward St., Cassopolis, Mich. 40931.

Habitrail

Small animals adapt more easily than large ones to new natural conditions, and the most natural condition you can create for your hamster or gerbil is the Habitrail, a whole world of clear plastic houses connected by tube tunnels which you can rearrange or add on to. This is the most interesting concept in home animal care to come along in years; it is much neater than an open wire cage. Enjoy watching your pet have fun running around the exercise wheel, through the tunnel, or up into the sky house. Habitrail is made by Living World®; you can get sets and accessories from most pet shops or pet departments.

Hamsters

Golden hamsters were almost unknown until 1930, when a zoologist brought back a litter of baby hamsters from the Syrian desert. Since then, these charming little creatures have become one of the most popular of all pets.

Hamsters: All About Them is a complete manual for the young owner of hamsters. It provides extensive information on the care, feeding, and housing of these agreeable pets, whose friendliness and curiosity make them great fun to play with. Hamsters are easy to breed; Alvin and Virginia Silverstein point out some of the fascinating possibilities of crossbreeding and mutations. Fully illustrated by Frederick Breda's photographs, this book will be treasured by everyone who has ever been captivated by a hamster. Lothrop, Lee & Shepard Co. $5.50. Also ask at your bookstore for *Rabbits: All About Them* and *Guinea Pigs: All About Them.* (Same authors, same publisher.)

A Special Book on Gerbils

This book was written by Larry Woods when he was 9 years old and first kept gerbils. He paid close attention to them and wrote observantly of what he saw. A useful book for anyone responsible for the care of gerbils. For your copy of *Gerbils* SEND $1.00 to Workshop for Learning Things, 5 Bridge St., Watertown, Mass. 02172.

Five Guides to Horses and Horsemanship

Meet the Horse, by Pat Johnson, is a complete history of the horse with detailed discussions of the most popular breeds, both light and heavy. Fully illustrated with Walter D. Osborne's excellent photographs, this is an informative and entertaining guide. $4.95.

The Quarter Horse, by Pat Johnson, is the complete story of America's favorite breed: how it was developed and why its popularity is increasing. Fully illustrated with stunning photographs and rare drawings and engravings. $4.95.

Select, Buy, and Care for Your Own Horse, written by Barbara Van Tuyl and edited by Patricia H. Johnson, is a complete guide for new and potential owners. Fully illustrated with exciting photographs by Walter D. Osborne. $4.95.

How to Ride and Jump Your Best, by Barbara Van Tuyl, is a complete guide for the beginning rider and jumper, including helpful hints on technique for the novice as well as guidelines on how to avoid picking up poor riding habits. With photographs by Walter D. Osborne. $5.95.

Winning Ways at Horse Shows, by Barbara Van Tuyl, is the complete guide for the rider who wishes to compete at horse shows. Kinds of shows, prizes and the divisions in competition, preparing the horse to be outstanding are all clearly described; photographs by Budd make this an excellent manual for the horse-show enthusiast. $6.95.

Grosset & Dunlap.

Learn to Ride

This happy book, with its informative text (by Robert Owen) and wealth of photographs and drawings (by Tony Streek and Peter Kesteven), is for young people who ride horses or dream of doing so. You learn how to choose a healthy horse and care for it, how to exercise and train it, how to manage the stable, and how to ride and jump. Above all, as the child progresses, this excellent book encourages the growth of love and sense of responsibility for the horse. Golden Press. $3.95

Horses

A comprehensive guide to the fascinating world of horses. Moira Duggan explains their origins, evolution, physiology, and importance in history; major breeds; and the modern activities performed by horses. Nearly 100 photographs, many in vivid full color, plus drawings and glossaries complement the text. *Horses* is a Golden Handbook Guide. Golden Press. $1.95.

FREE Horse Booklets

The American Quarter Horse

The quarter horse is the most versatile, best all-around horse in the world. It is gentle and has an easygoing disposition which makes it an ideal mount for the whole family. The American Quarter Horse Association is an organization dedicated to people who love horses, and to the advancement of the breed of quarter horses; it publishes free, informative pamphlets and booklets such as *Judging American Quarter Horses; Quarter Horse Racing; Training Riding Horses; The Now Thing: The American Junior Quarter Horse Association; Ride a Quarter Horse;* and *Youth Activities.* You can also get a full-color reproduction of a majestic American quarter horse, either 9 x 12 or 18 x 24 inches, as well as a detailed chart of the anatomical parts of the horse. For any one or all of these publications WRITE TO American Quarter Horse Association, Amarillo, Tex. 79168.

Anatomy of a Horse

If you like horses and want to learn more about them, here are two charts that will help. One labels the parts of the body from forelock to fetlock. The other names the sections of the body in terms of the equipment used on a horse, such as halters, martingale, stirrup, and saddle. For your free charts WRITE TO Horse Drawings, Education Department of the ASPCA, 441 E. 92d St. New York, N.Y. 10028. Enclose 10 cents for mailing.

FREE

Care of the Horse

This general guide, *Care of The Horse in Health and Disease,* contains important basic information for horse lovers about stalls, feeding, foaling, grooming, equipment, and first aid. For this free pamphlet WRITE TO American Humane Education Society, 180 Longwood Ave., Boston, Mass. 02115. Enclose 15 cents for mailing and handling.

Knowledge Through Color

Here are books about pets from the Knowledge Through Color series. Each book is magnificently illustrated in full color and written by a noted authority. Collect your favorite subjects from this series and build your own encyclopedia. Knowledge Through Color is published by Bantam Books.

Dogs: Selection, Care, Training gives advice on the choice and rearing of puppies, sections on dog senses and canine behavior, and basic training techniques for pet or show dogs. By Wendy Boorer. $1.95.

Cats: History, Care, Breeds tells you how to select a kitten, how to look after your pet, and what procedures to follow when exhibiting your cat. The Cat Fanciers Association has provided the standards and scales of points for the many breeds described and illustrated in the book. By Christine Metcalf. $1.95.

Tropical Freshwater Aquaria has complete suggestions on maintaining a home aquarium, choosing the best equipment and plants, treating fish diseases, and details on simulating natural environment. By George Cust and Peter Bird. $1.45.

Tropical Marine Aquariums describes how to set up a marine aquarium, including specifications for proper conditions, equipment, species, and care of available tropical fish. By Graham F. Fox. $1.45.

Horses and Ponies has delightful, instructive histories of horses and ponies, with pointers on how to care for them and tips on the sport of riding. By Judith Campbell. $1.95.

Selecting and Training Your Dog

Ready for a dog? Here are two good, inexpensive books especially written for children.

Both Ends of the Leash, by Kurt Unkelbach, helps you select your dog, tells you the ten best breeds for boys and girls, discusses early care, and explodes a lot of myths about our canine friends. Illustrated by Haris Petie. Prentice-Hall (Treehouse Paperback). 95 cents.

Training a Companion Dog, by Dorothy M. Broderick, answers the question: What is obedience training? Also, it tells you how a dog learns and goes over all the lessons you'll need to teach your dog to heel, stand-stay, sit, and down-stay, as well as recall and go-to-heel. You'll learn how to officially qualify your dog as a CD (Companion Dog), CDX (Companion Dog Excellent), and UD (Utility Dog). Illustrated by Haris Petie. Harvey House. $1.25.

FREE

Handbooks on Dog Care

If you have questions about what dog to choose, and how to care for the dog you've chosen, here are some booklets that are brimming with the pertinent information you need to make your pet's life healthy and happy. You'll learn about the different breeds, registration, housebreaking, feeding and exercise, and much more.

Handbook of Dog Care. P.O. Box 9475, St. Paul, Minn. 55197. Enclose 25 cents for postage and handling.

Care of the Dog. American Humane Education Society, 180 Longwood Ave., Boston, Mass. 02115. Enclose 15 cents for mailing and handling.

Tips on Feeding Your Dog and Cat. Education Department of the ASPCA, 441 E. 92d St., New York, N.Y. 10028. Enclose 10 cents for mailing.

You and Your Dog. Animal Welfare Institute, P.O. Box 3650, Washington, D.C. 20007. Enclose 10 cents for mailing.

The Battle of the Puddle. Education Department of the ASPCA, 441 E. 92d St., New York, N.Y. 10028. This leaflet is also available in Spanish—ask for *La Batalla del Charco.* Enclose 10 cents for mailing.

How to Care for Your Dog

As many a puppy may ask:

Do you know what to do
When I catch cold and sneeze?
When I bark too much
Or have too many fleas?

This practical book for young pet owners answers these questions and dozens more. It gives simple instructions about how to help your dog live happily and healthily.

From the day you pick out your puppy and bring him home, there are easy rules to follow (and have him follow) for feeding, paper training and housebreaking, curing his homesickness and first-night blues, teaching him good manners and tricks, taking him on trips, and caring for him when he cries or gets sick. Delightful drawings by Norman Bridwell on almost every page make it all easier to understand.

So, young dog lovers of the world, rejoice! *How to Care for Your Dog* was written by Jean Bethell so you—not your parents instead of you—can give the best care possible to the pet you love the most. Four Winds Press (a division of Scholastic Magazines). $3.95.

The Common Sense Book of Puppy and Dog Care

If you already have a puppy or a dog, or if you are thinking about getting one, *The Common Sense Book of Puppy and Dog Care* will tell you all about the care, feeding, and training of your pet. What should you look for when choosing a collar and leash? Should your dog be allowed to sleep in your bed? How much exercise does your dog need? What are the main points to keep in mind when selecting a puppy? Harry Miller has drawn upon his extensive knowledge of dogs and his close association to many research projects devoted to their welfare to provide answers to these questions and hundreds more. Bantam Books. $1.25.

FREE Beware of Distemper

Distemper is a dread disease of dogs which can be long, costly, extremely uncomfortable, and often fatal. It is essential that your dog receives the proper vaccinations to guard against the onset of this illness. For a free leaflet about distemper vaccinations WRITE TO Education Department of the ASPCA, 441 E. 92d St., New York, N.Y. 10028. Enclose a 10-cent stamp for postage.

Small-Pet Care

Curious, comical, and cuddly—that's a gerbil. Gerbils are delightful pets, small enough to keep in your room. Dorothy E. Shuttlesworth tells you all you need to know about housing, feeding, and raising healthy, happy gerbils and other small animals. Squirrels, if you have a lot of patience, make wonderful outdoor pets. Eventually you can train them to eat out of your hand.

Caring for Gerbils and Other Small Pets also tells you about mice, rats, guinea pigs, and all kinds of rabbits. Scholastic Book Services. 75 cents.

Write for These

Where to Buy, Board or Train a Dog

If you're thinking of getting a dog, here's help in deciding what kind of a dog. Chapters in the 65-page *Where to Buy, Board or Train a Dog* include "Which Dog for You?"; "Large Versus Small"; "Short Versus Long Haired"; and "Male Versus Female." Most important is a special section entitled "A Dog for Your Child." 50 cents.

Dog Wall Chart

This beautiful wall chart of the main breeds of dogs in America shows 127 dogs in full color. A folded copy of the chart is 50 cents, and a rolled chart suitable for framing is $1.25.

Touring with Towser

Touring with Towser lists over 2000 hotels and motels in the United States and Canada where you can take your pet. If you plan your trip with this guidebook, the "whole family" can take a vacation together. 50 cents. To get the booklets or chart SEND your order to Gaines, P.O. Box 1007, Kankakee, Ill. 60901.

Know Your Cat

There are many books on cat care. *Catnip: Selecting and Training Your Cat* is a particularly good one because it lets you get to know your cat and the cat world he comes from. Did you know that cats were very important in Egypt over 5000 years ago? Do you know why a black cat is thought to be bad luck? Are you aware that the cat's whiskers and eyebrows provide him with the most delicate sense of touch in the whole animal kingdom?

This delightful, informative book, by Kurt Unkelbach, has lots of information about cats. The care of the cat is dealt with fully, from grooming to breeding and training. There is a most interesting section on the myths and superstitions that have grown up around the cat. Prentice-Hall (Treehouse Paperback). 95 cents.

FREE

Plays About Animals

Two plays about animals which you will have fun acting in are *The Kindness Train* and *One Morning Long Ago* (about St. Francis of Assisi and the birds). They make excellent school, community, or rainy day projects. For free copies WRITE TO American Humane Education Society, 180 Longwood Ave., Boston, Mass. 02115. Enclose 15 cents for mailing and handling.

FREE

Good Kind Lion

This delightful story begins:

Suppose an enormous lion suddenly appeared right in front of you, what would you do? Well, there probably wouldn't be much you *could* do, except sit tight and hope that he was a good kind lion. And that's pretty well what all smaller creatures have to do when YOU suddenly appear in front of them—they all just hope YOU are a good kind lion, too.

For a free copy of *Good Kind Lion* WRITE TO Animal Welfare Institute, P.O. Box 3650, Washington, D.C. 20007.

Birds as Pets

Finches, canaries, budgerigars, parrots, lovebirds: How do you care for them? How do you feed them? How do you breed them? How do you train them? All these questions are answered by Paul Villiard in *Birds as Pets,* an easy-to-read, scientifically factual book. Not only do you learn about the care, feeding, housing, breeding, and training of various birds, but the author covers the care and treatment of the more common bird diseases. Lavishly illustrated, this is a fine book for the amateur zoologist. Doubleday & Co. $5.95.

FREE

Care and Feeding of Cage Birds

If you have a parakeet or a canary, or are thinking of getting one, here is a booklet that will help your bird have a happy, healthy life. It shows you how to recognize when your bird isn't feeling well, and gives you tips on proper nutrition and environment. For your copy of *Care and Feeding of Cage Birds* WRITE TO American Humane Education Society, 180 Longwood Ave., Boston, Mass. 02115. Enclose 15 cents for mailing and handling.

FREE

Start an Animal Club

Few things hold a child's interest as deeply as pets and other animals do; but because children know little of their ways and needs, they sometimes cause animals much discomfort and thoughtlessly mistreat them. To overcome this, the American Humane Education Society started animal clubs (or "bands of mercy") about 60 years ago to help youngsters acquire a greater knowledge, understanding, and sympathy for pets, farm animals, and wildlife.

For information on how to start an animal club, band of mercy, and junior humane league, plus program activities, WRITE FOR a free copy of *What Do You Know About Animals?* to American Humane Education Society, 180 Longwood Ave., Boston, Mass. 02115. Enclose 15 cents for mailing and handling.

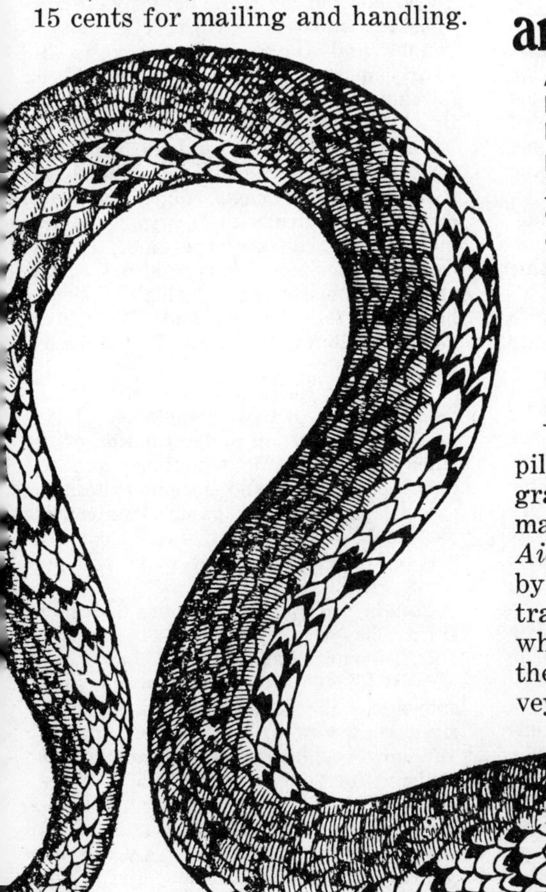

Your Insect Pet

Think of a pet and a dog or cat will probably come to mind. But since a pet is defined as an animal kept for our enjoyment, we can also include birds, fish, and even insects. Millions of youngsters collect a wide variety of insects, from field crickets and dragonflies to whirligig beetles and water striders.

Insects are easy to obtain, require little care, and take up little room; moreover, they are extremely interesting animals to watch and study. *Your Insect Pet* introduces the beginning entomologist to one of the most fascinating groups of animals. Carefully explaining how to find and keep insects, Richard Headstrom covers the subject simply enough for the beginner, yet thoroughly enough to satisfy the serious student. David McKay Co. $4.95.

First Aid for Insects and Much More

An insect can be an interesting pet. Perhaps you know this already. You have caught a cricket or a caterpillar and tried to keep it in a glass jar or a box. Then what? Your pet caterpillar may refuse to eat. Your cricket may stop chirping. If that happens, your pet needs First Aid! This First Aid book will tell you how to keep insects well fed and lively. It will give you hints on what to do about problems. It will help you to choose insects that make good pets.

You'll learn about ants, caterpillars, cocoons, crickets, fireflies, grasshoppers, beetles, praying mantises, and water insects. *First Aid for Insects and Much More,* by Arthur A. Mitchell with illustrations by Wendy Worth, tells you where to find insects, where to keep them, and what to feed them. Harvey House. $5.35.

FREE

First Aid for Animals

If your pet were hurt, would you know what to do? Have you ever found an injured or sick animal near your home and wanted to help? Would you like to know how to save the life of an animal in an emergency? Here are two free pamphlets and a booklet that can give you the information you need on symptoms and remedies, shock, broken bones, bleeding, heatstroke, and many other emergency situations.

First Aid Hints. Educational Department of the ASPCA, 441 E. 92d St., New York, N.Y. 10028. Enclose 10 cents for mailing.

First Aid to Animals. American Humane Education Society, 180 Longwood Ave., Boston, Mass. 02115. Enclose 15 cents for mailing and handling.

First Aid and Care of Small Animals. Animal Welfare Institute, P.O. Box 3650, Washington, D.C. 20007. Enclose 50 cents for mailing and handling.

See What I Caught

What do crickets like to eat? Do fireflies eat at all? When is the best time to let a salamander free? Why do you have to wash your hands after you play with your turtle? The answers are in *See What I Caught!* an illustrated guide to catching, caring for, and setting free eight cold-blooded animals. Ann Thomas Piecewicz, a teacher and nature guide, shows how to catch frogs and tadpoles, land and water turtles, toads, fireflies, salamanders, crickets, and tree frogs. Instructions for making an insect net, a critter carrier, and special cages are included in this loving guide to woodland creatures. Illustrations by Perf Coxeter. Prentice-Hall. $4.95.

SPORTS

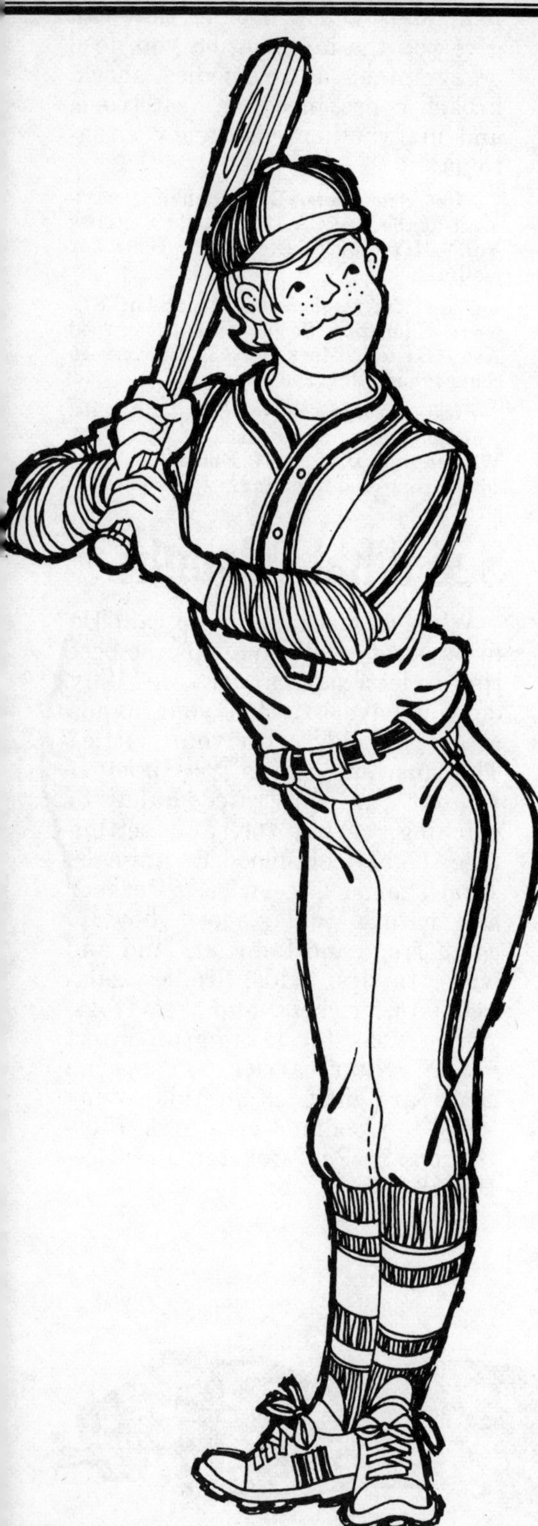

A Library of Pictorial Sport Books

Here are some large pictorial handbooks (8½ x 11 inches) on 14 sports. These make-you-want-to-learn books are filled with photographs, diagrams, and action drawings, which makes them just right for beginners; the instructions are written by sports stars like Bob Toski and Dick Button—all experts in their field. Grosset & Dunlap.

Beginner's Guide to Golf demonstrates the fundamentals of the game in action photographs. A book for beginners—and all golfers who plan to break 90. One of America's top pros, Bob Toski, shows you how to develop a sound golf swing and how to use every club. $1.25.

Golf Rules in Pictures. This official publication of the United States Golf Association makes the rules come alive in graphic drawings. Edited by Joseph C. Dey, Jr., it is the perfect antidote for the inevitable controversies about rules that come up during a game. $1.95.

The Book of Tennis. The world's great players have written, with photographs by Jeff Chapleau, a step-by-step method for teaching yourself tennis. You are instructed in choosing the racket that's right for you and how to grip it. You are taught the vital "ready position" and how to hit with the forehand, backhand, volley, and dropshot, as well as serve, overhead, and lob. Here, in the best book yet on beginning tennis, is all you need to know as you step on the court the first time. It is written by Cornel Lumiere and the editors of *World Tennis* magazine. $1.95.

Ski in a Day! The famous instant skiing method on shortee skis is demonstrated in magnificent sequence photographs. Clif Taylor shows how you can start on short skis and move up to long skis without ever having to do a "snowplow." Learning to ski becomes fun this sensible new way. $2.95.

Instant Skating. Packed with action sequence photographs and drawings, this book by Olympic gold-medal champion Dick Button shows the beginner how to dress correctly for skating, how to choose the right skates and lace them properly to avoid weak ankles, how to step out on the ice for the first time, and how to skate forward and backward without fear of falling. The book goes on to basic edges and figures. $1.95.

Ice Hockey Rules in Pictures. This game of rousing speed, spectator appeal, and color is the fastest team sport in the world. An understanding of the rules is essential to its appreciation. In addition to illustrating the basic rules, this handy volume contains special sections on scoring goals and on defensive hockey. By Robert Scharff with illustrations by John McDermott. $1.95.

Pro Football Plays in Pictures. Offensive formations, running plays, passing plays, pass patterns, and defensive formations are illustrated with action photographs and diagrams accompanied by clear, concise explanations. The ideal book for any youngster who plays or watches. By George Sullivan. $2.95.

Football Rules in Pictures. New and revised edition containing the official National Football League digest of rules, clearly illustrated with drawings explaining the basic rules of the game. Professional and college interpretations of the rules are covered, making this the perfect handbook for the weekend TV viewer. Edited by Don Schiffer and Lud Duroska. $1.95.

Basketball Rules in Pictures. With the tremendous growth in popularity of this fast-paced sport, an understanding of the rules and of officials' techniques and signals becomes more and more important for the enjoyment of the game. This concise handbook, edited by A. G. Jacobs, contains a special section on basic plays and patterns. $1.95.

Sail in a Day. The author, George D. O'Day, has won more than 30 American and international sailing honors and awards. This easy-to-understand guide shows you in pictures how to select the right boat, how to handle tiller and sail, and how to plan weekend and vacation cruising for your family. $1.95.

Get in the Swim with Esther Williams. Hollywood's great aquatic star takes you for your first swim and shows you how to improve your swimming and diving techniques. Action photographs cover everything from treading water to the Australian crawl, from poolside dive to jackknife. There are instructions for begin-

Sports fans! How to get you in the swim. Water skiing, scuba diving, snow skiing, tennis, hockey, golf, football, basketball, baseball, sailing, bowling, track, karate, archery, judo, back-packing. Fun, excitement, and sports, sports, sports!

ners, tips on teaching kids to swim, and the vital rules of water safety. By Bob Thomas. $1.25.

Outdoor Life Fishing Book. P. Allen Parsons, well-known former editor of *Outdoor Life,* helps you catch your limit of trout, bass, salmon, pike, muskie, and lots of other fishes. There are tips on wet- and dry-fly fishing, bait casting, spinning, trolling, and nymph fishing; the book also covers special baits, rods, lines, leaders, reels, flies, and illustrated descriptions of fishes and their habits. $1.25.

Billiards for Everyone. A complete picture handbook: in fact, the first billiards instruction book to teach entirely by photographs. The man behind the cue in every easily understood picture is World Pocket Billiards Champion Luther Lassiter. You learn how to hold the cue; how to stroke correctly; how to control the cue ball; how to master advanced shooting—bank shots, reverse banks, frozen combinations; and how to practice to become a real pool shark. $1.95.

Instant Bowling. Star bowlers Harry Smith and Steve Cruchon show you how to become a good bowler yourself, whether you are a beginner or want to improve your game. The book gives you the fundamentals: how the bowling ball should fit your hand; how to do the pushaway, the approach, the release, and the follow-through; "spot" and "pin" bowling; how to make spares and splits; and how to score. $1.95.

Warm Up for Little League Baseball

So you want to play baseball? Who is going to teach you: your dad, the kids at school—or will you pick it up as best you can watching other boys playing? How will you know you are learning it right?

It's just as easy to learn it right as to learn it wrong, and that is what *Warm Up for Little League Baseball,* by Morris A. Shirts, is all about. It's a show-and-tell in words and pictures about throwing, hitting, catching, baserunning, sliding, playing the infield and outfield. Excellent photographs of boys playing the game plus an approval by Little League Baseball make this book a must for the beginner at baseball. Sterling Publishing Co., cloth $3.95. Pocket Books, paper 95 cents.

FREE Little League Baseball

Did you know that Little League baseball is the world's fastest growing youth movement? There are more than 5700 leagues with more than 29,000 teams in Little League baseball. Little League is played in every state, every province of Canada, and 24 foreign countries. If you are interested in starting a Little League team in your community, you can get free information; WRITE TO Little League Baseball, P.O. Box 1127, Williamsport, Pa. 17701.

Famous Slugger Yearbook

Famous Slugger Yearbook is an annual 64-page publication containing baseball records in batting, fielding, and so on, photographs, and statistics along with interesting sidelights. There's a special article on hitting by Dick Allen, outstanding slugger for the Chicago White Sox, and dozens of photographs of famous baseball players and teams, and of the Little League champions. If you want to know how a baseball bat is made, there's a picture story that will give you all the answers. For your copy SEND 25 cents (mailing and handling charge) to Hillerich & Bradsby Promotions, P.O. Box 18554, Louisville, Ky. 40218.

FREE

Amateur Softball Association

Membership is free. Boys and girls ages 9 to 17 may join. WRITE TO Amateur Softball Association, P.O. Box 11437, Oklahoma City, Okla. 73111.

How to Play Better

If you are a beginner, a member of a team, or just like to watch football on television, *How to Play Better Football* is for you. It is written clearly and simply by C. Paul Jackson with cheerful illustrations by Leonard Kessler. Thomas Y. Crowell Co. $4.95. Also try *How to Play Better Baseball* and *How to Play Better Basketball* (same author, illustrator, publisher, and price).

FREE

Basketball Was Born Here

This free booklet tells you the history of basketball, how it was invented by Dr. James Naismith at Springfield College, Springfield, Massachusetts. There's a picture of the world's first basketball team and drawings of the first game: the baskets were half-bushel peach baskets, and a man was stationed behind each goal in the balcony to pick the ball out of the basket and put it back into play. For a free copy of *Basketball Was Born Here* and a pamphlet on the Naismith Memorial Basketball Hall of Fame SEND a self-addressed stamped envelope to Basketball Hall of Fame, P.O. Box 175, Highland Sta., 460 Alden St., Springfield, Mass. 01109.

How to Star in Swimming and Diving

You can be a better swimmer and more skillful diver. This handbook will show you how: simple instructions plus diagrams and step-by-step action photographs demonstrate the four basic strokes: crawl, backstroke, breaststroke, and butterfly.

You'll find out how you can master many kinds of dives, too: the racing dive, jackknife, half gainer, half twist, back dive, back jackknife, and other exhibition dives. Whether you're interested in swimming and diving for competition or just want to improve your style, *How to Star in Swimming and Diving* by Charles Batterman is the book for you. Four Winds Press. Available at your local library.

Other "How to" books for young athletes from the same publisher are: *How to Star in Basketball* by Herman L. Masin; *How to Star in Football* by Herman L. Masin; *How to Star in Baseball* by Herman L. Masin. Four Winds Press. Cloth $3.95 each. Starline/Scholastic Books Services. Paper 75 cents each.

Fun and Fitness

Bonnie Prudden, who has won nationwide acclaim for her work in alerting people of all ages to the need for fitness, believes in exercise for everyone, and now she presents *Fun and Fitness,* which is not just a booklet about bowling but a booklet about exercise with a bowling ball. Of course, the fitness you gain will help your game. Send for this new look at exercise. WRITE TO Fun and Fitness, AMF Bowling Products Group, Advertising Dept., Jericho Turnpike, Westbury, N.Y. 11590.

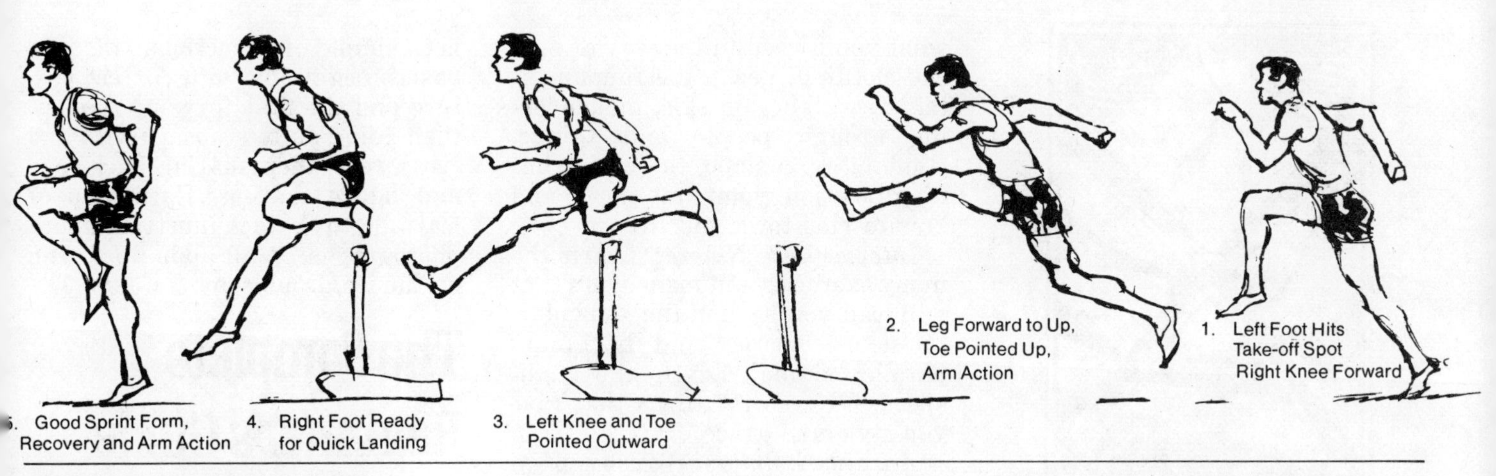

Track and Field for Young Champions

The sport of championship track, like a circus, is made up of many events. Some meets are held outdoors, others indoors. Robert Antonacci and Gene Schoor cover the spectrum: distance running; relay and hurdle racing; jumping for distance and height; the shotput, discus, javelin, hammer, and weight throws; and walking, jogging, and hiking.

The rules are explained in detail, and the history of each event is related. Illustrated throughout with graphic drawings, *Track and Field for Young Champions* makes track and field more exciting and enjoyable for everyone: those who participate, those who coach, and those who cheer. McGraw-Hill Book Co. $5.95.

Junior Tennis

A tennis book that doesn't start with a tennis racket? Yes. Instead of starting on a court, you begin by "bumping" the ball with your open palm. When you get a sense of how the ball bounces you start using a practice paddle, and with it you develop the skills you need for many tennis strokes.

When you pick up a tennis racket and walk out on a court you know how to hold the racket, how to stand, and how the serve, forehand, and backhand feel. You will learn how to hit a slice serve, and what footwork you need to use to reach and hit the ball into every part of the court.

Each step is illustrated with action photographs and diagrams. You'll find the rules of the game, basic strategy, and answers to questions which you thought were silly but which are not.

A tennis book where the pictures tell the story? Yes: *Junior Tennis*, by Harry Leighton. Sterling Publishing Co. $3.95.

Table Tennis

Si Wasserman, coaching chairman of the United States Table Tennis Association, has written the step-by-step story that accompanies the diagrams and more than 80 photographs in *Table Tennis*. Here are the fundamentals of the game, how to score, and many tips on techniques. If you study the illustrations with the text you will learn not only why the ball bounces or spins as it does when you serve or return it, but how you should stroke or chop the ball to win the point. Sterling Publishing Co. $3.95.

American Badminton Association

Junior membership ages 7 to 18 costs $1.00 per year; members receive a subscription to *Badminton USA*. For the name of the nearest club WRITE TO American Badminton Association, 1330 Alexandria Drive, San Diego, Calif. 92107.

Karate for Young People

For all young people age 10 and up this book teaches the basics of sport karate. 124 large photographs of young people demonstrate, step by step, how to position the body and carry out each movement. Russell Kozuki, author of *Junior Karate* and formerly head of a judo and karate school, shows you how to execute each punch, kick, and block basic to the sport. By following the clear instructions in this book you gain the satisfaction of knowing you have complete control over yourself and can react swiftly and correctly to any challenge. *Karate for Young People* is a fine beginner's book. Sterling Publishing Co. $3.95.

Skiing for Beginners

Can you really learn to ski from a book? Haystack Ski School Director Bruce Gavett emphatically says yes. The pictures in *Skiing for Beginners* demonstrate every technique from the first time walking on skiis to the parallel christie. The book describes each maneuver in the terms ski instructors use all across America. The children in the pictures tell in their own words exactly what they are doing as they demonstrate each maneuver. It's like having an instructor explain it to you while you are on the slope. The book is a most effective self-teacher before and after your ski class or practice session; it's a way to make sure you've got it all together in your head, your muscles, and your reflexes. If you follow the sequence of steps faithfully while practicing you will experience the great thrill of swinging down a mountain with perfectly controlled turns, whipping up a sparkling plume of snow. Written by Bruce Gavett and Conrad Brown; photos by Kim Massie. Charles Scribner's Sons. $5.95.

Skiing with Control

This book will teach you the basics of every ski maneuver from the herringbone to parallel skiing and wedeln.

Beginners: Learn what you should know before going out on the slopes: how to get in condition, how to choose your equipment, what you'll need in the way of special clothing. Learn the fundamentals of walking on skis, maneuvering around people and things, climbing up a slope, the basic running position, your first turns, and how to ride tows and lifts.

Intermediate Skiers: Learn the many exercises and maneuvers that will lead you beyond the snowplow to stem christies and beginning parallel skiing. Learn the high-speed "less effort" turns that give you style and grace.

Advanced Skiers: Brush up on the forms of parallel skiing that give control. Learn some of the fun maneuvers that make the real experts stand out on any hill.

Skiing with Control, by Rick Shambroom and Betty Slater, is a self-instruction guide to every ski maneuver, basic to advanced, with 300 photographs of ski pros in action demonstrations. Collier Books (imprint of Macmillan Publishing Co.). $1.95.

The Complete Beginner's Guide to Ice Skating

Have you ever wished you could achieve the rhythmic movements and graceful glides of the skilled ice skater? You can. The only requirements are that you have a normal sense of balance and can walk.

The relative simplicity of equipment and many ways to use skating (hockey, speed skating, competitive figure skating, ice dancing, to name a few) make its appeal widespread. In *The Complete Beginner's Guide to Ice Skating,* Edward F. Dolan, Jr., gives complete yet simple instructions on the basics, beginning with correct posture and the first forward strokes, then going on to stops, curves, and crossovers, then skating backward and figure skating. Each step of learning to skate is illustrated with photographs. It all adds up to fun on the ice. Doubleday & Co. $4.95.

The Complete Beginner's Guide to Bowling

Can you pick up a rubber or plastic ball weighing 16 pounds, walk with it in a straight line for four steps, and swing your arm in an easy pendulum arc as you walk?

If so, you can bowl.

Here's a guide to a sport that requires no special physical qualifications and that anyone of any age can participate in. Bowling does, however, require accuracy, coordination, and concentration. An avid bowler himself, Edward F. Dolan, Jr., explains the fundamentals for developing this skill, from the lanes, equipment, and rules to the actual steps involved, in a lively text supplemented by photographs and diagrams. A practical, concise guide to a sport that has attracted over 40 million bowlers and attracts new ones every day. *The Complete Beginner's Guide to Bowling* is perfect for the teenage reader. Doubleday & Co. $4.95.

FREE

Learn to Bowl

Bowling is America's most popular participant sport—and you're never too young to start. Included in the 24-page booklet *Learn to Bowl* are all sorts of tips which will help put more fun in your bowling. Besides supplying hints on ball care, the booklet gives valuable help on improving bowling scores, scoring methods, and bowling sportsmanship. For your free copy WRITE TO Manhattan Leisure Products Co., P.O. Box 1526, La Grange, Ga. 30240.

Junior Judo

This is a great book for the beginner. Judo is the famous art of self-defense without weapons, devised by a Japanese educator more than 75 years ago. Judo should appeal to every person eager to become healthy, strong, skillful, and capable of self-defense against a bully. With the aid of judo you can do this. In *Junior Judo,* by E. J. Harrison, step-by-step fundamentals are simply described and pictured; all instruction is directed toward the young learner. The positions, holds, throws, locks, bends, and twists which are most useful and not harmful to the young judoka (judo fans) are explained. Sterling Publishing Co. $3.95.

Better Bowling Tips

Here is a pamphlet that is sure to improve your game. *Better Bowling Tips* gives you step-by-step pictures of the four-step approach with explanation from stance to follow-through. There are tips on equipment and attire, bowling etiquette. spot bowling, how to make spares, and how to score. You get a personal bowling score record with this pamphlet, both free. WRITE TO Better Bowling Tips, AMF Bowling Products Group, Advertising Dept., Jericho Turnpike, Westbury, N.Y. 11590.

Bowling Talk

As the *anchor man* used the *four-step approach* down the *alley* he *balked,* went back to the *foul line,* and delivered a *hook* that left a *baby split* standing on the *lane.* The next throw knocked down the two remaining pins, so he got a *spare.* The next bowler threw a *powerhouse* that hit the *pocket* squarely causing a *pocket split.* Unfortunately, it wasn't a *cheesecake,* and there certainly weren't many *strike artists* in this game.

If you have trouble following this paragraph, *Bowling Talk for Beginners,* by Howard Liss, can help you develop new insights into the popular sport of bowling. Learn the slang terms in common use for the intricate plays in bowling with this cross-referenced dictionary, which also helps you learn scoring. A valuable tool to improve your confidence as a bowler. Pocket Books. 75 cents.

American Junior Bowling Congress

Why not organize a junior bowling league in your community for yourself and your friends? You can get emblems and certificates to help measure your progress. Bowling helps develop poise, skill, and sportsmanship and is a regular part of the physical education program in thousands of schools throughout the country—a tremendous help in combating juvenile delinquency. Over 700,000 boys and girls under 19 participated in the 1972–73 season offered by the American Junior Bowling Congress. To get your free copy of the current program WRITE TO American Junior Bowling Congress, 5301 S. 76th St., Greendale, Wis. 53129.

FREE

Learn to Roller Skate

Everything you need to know about roller skating is told in these two illustrated booklets.

How to Roller Skate includes "How to Start"; "How to Stroke"; "How to Steer"; "How to Stop"; "Turns, Tricks and Spins"; and seven fun games you can play on roller skates: cap tag, obstacle race, tin-can rolling, wood tag, Japanese tag, follow the leader, and cross tag.

Skating Skills is a 10-page comic-style book which tells the story of how four beginner roller skaters became artists of the rink. You will, too, after you put the skating tips in this booklet to use. Everything's demonstrated in pictures, so it's easy to learn.

For your free copies WRITE TO Chicago Roller Skate Co., 4450 W. Lake St., Chicago, Ill. 60624. Enclose 25 cents for each booklet to cover postage and handling.

Archery Lessons

Free lessons in archery are yours with a copy of *How to Shoot a Bow and Arrow.* A sport that the family can enjoy together, archery's popularity is growing rapidly. This excellent illustrated booklet (12 pages) tells you all you need to get started: arrow length, bow length, bow weight, how to brace or string the bow, and proper shooting technique (stance, drawing, anchoring, aiming, and releasing). If you're a junior Robin Hood, to get your free copy WRITE TO Ben Pearson Archery, P.O. Box 270, Tulsa, Okla. 74101.

The Complete Beginner's Guide to Motorcycling

Types of cycles; you and the law; helmet hints; careers for cyclists; contest riders; accessory tips; wind factors; and much more are included in this comprehensive, easy-to-understand guide to motorcycling for the novice.

Over 3 million American families have recently taken to motorcycling as a life-style, and millions more are on the verge. Bernhard A. Roth's book *The Complete Beginner's Guide to Motorcycling* gives up-to-date information on all aspects of buying, licensing, riding, maintaining, and using powered two-wheelers for maximum fun and utility. He offers extensive advice on motorcycle safety and sensible protection for riders of all ages. Lavishly illustrated with almost 100 photographs, this book is a must for motorcycle enthusiasts. Doubleday & Co. $4.95.

Auto Racing

Do you like racing cars? Are you interested in all the different kinds of races? Do you like pictures of racing cars? If so, here is a coloring book that will please you. Not only are there 16 large pictures of races (by Albert Nebeker) for you to color with all the action that goes with racing, but each race is explained by James Eckman and Terry Wilson. You'll learn about the world land speed record (and color the *Blue Flame*), stock-car racing, top fuel dragsters, the pit stop, the Indianapolis 500, Can-Am racing, and the Grand Prix.

Get out your colors and have fun with *Auto Racing*. Troubador Press. $2.00.

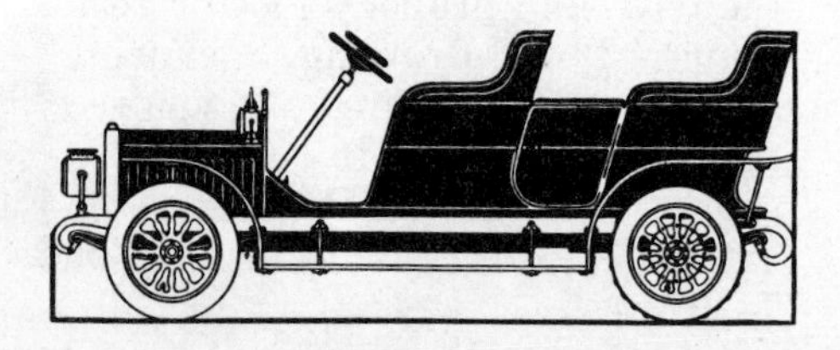

Invitation to Sailing

This book will teach you how to sail. It doesn't matter if you've never been in a boat in your life: Alan Brown's instructions will teach you what you need to know. He begins with fundamentals, telling you what makes a boat go, how to tack and jibe, how to set sail, drop the anchor, make a proper (and graceful) entry to a crowded dock—and what the name is for everything in, on, and around a small boat.

Then go out on the water and learn how to sail in light winds, in heavy weather, how to handle a spinnaker. You learn about racing, from basics to fine points on how to outsmart the field (and keep it from outsmarting you). Most important, you learn to develop the indispensable instincts that are the hallmark of a good sailor.

It's all taught in *Invitation to Sailing* by techniques the author has perfected over years of sailing and teaching sailing. Simon and Schuster (Fireside Book). $3.95.

Dinghy Sailing Illustrated

A sailor once said, "When I first started sailing some years ago, one item that always went into the boat was a plastic-wrapped copy of Patrick M. Royce's *Sailing Illustrated*. It didn't always save me from stupidity, but it explained afterwards what I'd done wrong."

There's no question that *Dinghy Sailing Illustrated* will provide the same kind of answers to a new generation of sailors. It offers a comprehensive look at the world of the small boat from basic sailing theory to racing rules. To the thousands of sailors who were introduced to correct terminology and techniques by Pat Royce, his books are truly the sailor's Bible.

This is an ideal book for the novice or for the experienced sailor who wants to get into racing. A number of yacht-club and college sailing programs have adopted this book as their text, with good reason.

Each page has good illustrations of typical small boats and the book is held together with Pat Royce's private brand of wry humor, because, after all, sailing is really fun. Japan Publications Trading Co. $2.00.

The Craft of Sail

This book, written, designed, and illustrated by Jan Adkins, belongs in the hands of everyone who sails a boat or wants to learn how. If you only dream about boats, this book will help make your dream come true. The drawings are so beautiful you may want two copies of the book, one to read and learn from, and the other to cut out the pages from and frame.

Here are a few words from the introduction, so you can get a feel of the man:

> To be in harmony with the forces of nature, you must know them intimately. A truthful book may help, but a real acquaintance with sea and moving air is indispensible. Here is where this book may fail at the whole truth: I cannot build spray into page 22, I cannot have the foghorn start its phomphing as you open page 59, I cannot write and draw the funky smell of low tide on page 52. You yourself must find those things and more. . . . I am a learner, as we are all learners on the water, and though I know this cannot be all of your sea lesson, I hope it can begin it.

The Craft of Sail is a very special book. To have your own copy SEND $6.95 plus 25 cents for postage and handling to Walker & Co.

Jumbo Wall Charts

"Know Your Knots and How to Use Them" is a small reproduction of one of six jumbo wall charts you can get as a *Boys' Life* reprint. Each chart opens up to almost 2 by 3 feet, and the price for all six is only 60 cents. The other charts are "Boat Safety"; "Physical Fitness"; "Firebuilding"; "Outdoor Code"; and "Outdoor Hazards." These wall charts will make handsome and practical additions to your room or clubhouse. Ask for No. 26–061 and send 60 cents to Boys' Life, North Brunswick, N.J. 08902.

FREE

Tips About Boating ...So the Boat Won't Tip

(Almost) Everything You Ever Wanted to Know About Boating . . . But Were Ashamed to Ask is a comically written but seriously intended booklet about boating. It is a short, self-learning text on boating for fun and health. A safety test for all boaters is included. Some of the topics covered are personal flotation devices; boat loading commandments; staying with the boat; the float plan; preventing accidents; and fueling commandments. For your free copy of this amusing and informative booklet comprised of reprints from a Coast Guard booklet WRITE TO Outboard Boating Club of America, 401 N. Michigan Ave., Chicago, Ill. 60611.

FREE

Boat and Canoe Catalog

Is yours an on-the-water family? Then you'll be delighted with this exciting catalog of the finest in canoes, kayaks, and powerboats in 28 pages of full-color water pictures. It begins with the wooden and canvas canoes made many years ago and brings you up to date on the latest in boats and canoes. For a free copy WRITE TO Old Town Canoe Co., Old Town, Me. 04468.

FREE

Water Skiing FUNdamentals

Water skiing is fun if you know how to do it properly. Here are three booklets which will introduce you to important aspects of water skiing that every beginning (and advanced) skier should be aware of.

Water Skiing FUNdamentals, the most basic pamphlet, teaches water skiing step by step, complete with photographs so you can see how you should look as you progress. There is a section on the most common errors of beginners, as well as tips on starting in deep water, from the beach, or from a dock.

Safety in Water Skiing outlines some basic commonsense precautions and safety pointers that every water-skier should know. Five topics are discussed in detail: the water, the boat and motor, the skier's gear, the driver, and the skier in action.

Safe Boat Driving for Water Skiing, which begins before you get in the boat to take your first skier for a ride, covers a vitally important aspect of safe water skiing. No matter how much the skier knows, he or she can't be safe without a safe boat driver. Subjects include boat-motor combination, speeds, accessories, ski area, signals, passengers, underway, picking up a fallen skier, return, dock starts, safety tips, and tournament driving.

For free copies of these excellent pamphlets WRITE TO American Water Ski Association. P.O. Box 191, Winter Haven, Fla. 33880. Be sure to enclose 10 cents for mailing and handling charges for each pamphlet.

FREE

American Water Skiing Association

Junior membership for boys and girls age 16 and under costs $10.00 for the first year, $7.50 for renewals. Members receive *Water Skier* magazine seven times a year and a book of Official Tournament Rules is available free. For information WRITE TO American Water Ski Association, P.O. Box 191, Winter Haven, Fla. 33880.

Scuba Diving

The basic techniques and knowledge needed to explore the underwater world are in *Scuba Diving: Handbook of Underwater Activities,* by Wheeler J. North. You'll read about training; safety; equipment; navigating underwater; and understanding waves, tides, and currents. There are chapters on observing and collecting marine creatures; hunting and spearing; and search and salvage. The book is fully illustrated, Golden Press. $1.95.

Guide to Fishing

Here in one clearly written, comprehensive book is all a beginner needs to start fishing. George X. Sand goes from the basic hook to elaborate new inventions for the fisherman, including the electronic depth indicator. He shows the novice where to look for fishes, what the proper bait is, what tackle to use, and what to wear. There are detailed chapters on the rod and tackle; lures and bait; the most popular game fishes in America; casting; and different types of fishing, including everything from wading and pier fishing to deep-sea fishing and old-fashioned cane-pole fishing.

The Complete Beginner's Guide to Fishing gives the reader the feel of fishing with its endless expectation followed by sudden high excitement, whether the fisherman is after a catfish in a muddy river or marlin in mid-ocean. Written for beginners of all ages, this immensely useful guide is illustrated with 30 photographs and over 60 drawings. Doubleday & Co. $4.95.

Let's Fish

Here's an excellent guide to fresh- and saltwater fishing, written and illustrated by Harry Zarchy. *Let's Fish* is for beginners; it will be of great help to people who would like to fish, but don't know how, where, or when. Fishing is open to everyone, regardless of age: no matter where you live, there is bound to be some form of fishing you can enjoy.

Chapters on freshwater fishing include "Bait Casting"; "Fly Rod Fishing"; "Spinning"; "Fishing Lakes and Ponds"; "Stream Fishing"; "Live Bait"; and "Artificial Bait." Saltwater fishing chapters include "Bay Fishing"; "Surf Fishing"; "Deep Sea Fishing"; "Live Bait"; "Artificial Lures"; and "Knots for Fishermen." Fishing teaches youngsters how to get along with other people. It teaches good sportsmanship and a consideration of the rights of others. Alfred A. Knopf. Available at your local library.

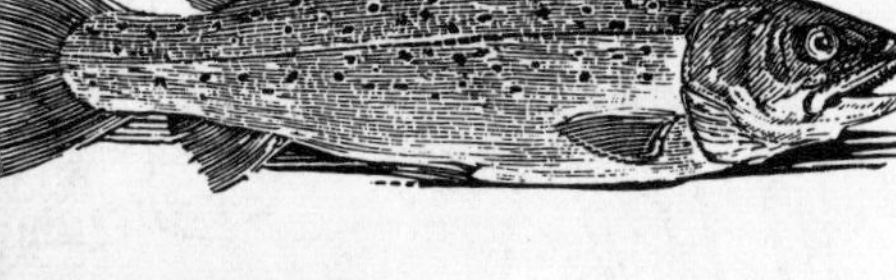

Fishing

This is a *Boys' Life* reprint—24 pages with pictures and text to help you catch the big ones! It includes articles on fishing with natural bait; a pocket fishing kit; a lightweight tackle box; fishermen spinning; bringing back the trout; making your own spinning lures; flytying; outboard fishermen; fish bait for the future; and finding fishes. To make sure they don't get away SEND 50 cents to Boys' Life, North Brunswick, N.J. 08902, for a copy of *Fishing* (No. 26-045).

Want to Go Backpacking?

Walking in the Wild is a complete guide to hiking and backpacking. This complete treatment of the subject is written by Robert J. Kelsey out of many years of experience and much love for the out-of-doors and how it is enjoyed by the young. The author started backpacking when he was 5 years old in the Sierra Madre mountains; took some of his children backpacking when they were 4—one of them, when he was 2!

If you are in, or want to be in a backpacking family, this is a book you should get for your house. It covers everything from clothing and equipment to trail techniques, setting up camp, and cooking. The author points out how to make do with what is readily and cheaply available, without buying expensive equipment. There is a long list of outdoor books, camping food producers, and equipment suppliers.

The information in this book will be useful as you grow up, but note the author believes in children in the out-of-doors:

> Start simply, both in equipment and ambition. It takes very little gear to enjoy warm-weather camping, as we shall shortly see. And some of the most beautiful spots on earth are those you see for the first time with a close friend, the most enjoyable times those quiet moments shoulder to shoulder in front of a snug campfire—no matter where it is built.
>
> As you gain in experience, your awakening sense of adventure should lead you to sample the joys of crisp spring and fall days and even the exhilaration of a backpack over deep snow in the crackling cold of winter.
>
> And it doesn't take a remote peak in a far country to achieve this state of mind either. For the very young and the beginner of any age, the attainment of a local tor may hold the same thrill as experienced by a mountaineer in his struggle up one of the world's giants.

This is a great book on an exciting subject. Funk & Wagnalls. $6.95 cloth, $2.50 paper.

The High Adventure of Eric Ryback

If you like backpacking, here is one of the truly great outdoor sagas of our time: *The High Adventure of Eric Ryback,* the extraordinary true story of how an 18-year-old boy accomplished what no man had done before. The adventure began in June 1970 when Eric Ryback set out to traverse the nation's longest hiking trail, from Canada to Mexico, alone on foot. Facing him were lingering winter snows as deep as 16 feet, towering mountain ranges, burning deserts, hundreds of miles without meeting another human being. Awaiting him were rattlesnakes, bears, coyotes, loneliness, sometimes near despair. Ahead of him too was some of the most spectacular scenery in the world, as well as a sense of mastery and achievement that marked an epic 2500-mile trek to maturity and manhood. Bantam Books. 95 cents.

Careers in the Outdoors

Increasing numbers of young people are turning to outdoor work for fulfillment and satisfaction in living. Whether your interest lies in helping to put out a raging forest fire, creating a better habitat for an endangered species of wildlife, or helping to stop soil erosion, you may find a niche in this challenging and truly rewarding field of endeavor. In *Careers in the Outdoors,* Mark Boesch, a veteran Forest Service professional, tells how you can find fulfilling work caring for our natural resources. E. P. Dutton & Co. Paper $4.95.

INDIANS

American Indian Handicrafts

Want to make your own Sioux warbonnet, a genuine leather Indian vest, an eagle feather back bustle, or a black buffalo headdress? If you choose Indian handicrafts as a hobby you can make all of these and many more with authentic, colorful kits. If you need a book about Indians (their history, lore, customs, legends, dances, and so on), maps that show tribe locations, or phonograph records with war, hunting, and love songs, Grey Owl Indian Craft Manufacturing Co. can supply it to you.

Grey Owl is continually working to preserve the fascinating heritage of the Indian. The owner, Don Miller, took up Indian lore as a hobby when he was a boy scout: he still remembers the first warbonnet he made, spending hours in Indian museums, never missing a movie about Indians, and putting his face into every book on the subject. Indian Lore is still his hobby.

If you want to take up this interesting hobby, SEND 35 cents to Grey Owl Indian Craft Manufacturing Co., 150–02 Beaver Road, Jamaica, N.Y. 11433, for the latest catalog which lists hundreds of kits, supplies, and books.

Here are Indian costumes and handicrafts that you can make with do-it-yourself kits.

Kachina Doll Coloring Books

The purpose of these coloring books is similar to the purpose of the kachina dolls themselves: to explain the tribal religious heritage of the Hopi Indians of the Southwest. There are about 250 kachinas; they play an important role in various religious ceremonies, many of which take the form of dances and chants. When a Hopi dresses himself in the costume of a particular kachina, he believes the spirit of that kachina has replaced his personal identity. The ceremonies ask the kachinas to bring plentiful crops, give the people of the village good growth, control the seasons, aid the warriors, or make miraculous things happen in the night.

Kachina Doll Cut-Outs written by Julie West Staheli, not only has magnificent Hopi dolls to color, but also a Tawa mask which you can cut out and wear. Troubador Press. $2.00.

Kachina Doll Coloring Books 1 and 2, created by Donna Greenlee, contain large kachina dolls for you to color, with a description of each. Fun Publishing. $1.95 each.

From tepee to tom-tom, from totem pole to tomahawk, it's all here: Indian craft and Indian lore and Indian life and Indian magic. Make a Sioux warbonnet, a leather vest, a peace pipe, bows and arrows, snowshoes, moccasins, and hundreds more!

Akwesasne Notes

Interested in learning about the Indians of today? *Akwesasne Notes* is the official publication of the Mohawk Nation at Akwesasne (People of the Longhouse); it contains *Longhouse News*, the official publication of the Mohawk Nation at Kanawake. The notes (48 pages, tabloid size: 11½ x 17 inches) are published in March, May, July, September, and December; the newsstand price is 50 cents per issue. Says the publisher:

> There is no fixed subscription price. But that does not mean this paper is free. Some people have lots of money; others have none. If you want the paper, we'll be glad to send it to you. If you want to help with costs, we will appreciate that—that's the Indian way. And since we have no grants, no other sources for finances, we must depend on you, the readers for survival.

To subscribe to this interesting Indian newspaper WRITE TO Akwesasne Notes, Mohawk Nation, via Rooseveltown, N.Y. 13683.

Akwesasne Notes Posters

In each issue of *Akwesasne Notes* there is a centerfold poster. The series is available as separate posters printed on heavy poster paper of various colors, size 17 x 22 inches. Posters are available at 3 for $1.00 or 50 cents each, plus 25 cents for a cardboard mailing tube. The right is reserved to substitute when out of stock. Posters currently in stock are "South America"; "Chief Joseph"; "Wounded Knee"; "Longhouse to Kiva"; "Zuni Governor"; "Sitting Bull"; "Grandma Hunter"; "Poundmaker"; "Shackled Native"; "Fire Carrier"; "La Raza"; "Dan Katchongva"; "Our Ideas"; "Statue of Liberty"; "Signs of the Times"; and "Flathead Chief." ORDER FROM Akwesasne Notes, Mohawk Nation, via Rooseveltown, N.Y. 13683.

Native American Calendar

These calendars (17 x 22 inches) are published by *Akwesasne Notes*. The 1975 theme was "Original Instruction for Human Beings."

> The calendar included each day's major events on the history of native peoples of North America, plus the cycles of our Grandmother, the moon, names of the month in native languages. It was an attractive and easy way to help friends come into greater harmony with historical realities. In our homes and offices, it was a means of educating ourselves on our past—and our future as human beings in this Creation.

The Indian calendars, published each year, cost $3.00 shipped in a cardboard mailing tube. To order WRITE TO Akwesasne Notes, Mohawk Nation, via Rooseveltown, New York, N.Y. 13683.

From a painting by Karl Bodmer of Pehriska-Rupa, a Minnetaree Warrior, 1834; Northern Natural Gas Collection, Joslyn Art Museum.

A PLAINS WARRIOR
DOING THE DOG DANCE

Coloring Book of American Indians

2000 years of Indian art from many tribes: this is a fine survey of art from prehistoric to Iroquois, Sioux, Cree, Navaho, Cheyenne, and Hopi. There are beautiful illustrations of Indians by artists George Catlin and Karl Bodmer. The drawings are simple, allowing younger children to be introduced to Indian art. A great book for Indian projects at school. *A Coloring Book of American Indians* is available at your bookstore or SEND $1.95 plus 25 cents for postage and handling to Bellerophon Books.

The World of the American Indian

In this excellent, profusely illustrated book you will meet the original Americans and follow their footsteps from 25,000 B.C. to the mid-1970s. Savor 400 pages about the richly varied cultures and contributions of arctic nomads, fishermen, foragers, woodsmen, pueblo dwellers, farmers, and horsemen. To order your copy of *The World of the American Indian* WRITE TO National Geographic Society, 17th and M Sts., NW, Washington, D.C. 20036. The Society will bill you $9.95 at time of shipment, plus postage and handling.

Catlin's Indian Gallery Coloring Book

George Catlin lived with the Indians for many years in the 19th century. He painted their portraits, recorded their customs, and learned their languages. His paintings and notes are the first important records of Indians west of the Mississippi River. In this excellent coloring book containing line illustrations by Warren Cutler based on Catlin's Indian portraits, you will meet Wah-Pe-Ke-Suck, the White Cloud; Ha-Won-Je-Tah, the One Horn; Tah-Teck-A-Da-Hair, the Steep Wind; Sky-Se-Ro-Ka, Second Chief of the Pawnee Picts; Tal-Lee (a well-known Osage warrior); and many others. You will learn about each Indian from a short history (by Susan and Warren Cutler). For example, Ta-Wah-Que-Nah, the Mountain of Rocks, told George Catlin he had earned his name by successfully leading his fellow warriors through a secret underground passage in the Mount of Rocks after they had been surrounded by an enemy tribe. Color these in *Catlin's Indian Gallery Coloring Book* and you'll feel you're bringing these historic Indians to life again. Grosset & Dunlap. $1.95.

W. Ben Hunt

W. Ben Hunt was born in Wisconsin about 90 years ago. He lived and worked in a log cabin not far from the site of the last Sioux uprising, and traveled and stayed with several Indian tribes. Most of his life was spent writing and lecturing about Indiancraft. Although most of us will probably never be forced to use our own hands, wits, and woodcraft to survive on a camping trip, it is nice to be able to do it. Everyone should be able to use a knife, ax, or thongs to make simple tools and equipment. Being able to survive by using Indian woodcraft skills makes you feel so confident it is worth acquiring the knowledge. When using his book, try to be like the Indian. Select what you can from nature, adapt it for your own use, make your own tools, and always be aware of the bounty of materials that is part of the American Indian heritage. When you use the woodcraft and campcraft techniques of the American Indian you are fashioning objects that recreate the time and life of a great people. Most of all, you enjoy the forests and live for and with the land the Indians have loved so well.

Indian Lore

Along with a Ben Hunt article, "Get Acquainted with Our Oldest Inhabitants: The Indians of the United States and Canada," there are many pages of famous Ben Hunt instruction-drawings. Subjects include Indian games; Chippewa dance drums; the flat roach; beaded necklaces; chokers and ties; beaded Indian belts; Indian vests; breechclouts and leggings; and Indian wigs and facial makeup. This *Boys' Life* reprint is a real prize. For your copy of *Indian Lore* (No. 26-084) SEND 50 cents to Boys' Life, North Brunswick, N.J. 08902.

Indiancraft

W. Ben Hunt's *The Complete How-to Book of Indiancraft* includes 68 projects for making authentic Indian articles, from tepee to tom-tom. His readable, easy-to-follow text and hand-drawn diagrams show how to use techniques and designs developed and perfected by the Indians. No special tools or prior skills are required. From making rawhide to putting the finishing touches on a Sioux shirt, beginners and seasoned woodsmen alike will enjoy making the clothing and other objects that American Indians have used for years. Projects include buckskin shirts, peace pipes, bows and arrows, totem poles, moccasins, breechclouts, warbonnets, headdresses, snowshoes, sleds, and wigwams. Collier Books (imprint of Macmillan Publishing Co.). $2.95.

Drums, Tomtoms & Rattles

Drums to have fun with, drums with magical powers, dance drums, tomtoms, water drums—just about any type of drum you can wish to know about or make is described in this book by Bernard S. Mason, a well-known authority on Indians. Complete directions for making each drum are given; you may be surprised to find that many of them can be constructed from simple, everyday materials such as wooden kegs, flowerpots, coffee cans, buckets, old inner tubes, and airplane cloth. Many authentic decoration ideas are included that capture the flavor and appearance of American Indian drums. For your copy of *Drums, Tomtoms & Rattles* SEND $2.50 plus 35 cents for postage and handling to Dover Publications.

FREE Indian Reservation Map

Are you studying the American Indian? Here's a map of the United States that shows all the Indian tribes, reservations, and settlements. Also available is an informative pamphlet, *American Indians Today,* which describes the Indian Rights Association as well as the education, health, and goals of the 800,000 American Indians. The map and pamphlet are free. WRITE TO Indian Rights Association, 1505 Race St., Philadelphia, Pa. 19102. Enclose 10 cents in stamps to cover mailing and handling.

Indian Harvests

Everywhere in America you can find wild plants. The American Indians harvested many of them for food, and the bounty of the land provided rich feasts for all the tribes. Today you can follow in the footsteps of the Indians and find the same foods they used, from prickly pears in the Southwest to cranberries in New England. Some plants are familiar to us—strawberries, crab apples, and chestnuts; some are less well known—yampa, arrowroot, and piñon nuts. The major plants the Indians used are described, along with the ways the Indians prepared and used them. You can find these plants easily, and prepare them the way the Indians did. Accurate line drawings by Ronald Himler will help you recognize the plants. William C. Grimm wrote this excellent book, *Indian Harvests.* McGraw-Hill Book Co. $5.72.

North American Indian Arts

North American Indian Arts, by Andrew Hunter Whiteford, is a colorful guide to Indian arts and crafts. Packed with historical information and authentic full-color illustrations, this handsome 160-page paperback book is a key to understanding the Indians and the rich diversity of their cultures. The book is a great show-and-tell, covering everything from pottery to baskets, from textiles and leatherwork to wood and stonework. Golden Press. $1.95.

Indian Beadwork

The ancient craft of beadwork was perfected by the American Indians, who used bones, shells, and seeds to make their jewelry. Author Marjorie Murphy covers basic handmade stripwork suitable for simple collars, bracelets, and necklaces, and explains how to use the bead loom to make more complex designs, including the traditional thunderbird. There are informative chapters on choosing needles, beads, preparing thread, setting up the loom, and making your own designs. *Beadwork: From American Indian Designs* is an easy-to-follow guide to a craft as contemporary as it is historical. Watson-Guptill Publications. $8.95.

Bury My Heart at Wounded Knee

This is an Indian history of the American West during the years when the culture and civilization of the American Indian were destroyed. You probably know about the cowboys, gamblers, gunmen, and homesteaders—all the great myths of the West—but you may know very little about the Indian side of things, since the voice of the Indian was heard only occasionally. But these Indian voices are not all lost.

Dee Brown, the author of *Bury My Heart at Wounded Knee,* has drawn from records of treaty councils and other meetings with civilian and military representatives of the U.S. government, along with newspaper interviews of the day with tribal leaders, to provide an authentic picture of how the "conquest of the West" was experienced by the victims. Bantam Books. $1.95.

FREE Know the Truth About the Indians

Almost every school in the United States teaches about the American Indian; but the published material about Indians is often sentimentally unrealistic or brutally untrue. The publications of the Bureau of Indian Affairs, which are written by people who live and work with the Indians, present a truer picture. There are books for children in elementary grades and for older children. Most are low priced; some are free. For *Know the Truth About Indians,* a free annotated list of publications published by the Bureau of Indian Affairs, WRITE TO Publications Service, Haskell Indian Junior College, Lawrence, Kans. 66044.

slit F
tab F
slit E
tab E
slit D
tab D
slit C
tab C
slit B
tab B
slit A
tab A
fold up here
fold up here
EOTOTO
(chief kachina)
FROM KACHINA DOLL CUT-OUTS

Eototo's mask and clothes are white. His mocassins are red. The powerful simplicity of his appearance indicates the importance of this great kachina.

He is chief of all kachinas. He knows the ceremonies and controls the seasons.

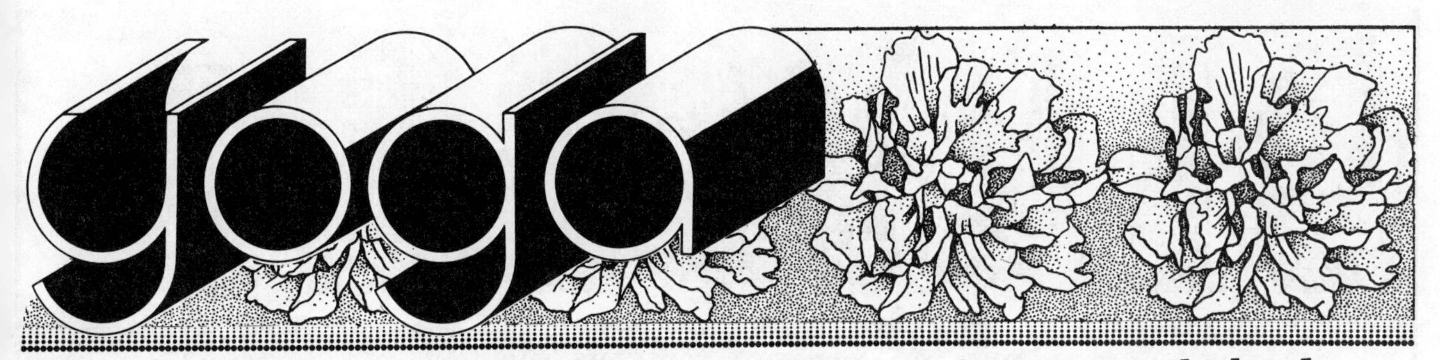

Discover the ancient Indian art of Yoga for a supple body and a sound mind. Enjoy the peace that comes from self-confidence and a sense of well-being.

Be a Frog, a Bird, or a Tree

Be a Bird, Be a Frog, Be a Tree, Be a Bridge, Be a Swan, Be a Bow. These 30 exercises make use of every child's most powerful creative tool—his or her imagination. They also teach balance and muscle control. The movements are called yoga exercises and they are fun to do, as children all over the world have discovered.

The author of *Be a Frog, a Bird, or a Tree,* Rachel Carr, says: "Children are easily bored by the suspended motion required of the traditional yoga postures which adults find so beneficial and relaxing. To overcome this barrier, I encouraged children to explore the sensation of movement through mimicry, such as pretending to be a jumping frog, a flying bird, a shooting arrow, or a hopping crow. They did not mind the stillness of some of the exercises when they could see themselves as a sturdy tree growing in a forest, or a strong bridge with cars traveling over them and ships passing under them."

Doubleday & Co. $5.95.

Lets Do Yoga

Yoga, the ancient Indian system for promoting physical and mental health, has become almost as commonplace in the U.S. as apple pie. One reason for the yoga's popularity is its beneficial effects; it is also fun and easy to do, and can be done almost anywhere. No paraphernalia is needed—just a place to sit and lie down.

Here then, to encourage good health (yoga promotes self-confidence and a sense of well-being) and to keep young bodies supple, is *Let's Do Yoga,* by Ruth Richards and Joy Abrams—a complete yoga book for young children. Clear and easy instructions take you step by step through the classic postures and some simple breathing techniques. The book is beautifully illustrated by Sandra Case. Holt, Rinehart and Winston. $5.95.

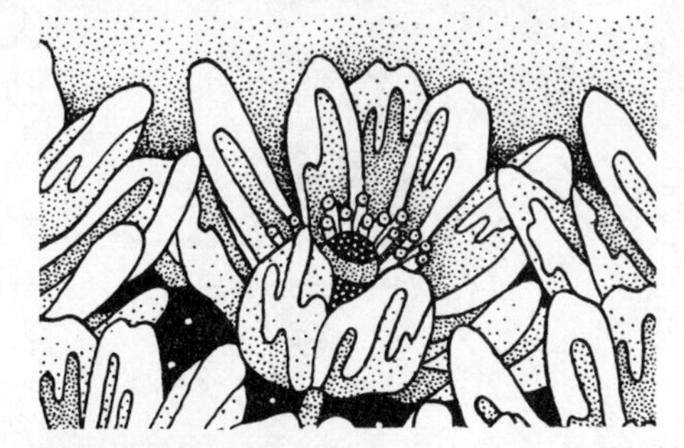

Yoga for Young People

Health is beauty. A healthy body, a sound mind—that is the purpose of yoga.

What is yoga? It is more than standing on your head, or sitting in the lotus position, or talking about meditation. It is an ancient system of exercise designed to promote physical and mental health. From it come almost all exercises for sports, the dance, and gymnastics. It keeps the body alive and clears the mind. Yoga will put you in touch with nature—and with yourself. It will strengthen your body and shape it.

Look good. Feel good. Learn to concentrate. Try yoga.

In the book *Yoga for Young People,* Michaeline Kiss gives you easy-to-follow instructions. Bobbs-Merrill Co. $4.95.

BICYCLES

snail race

All racers line up with their bikes at the starting line. On the signal, the racers go as slowly as possible to the finish line. The rider who gets there last wins. Anyone who puts a foot on the ground before the finish line is out.

Want a Great Bicycle Book?

That's the name of this one: *A Great Bicycle Book.* It was written because the author (Jane Sarnoff) and illustrator (Reynold Ruffins) got tired of having to take their bikes to the repair shop all the time. They say yes, even the non-mechanical boy or girl bicycle rider, age 8 and up, can understand, maintain, and repair a bicycle.

"Look, ma, no hands!" Ride a bike. Take a bike trip—to the end of town or overnight. The rules of the road. The anatomy of a bike: wheels, tires, brakes, chains, and derailleurs. How to buy a bike, take care of a bike. Get a great bicycle book.

Gears, wheels, tires, brakes, and chains lose their mystery as you learn exactly how they work. In clear and simple text and illustrations, this book should make three out of five trips to the repair shop unnecessary. It includes riding and racing suggestions, and silly and serious facts about bicycles and bicycle riding. You'll learn to be a better rider from the safety features woven throughout the book. Charles Scribner's Sons. $5.95.

The Complete Beginner's Guide to Bicycling

This book is a comprehensive bicycle manual for kids of all ages. In the first part of the book, author Richard B. Lyttle explains all about bicycles: parts, gears, and accessories. He tells you about the evolution of the bike and how you can join a bicycle club. The second part of *The Complete Beginner's Guide to Bicycling* is about using your bike; safety rules and habits you should get into; touring, camping, and bike competitions; and a complete guide to the care and repair of your bike. Doubleday & Co. $4.95.

Bicycling

Bicycling has more than 300 drawings, diagrams, and color photographs. One chapter, "Bicycling Through the Years," does a show-and-tell on the history of the bicycle and how it has influenced our way of life. A good book: comprehensive and concise, practical yet adventurous. You'll find out about bikeways, recreational trails, and how you can discover America by bicycle. The authors are George Fichter and Keith Kingbay; Mr. Kingbay is one of the foremost experts on bicycling in the United States—he rides 6000 to 8000 miles a year! Golden Press. $2.95.

Safety Fact Sheets

Here are two excellent fact sheets on the safety of bicycles and tricycles, with tips on how to operate and take care of these vehicles. You'll find out what to look for when choosing a bike. If you want more information, there is a list of 11 bicycle-oriented publications to send for. For your fact sheets WRITE TO U.S. Consumer Product Safety Commission, Washington, D.C. 20207. Ask for Fact Sheet No. 10, "Bicycles," and Fact Sheet No. 15, "Tricycles."

Bikes

You might call this book, written and illustrated by Stephen C. Henkel, the bicycle lover's Bible; it is a carry-along, how-to-do-it guide to everything about bikes. The chapters of *Bikes* include "The Anatomy of a Bike"; "How to Buy a Bike"; "How to Ride a Bike"; "Basic Care of Your Bike"; "How to Lubricate Your Bike"; and "Having Fun." If you don't know anything about bikes, this book will give you a proper start on the road. If you know a little, you'll learn more. And if you know everything about bicycles, you'll get a kick out of the pictures. Chatham Press. $5.95 cloth. Bantam Books. $1.25 paper.

The American Biking Atlas & Touring Guide

American families are bicycling in a big way. Did you know more new bicycles than new cars were sold in 1973? For all cyclists, adults and children, here is the most ambitious biking book to date, with 150 detailed trips in every region of the United States, from the accursed tombstone tour in Maine to the Kauai legend tour in Hawaii, from the 18th-century churches tour in Maryland to the Merrimack foliage tour in Massachusetts. There are city trips, for example, one for Sunday in New York. Two-color maps for each trip were drawn for the book. Sue Browder includes information on distance, traffic conditions, terrain, the best time to go, places to stay, sights to see, and clothes to wear. Each page is perforated for easy removal. For your copy of *The American Biking Atlas & Touring Guide* SEND $5.95 plus 50 cents for mailing and handling to Workman Publishing Co.

FREE Youth Hostels

The purpose of the AYH (American Youth Hostels) is to help all, especially young people, to a greater understanding of the world and its people through outdoor activities such as hiking, biking, and educational and recreational travel. It provides youth hostels (simple overnight accommodations in scenic, historic, and cultural areas) with supervising houseparents and local sponsorship. It's a group that helps you while you're traveling around the country: get to know the hostels. For a free copy of *What Is Hosteling, Highroad to Adventure the Hostel Way,* and *Youth Hosteling in the United States,* and information on membership, facilities, and specially organized trips, WRITE a postcard to Travel Dept., American Youth Hostels, Delaplane, Va. 22025.

Amateur Bicycle League of America

Registration costs $3.00 per year. Ages 8 and up. The league sponsors bicycle races. Midgets (ages 8 to 12), Intermediate (12 to 15). For information WRITE TO Amateur Bicycle League of America, 137 Brunswick Road, Cedar Grove, N.J. 07009.

FREE Bicycle Blue Book

The *Bicycle Blue Book* is a 16-page, two-color leaflet designed to help young bicycle riders learn the "rules of the road." The chapters include "Basic Bike Safety Code"; "Safety Steps You Can Take"; "Proper Size for Your Bicycle"; "More Things You Can Do"; "Jobs for Your Bicycle Serviceman"; "Special Tire Tips"; and "A New Dimension in Bicycle Safety." For your free copy of this helpful leaflet WRITE TO Pubic Relations Dept., Goodyear Tire and Rubber Co., Akron, Ohio 44316.

Back to the Bike

Illustrated with charming old-fashioned prints and modern technical drawings by Keith Halonen, *Back to the Bike,* by Clifford C. Humphrey, tells you how to buy, maintain, and use a bicycle as an alternative means of transportation. You'll find out about the basic types of bicycles and of riding, so you can be sure you are getting the right kind of bicycle for your needs. There are sections on "Maintenance of the Bicycle" and on "Maintenance of the Rider."

Philosophy in a book on bicycles? Yes; and it's good. In the opening, the author says:

> People today are trying to make peace with themselves, each other and the land. The bicycle is part of this revolution. We are living in a period of adjustment, re-evaluation and a redefinition of priorities. Along with the return of the bicycle, gardens are growing, labels are being scrutinized and people are becoming more involved in political activities. The growing realization that the earth is being clobbered by human activity has injected environmental issues into our daily lives. We have to ease up and learn to live within our means. The bicycle has an important role to play, as we move toward meeting our needs within the earth's limits. Riding a bicycle is healthy, fun and almost free. It produces neither noise nor pollution.

101 Productions. $2.95.

OUTDOOR FUN

Go Fly a Kite

Go Fly a Kite Store is the first in the world to specialize in kites and kite equipment; it carries the largest selection of kites in the world. The kites, whether made of brilliantly colored paper or shiny reflecting plastic, are beautiful to look at and wonderful to fly; they come with complete instructions and flying tips to add to your pleasure. With your purchase of a kite the store will send you a lifetime membership card in the Go Fly a Kite Association, and a button.

The store's catalog is 21 x 28 inches; on one side there is a poster; on the other side, photographs and descriptions of over 30 kites, including the Chinese Butterfly, the Thai Owl, the Star of India, and the Red Baron. The Indian Fighter is the most maneuverable of all kites: it glides, dips, dives, and twirls, and is made of bamboo and tissue paper in brilliant colors.

The poster side of the catalog is available unfolded and suitable for framing or mounting: free with a $25.00 order; $2.00 with an order under $25.00.

The Go Fly a Kite Store is at 1434 3d Ave. (between 81st and 82d streets), New York, N.Y. 10028. Telephone: (212) 988–8885. You are invited to visit the store to see its full line of kites, accessories, and materials—it's open Monday through Saturday, 11:00 a.m. to 7:00 p.m. If you can't get there in person, SEND 25 cents for a copy of the catalog.

More Kite Shops

Come Fly a Kite
900 North Point St.
San Francisco, Calif. 94109

A full range of kits from all over the world. Dinest Bahadur, proprietor.

The Kite Shop
542 St. Peter St.
New Orleans, La. 70116

A full range of imported and American kites. Sally Fontana, proprietor.

Windy City Kite Works
1750 North Clark St.
Chicago, Ill. 60614

Imported and domestic kites and accessories, including some designed by Carl and Peter DiDonato, proprietors.

Kites

This unique Golden Handbook Guide, *Kites,* by Wyatt Brummitt, explains the what and why of kites and describes different kinds of kites, where they came from, and how they are built and flown. It delves into the fascinating legend and folklore of kites and explains their place in ancient times and today. The great full-color drawings are by Enid Kotschnig. Golden Press. $1.95.

Kite Craft

This fine book, subtitled *The History and Processes of Kitemaking Throughout the World,* is by Lee Scott Newman and Jay Hartley Newman. It contains 357 photographs, 84 drawings, and 18 color plates; it is a book for older kite buffs.

Kite Craft tells you the history of kites and explains basic construction techniques. You will learn how to build many types of kites, including the standard diamond kite and Japanese lantern, bird, airplane, rectangular, and circle kites. There is a section on decorating the kite face using processes for creating designs on paper and plastic that include tape resist, batik, marbling, tie-dye, printing, and stamping. Crown Publishers. $8.95 cloth, $4.95 paper.

Kite Folio

If you like to fly kites you will want a copy of this exciting book, written and illustrated by Timothy Burkhart. It is a collection of illustrated anecdotes from the long history of flying kites. Facing each episode, detailed diagrams are joined with concise instructions to make kitebuilding and kiteflying understandable and easy. Says the author:

> The thrill of kite-flying is a tangible one. It comes from the pleasure of successfully launching and skillfully maneuvering a kite of your own creation, feeling the kite respond to the wind and the sun through the line in your own hands. Kite-flying puts you in touch with the birds and the sky and earth around you, and makes friends of curious strangers who gather to watch and question.

Historically, kites have been far more than toys. They have built bridges, saved lives, taken pictures, caught fish, and been worshiped as gods. For those who ask "Why fly a kite?" the book answers, "You will know only when you have done it," and proceeds to show you how.

For your copy of *Kite Folio* (11 x 14 inches) SEND $5.00 plus 30 cents for postage and handling to Ten Speed Press.

Go fly a kite! Shiny plastic or brilliantly colored paper. The Indian Fighter . . . it glides, it dips, it dives. Balloon kites. Box kites. Tetrahedral kites. Shoot marbles, throw a Frisbee, jump rope, play street games, fling a boomerang!

The Nantucket Kiteman

Betty and Al Hartig are very special people who make special kites. Al, who is an artist and ship model builder, has invented a new kind of hand-made kite, for which 13 design elements have been patented. While they were still living in New York City, Betty, who is a writer, set up a business to sell the kites to the many people who wanted them.

Now the Hartigs live on Nantucket Island, 30 miles off the Massachusetts coast, where they and other Nantucket residents still make the kites by hand. If you get to Nantucket during the summer, you'll find Betty and Al in their kite shop on Old South Wharf at the Nantucket Marina.

Each of his kites (the Ace, the Valkyrie, the National Eagle, and the Flying Fox) is a tailless kite of durable cloth that folds and reassembles in seconds. Struts may be removed and replaced. For a brochure, which tells you about all of his kites, prices, cord, handles, and parcel post charges, and features a picture of a whale flying a kite, SEND a stamped self-addressed envelope to Nantucket Kiteman, P.O. Box 1356, Nantucket, Mass. 02554.

25 Kites That Fly

Have you ever felt like flying a kite and feeling it tug at its string like a living creature? The joy of kiteflying is much greater when you make your own kites. And it's easy. With this book by Leslie L. Hunt, kitemaker for the U.S. Weather Bureau, you learn how to make personalized kites. All you need is some light wood, glue, wire, hammer, and nails—plus the clear, concise explanations in the book.

25 basic kites are covered in constructional detail: standard two-stick kites; six-point stars; figural kites, such as imps, fishermen, elephants, owls, and shields; balloon kites; tetrahedral kites; various box kites; really strong military kites; and so on. For your copy of *25 Kites That Fly* SEND $1.25 plus 35 cents for postage and handling to Dover Publications.

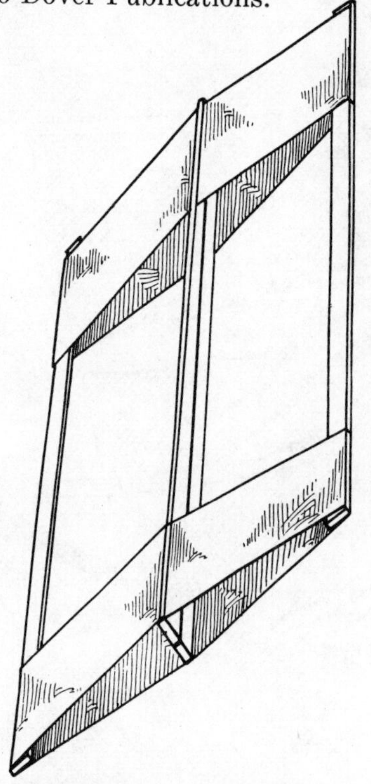

PERSPECTIVE VIEW OF THE SQUARE BOX KITE LET DOWN SHOWING HOW STICKS SHOULD BE SPACED

Tetrahedral Kites

Now you can fly with the tetrahedral kite that has revolutionized kiteflying. It's called the Amazing Classic Alexander Graham Bell TetraKite™ ($5.50). It combines scientific, historic, and hobby kit appeal with a new kind of kitefly-ing fun. Based on a principle developed by Alexander Graham Bell, its exclusive connectors, branches, and sails make it fit together easily and fly like a breeze. (You don't even need a tail.)

The size of these beautiful kites is as impressive as their lift. The 4-sail model is 33 inches on edge (4312 cubic inches); the 16-sail SuperTetraKite™ ($18.50) is a whopping 5½ feet on edge (34,496 cubic inches).

With the Skylinks I™ ($3.50) or Skylinks 4 Kite System™ ($8.50) you can build a train of kites, separate tetrahedrals flying from the same line, giant high-flying tandem kites that soar to incredible heights, and dual-line stunt kites that loop, dive, and figure-eight.

These kites are easy to assemble; they come with completely illustrated instructions plus excerpts from Alexander Graham Bell's 1903 article in the *National Geographic* on kite structure. At your toy or hobby shop.

Kiteflying Organizations

American Kitefliers Association
315 North Bayard St.,
P.O. Box 1511
Silver City, N.M. 88061

An organization of kitefliers interested in kite design, building and flying for sport, recreation, and scientific study of aerodynamics. Membership is $5.00 per year and includes a subscription to *Kite Tales Magazine,* which appears quarterly.

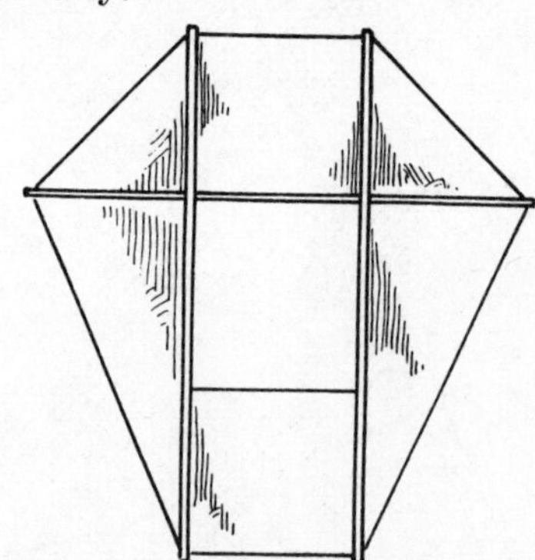

BACK FRAME FOR THE TRIANGULAR BOX KITE WITH WINGS

FREE

Make Your Own Kite

Go Fly a Kite is a reprint from *Seventeen's Make It!* magazine that shows you with diagrams, pictures and step-by-step instructions how you can make a colorful yellow target kite. A great summer project with tie-dyeing! For your copy SEND 10 cents for mailing and handling to Go Fly a Kite, P.O. Box 307, Coventry, Conn. 06238.

Boomerangs: How to Make and Throw Them

It takes only minutes to make a good, guaranteed-to-return boomerang. In this book—*the* book about boomerangs—you will learn how to throw a boomerang so it always returns to you. You can experiment with many types of boomerangs and fly them in new tricks and stunts. Author Bernard S. Mason has designed dozens of them that are easy to make and easy to throw, ranging in size from 14 inches to 3 feet. They have names such as boomabirds and tumblesticks. For your copy of *Boomerangs: How to Make and Throw Them* SEND $1.50 plus 35 cents for postage and handling to Dover Publications.

Frisbee

Frisbee, a Practitioner's Manual and Complete Treatise, is a huge work, carefully annotated and generously illustrated with photos and diagrams, with more information than you could have expected on platter trajectories. Dr. Stancil E. D. Johnson, the author, knows all the ways to play the game: bowling Frisbee, baseball Frisbee, circle Frisbee, field guts, football Frisbee, goalpost Frisbee, groove, hall hockey, dodge, conversion, lag, golf Frisbee, guts keepaway, indoors (and outdoors) Frisbee, posey pitch, innertube, and, for the daring only, ultimate Frisbee. Heads up! For your paperback copy SEND $4.95 plus 50 cents for mailing and handling to Workman Publishing Co.

American Marble Book

Need we say more? This current book is firmly established as the definitive treatise on over 50 variations of the historic game of mibs. Included are "Some History"; "Aggies—Alleys—and Others"; "A Lexicon of Mibology"; and all the games from corner-the-market to knuckle box. As the author, Fred Ferretti, says: "This book is for all former marble players . . . and for future marble players. Let their tribe increase!" For your copy of *The Great American Marble Book* SEND $2.50 plus 40 cents for postage and handling to Workman Publishing Co.

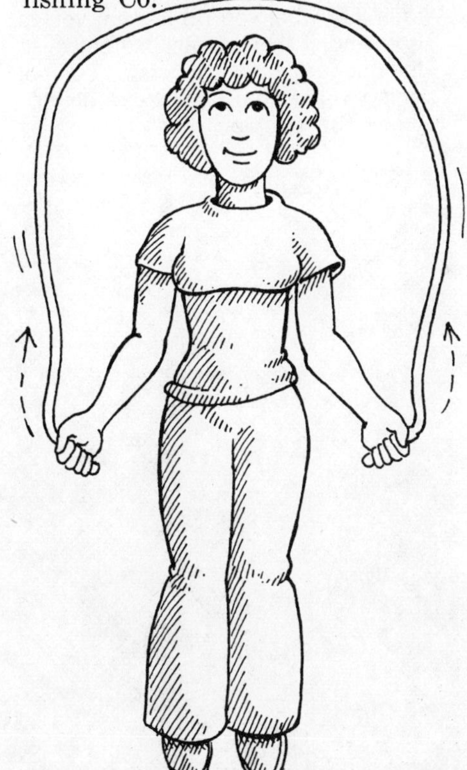

Jump Rope

Remember this one?

Charlie Chaplin went to France
To teach the pretty girls
The hula dance
First on heel,
Then the toe
Do the splits
And around you go.
Salute to the Captain,
Curtsey to the Queen,
Touch the bottom of the submarine

It's one of 250 chants and rhymes, old and new, to be found in *Jump Rope,* by Peter L. Skolnik, a whole world of jumping joy that includes jump-rope history, rope talk (a glossary of jump-rope terms), jump-rope technique, jump-rope games, and all the kinds of rope rhymes, including classic, counting, prediction, fortune, call-in, bump, and hot—you'll jump for joy! For your copy of this fun book SEND $2.95 plus 40 cents for mailing and handling to Workman Publishing Co.

Shimmy Shimmy Coke-Ca-Pop!

That's the name of a handclapping song and of a book with the subtitle *A Collection of City Children's Street Games and Rhymes.* For the children of big cities the sidewalk is "where the action is." From the secret society of city children come these songs, rhymes, and games. Some are very old ("London Bridge Is Falling Down"), some contemporary ("Rinsle, Tinsle, the Ordinary Soap"). There are sections on name-calling ("Eddie Spaghetti with the meatball eyes. Put him in the oven and make french fries"); ball bouncing; sidewalk drawing games; circle games; who's it?; tag games; jump rope; action games; follow-the-leader; handclapping; and dramatic play.

The authors, John and Carol Langstaff, write: "This collection is not in any way definitive, or to be followed rigidly for play or instruction. We have purposely made any bits of instruction sketchy to encourage improvisation." Photographs by Don Mac Sorley. Doubleday & Co. $4.95.

Build Your Own Playground

Children's Playgrounds

This is an exciting book: a sourcebook of play sculptures, designs, and concepts from the work of Jay Beckwith, written by Jeremy Joan Hewes. Children should show it to their parents, and vice versa.

When a Beckwith-designed playground is built, everybody pitches in—men, women, and children. The result is a play area that reflects the real will of the sponsoring community group; a play environment liberally drawn from the ideas of the children who will enjoy it.

> In a word, the thing that happens between people when they work together, particularly when they work together to create play, is magic. It's really invisible what happens, but people go away with a feeling of it.

Unorthodox materials and structures are used. The first step is awareness that a playground children will delight in and care for with pride need not be, *ought not* to be the usual sterile layout of asphalt, slides, sandbox, and swings.

Designer and teacher Jay Beckwith has developed more than 50 community and school playgrounds in California. His intense experience is distilled for use everywhere in *Build Your Own Playground!* a visually exciting, idea-packed book. Through text, 55 schematic drawings, and 215 sequential photographs, the reader is shown every detail of planning and building a creative, innovative playground with volunteer labor and amateur skills. San Francisco Book Co./Houghton Mifflin Co. $15.00 cloth, $7.95 paper.

A Belief About Playgrounds

Jay Beckwith, playground designer and builder, has written:

> It's my contention that playgrounds can only be built by community people. Creating an environment that stimulates play and supports kids' good feelings about themselves is so complex that the only way to handle it is intuitively. When we take our native intelligence and our feelings about ourselves and our kids and put them together intuitively, we make an expression that is both a conscious and unconscious act which reflects who we think we are, where we want to go, what our dreams are. You're not going to get that with a set of plans, a hard-hat crew, and a jackhammer; it just doesn't happen that way. We're going back to the "the medium is the message," the basic idea that what you do is what you've got. And if there is an alternative to the juggernaut of concrete, it's not going to be more bulldozers—it's going to be people.

Adventure Playgrounds

Robin Moore is a playground designer who has been working to incorporate the features of adventure playgrounds into children's environments. He has published three booklets on his ideas and experiences; the publications have a kidlike feeling reminiscent of the atmosphere of a child-centered playground. To get your copy of *Open Space Learning Place* ($1.00), *Washington Environmental Yard* ($1.00), or *Living Kid City* ($2.00) WRITE TO Dept. of Landscape Architecture, 202 Wurster Hall, University of California, Berkeley, Calif. 94702.

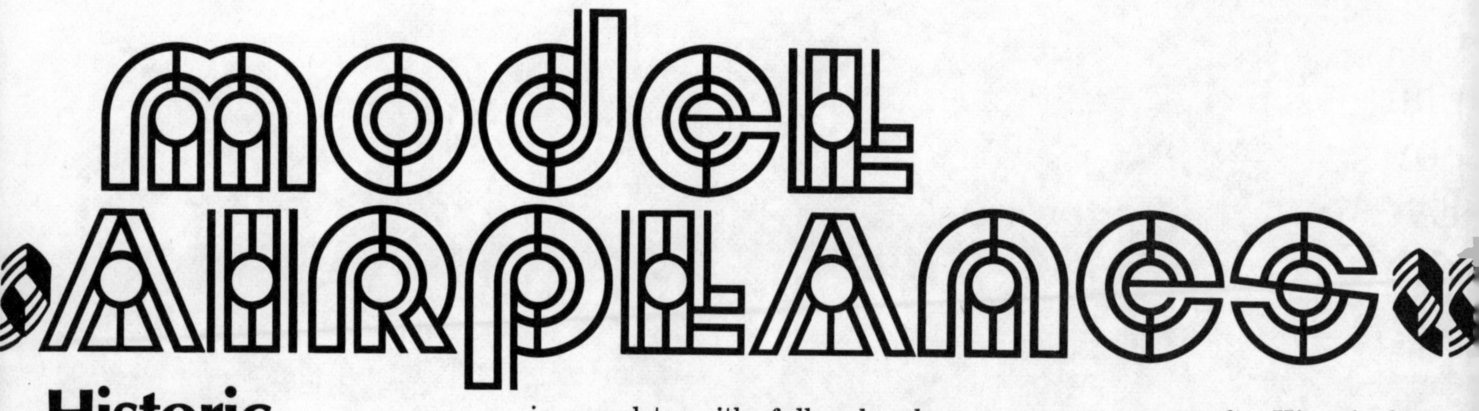

Historic Plane Models: Their Stories and How to Make Them

Here are the stories of seven history-making planes, with clear, detailed instructions and diagrams for making a model of each plane. Also included are photographs of the planes and maps showing the routes they took on their memorable journeys. Among the planes included are the Wright brothers' heavier-than-air flying machine; the *Spirit of St. Louis,* in which Lindbergh made the first solo flight across the Atlantic Ocean; and the *Floyd Bennett,* the plane in which Byrd made the first flight over the South Pole. A good mixture of history and model making, *Historic Plane Models* is written by Frank Ross, Jr. Lothrop, Lee & Shepard Co. $2.45.

Balsa Models

Making models from balsa wood is a fascinating pastime which can be tackled with confidence by anyone, thanks to the easy-working qualities of this lightest of all woods. No expensive tools: just some old razor blades, sandpaper, and a few other household odds and ends. *Bill Dean's Book of Balsa Models* has illustrated instructions for making 18 working models from balsa: 12 planes, 2 boats, a spaceship, and an automobile. Each is complete with full-scale plans and patterns. Dean has aimed everywhere at eliminating the usual stumbling blocks encountered by beginners. Arco Publishing Co. $2.95.

Airborne All-Stars

Here's a book of paper flying models (that you can make) of famous aircraft, including the Cessna 210, the Boeing 747, the International Concorde, the McDonnell-Douglas F-4, the Lockheed F-104, the Hawker Siddeley, and the Dassault Mirage III. You will learn about materials, patterns, construction, adjustments, and flying.

Author Yasuaki Ninomiya, winner of the grand prix at the First International Paper Airplanes Contest (Pacific Basin division) has a collection of sleek, high-performance paper craft. His airplanes hold the record for duration and distance in flight contests. *Airborne All-Stars* is one of the better books on paper airplanes. Japan Publications Trading Co. $4.25.

More Paper Airplanes

Paper Airplanes presents 20 aircraft to color, cut out, fold, and fly, including space shuttle, wing plane, X-wing racer, supersonic rocket ship, hi-altitude interceptor, and beautiful bird planes that actually fly: the "sparrow" and the "Bering sea gull." Designer-illustrator Marc Arcenaux includes complete instructions for folding and cutting, and there are special flying instructions for each plane. Troubador Press. $2.00.

Cut them and fold them and glue them and fly them. Paper planes with high-performance design. Planes that are rubber-powered or radio-controlled, gliders, hot-air balloons, a rocket. Everything that soars or roars . . . make it and fly it!

The Sig Catalog

If you're an old-timer at model airplanes, you know that more contest winners use Sig balsa than all other brands combined. Sig puts out one of the best model airplane catalogs in the country; its 256 pages are jam-packed with everything you could ever need whether you're launching a hand-held glider, putting up a rubber-powered airplane, or flying a radio-controlled plane. You name it and they have it, from balsa and spruce to a motor or a kit; there are dozens of full-color photographs of kit planes that you can build and hundreds of black-and-white photographs of everything from a jetstar with a 65 inch wingspan to a Kavan bell jet ranger helicopter. For your copy of this fabulous airplane catalog SEND $1.50 to Sig Manufacturing Co., 401 S. Front St., Montezuma, Iowa 50171.

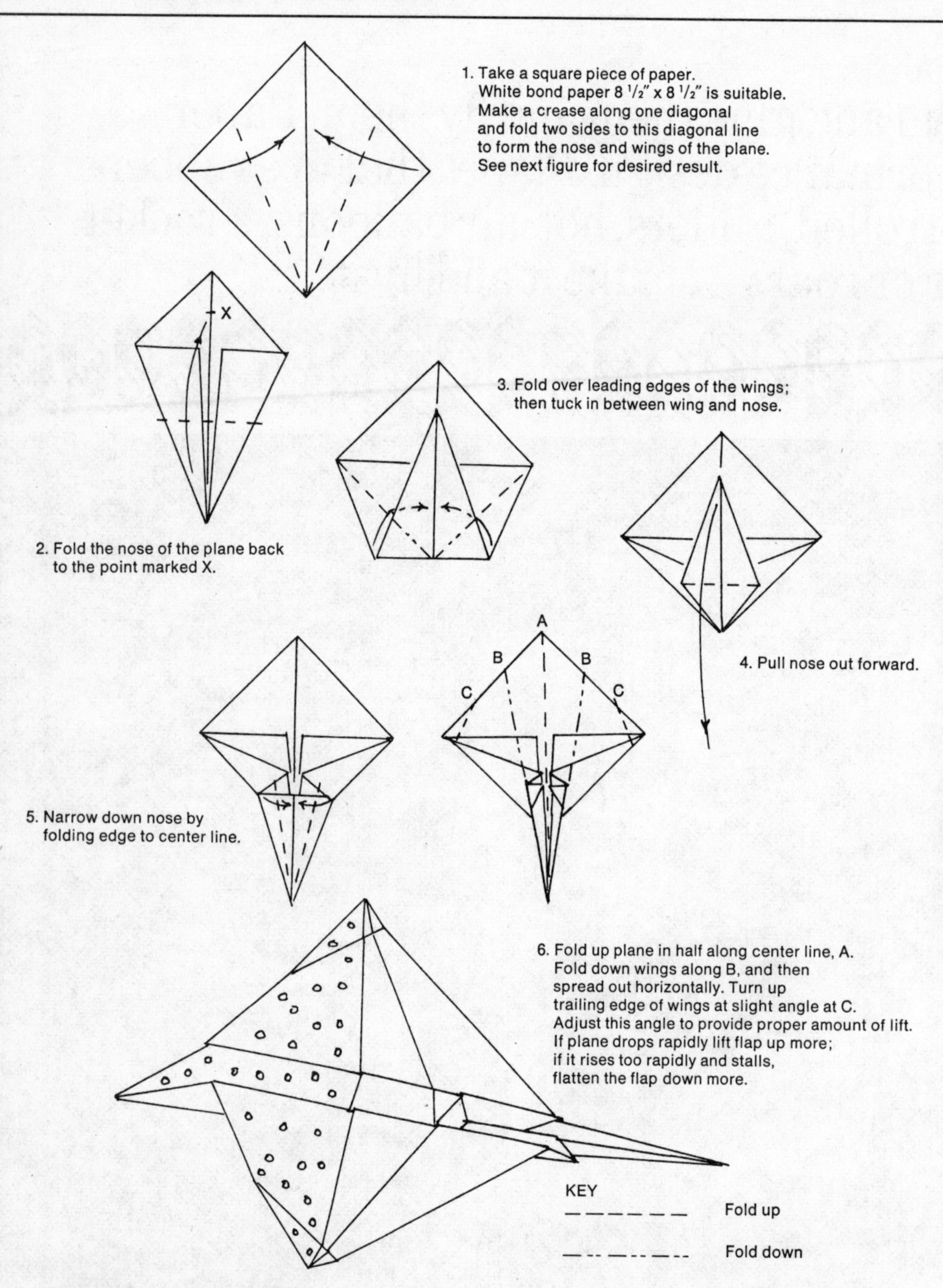

The Great International Paper Airplane Book

If you are into flying paper airplanes, you'll want a copy of *The Great International Paper Airplane Book,* by Jerry Mander, George Dippel, and Howard Gossage. Besides a fully illustrated story of the First International Paper Airplane Competition conducted by *Scientific American,* it contains 20 patterns of the seven winning planes and other notable entries; you can cut them out, fold them, and fly them. The instructions say:

> The success of each plane depends upon how faithfully you attend to minor details in folding and cutting and how patiently you experiment with various launching efforts. The pattern pages are perforated so as to more easily allow you to tear them out and fold. You may find in the case of some planes that your success improves with your own paper of different size and weight.

In the competition paper airplanes were submitted from 28 countries; almost half of the nearly 12,000 entries were from children, although the seven winners were all adults. The longest distance flown was 91 feet 6 inches. The longest time aloft was 10.2 seconds. *The International Paper Airplane Book.* Simon and Schuster. $2.95.

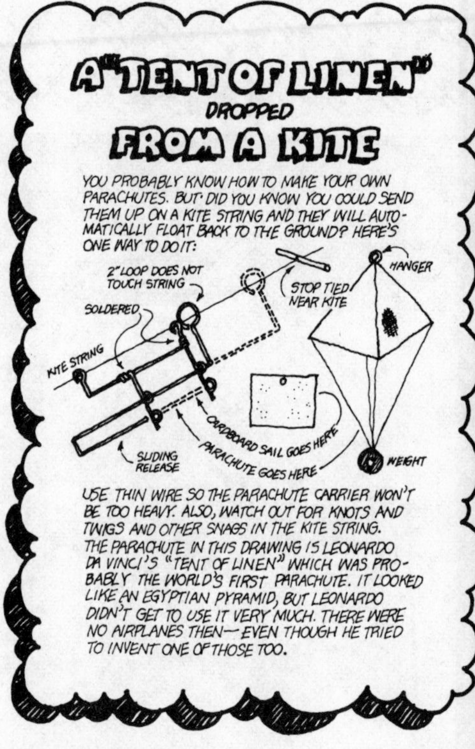

Fly Paper

If you like building and flying airplanes you'll want a poster (22 x 22 inches), called "Fly Paper," another amazing creation from the Amazing Life Games Company (of which we show one corner). You can pin it on your wall as a decoration and you can learn from it. How to make four kites, three paper planes, a helicopter, a parachute, a hot-air balloon, and a rocket. The step-by-step drawings make it easy. To get your copy of "Fly Paper" SEND $2.00 to The Amazing Life Games Co., P.O. Box 506, Sausalito, Calif. 94965.

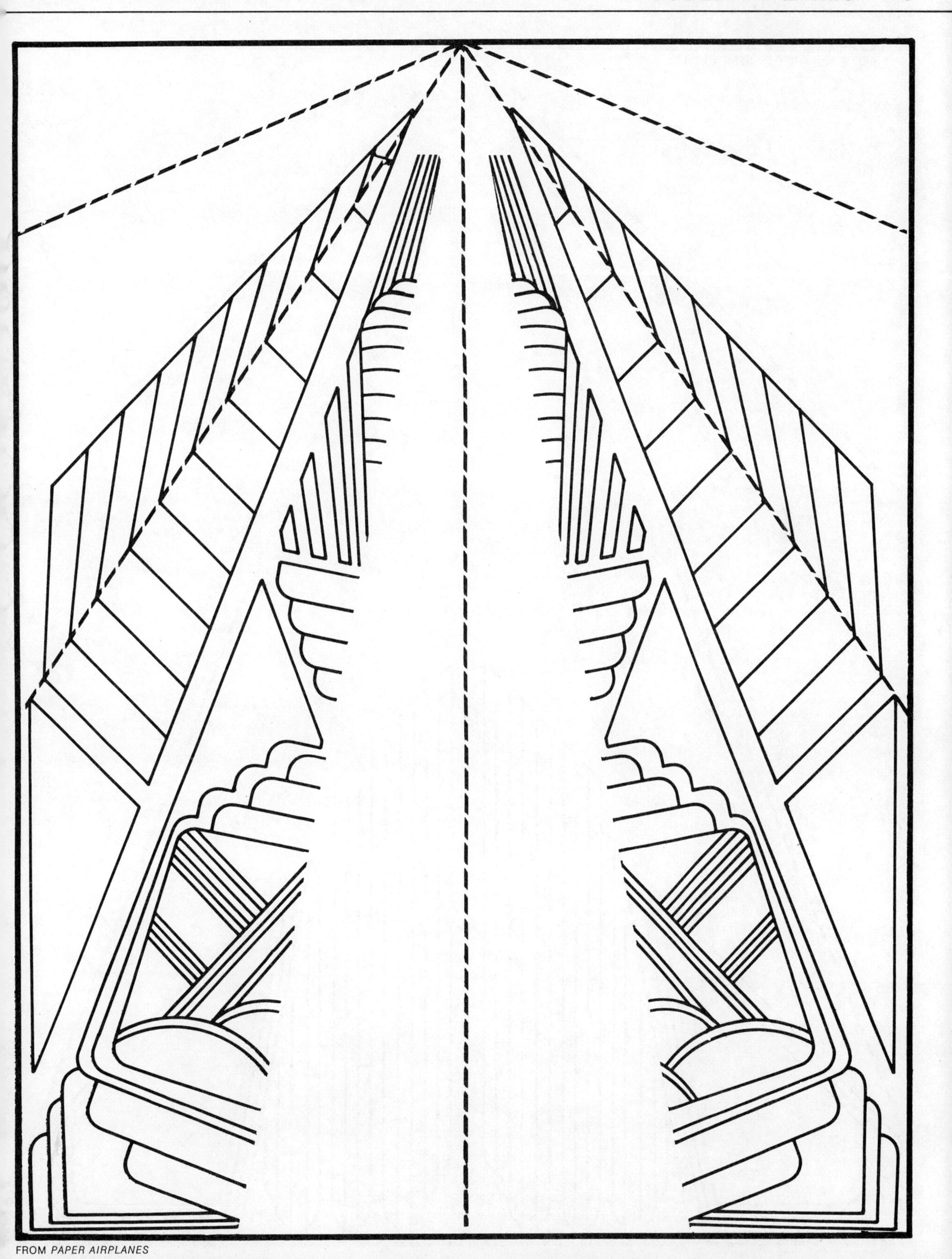

FROM *PAPER AIRPLANES*

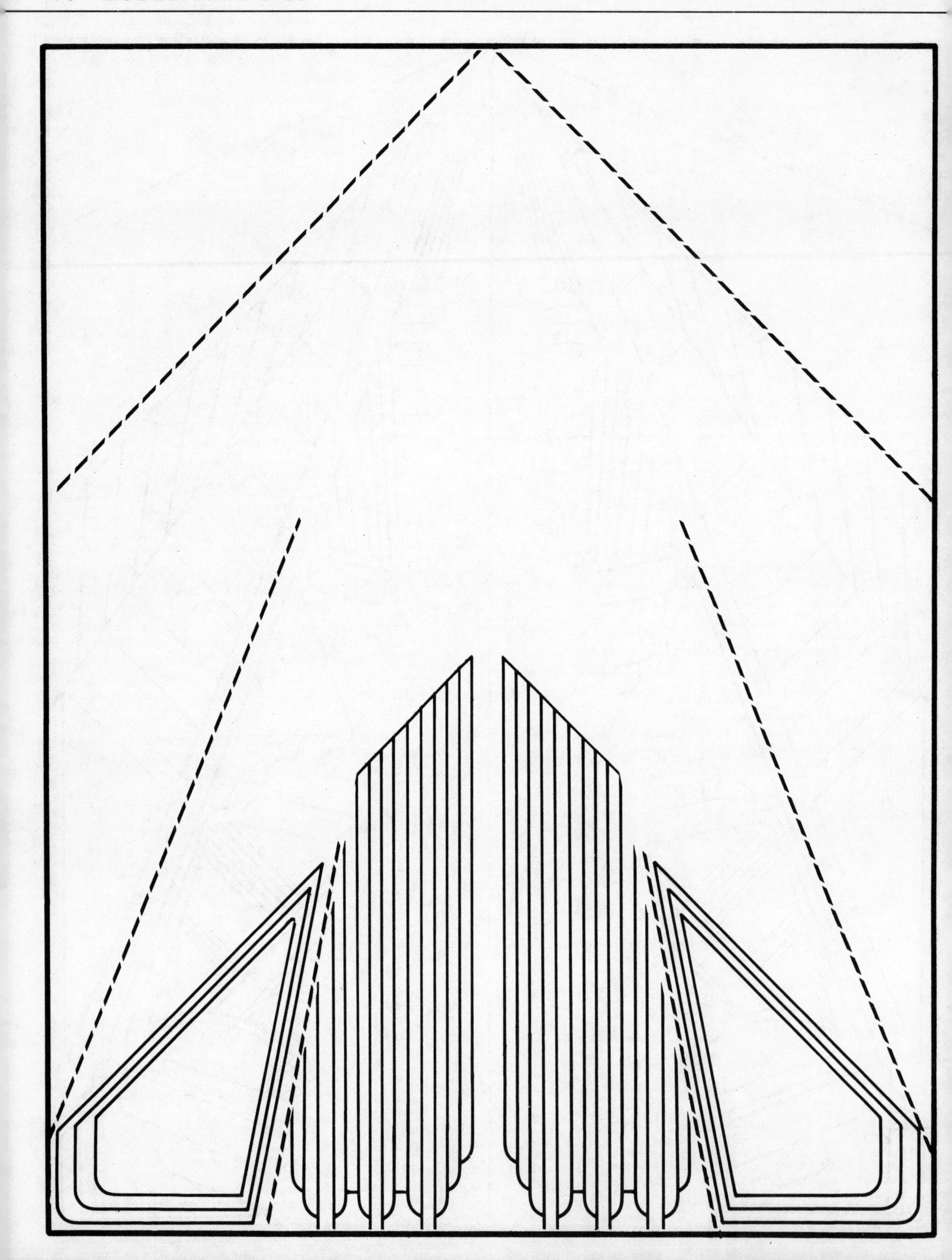

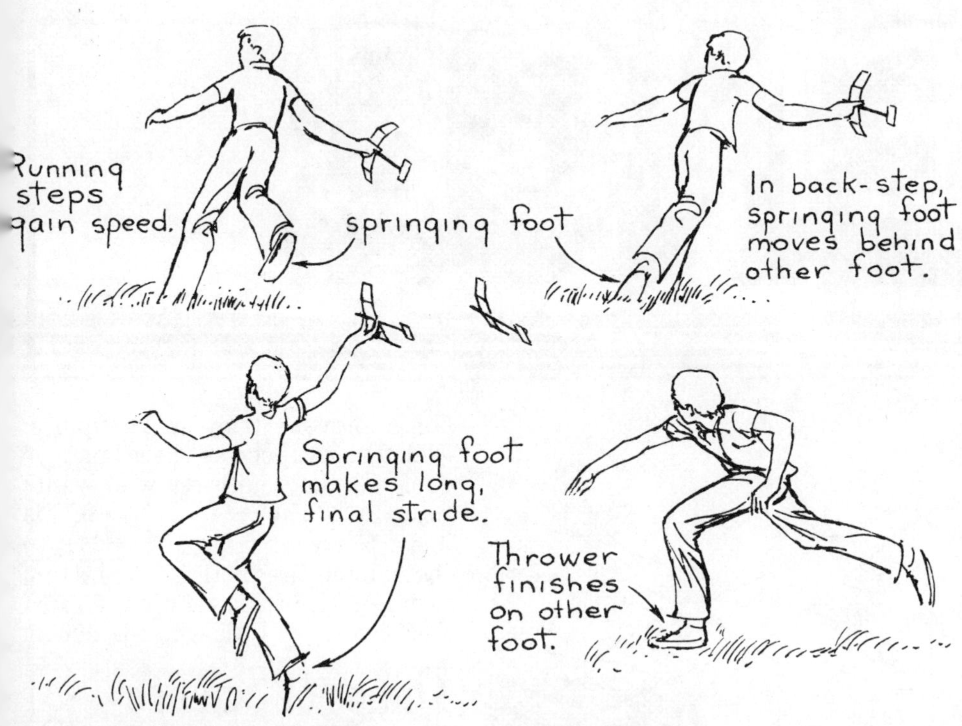

Flying Hand-Launched Gliders

There are simple directions with diagrams and drawings in this guide to building and flying hand-launched gliders. You will learn about selecting the right balsa, shaping the parts carefully, and assembling them accurately; and about the fine points of adjusting and flying a glider. If you want to compete, there is a chapter on flying in contests sponsored by the Academy of Model Aeronautics and the National Free Flight Society. *Flying Hand-Launched Gliders* is written and illustrated by John Kaufmann. William Morrow & Co. $5.50 cloth, $2.50 paper.

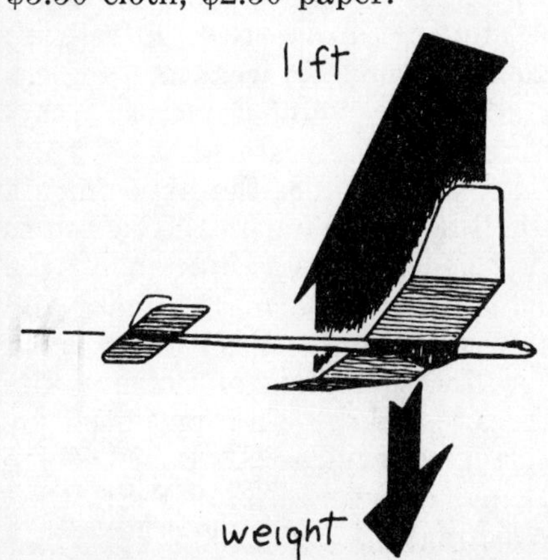

Model Satellites and Spacecraft

This book, illustrated with many photographs and drawings, tells you about 12 satellites and spacecraft, their purpose and history, and how to make models of them. The directions are fully diagramed; there are photographs of the space vehicle and the model. You can display your finished models on stands or make wonderful mobiles with them to hang in your room; you can also use the models for school science projects. *Model Satellites and Spacecraft* is by Frank Ross, Jr. Lothrop, Lee & Shepard Co. $2.45.

Flying Origami

This book gives full directions for making 30 origami aircraft that really fly. Some are easy to make, some more sophisticated; but all of them fly with speed and accuracy. There are two origami birds—probably the first of the kind—that move their wings in flight. Photographs, charts, and clear explanations in *Flying Origami: Origami from Pure Fun to True Science,* by Eiji Nakamura, tell the reader how to make all these exciting planes. Japan Publications Trading Co. $4.95.

How to Make & Fly Paper Airplanes

Captain Ralph S. Barnaby, U.S. Navy (retired) has written the definitive book about how to successfully design and fly a paper airplane. He blueprints six basic models, including the one that took a prize at the First International Paper Airplane Competition. Clear illustrations of the planes make each construction step easy to follow.

Captain Barnaby, who knew Orville and Wilbur Wright personally, is an aviation pioneer. He designed, built, and flew his first glider in 1909; he holds soaring certificate No. 1 from the National Aeronautic Association. You couldn't find a more qualified instructor. All you have to do now is find some paper and scissors to learn *How to Make & Fly Paper Airplanes.* Bantam Books. 75 cents.

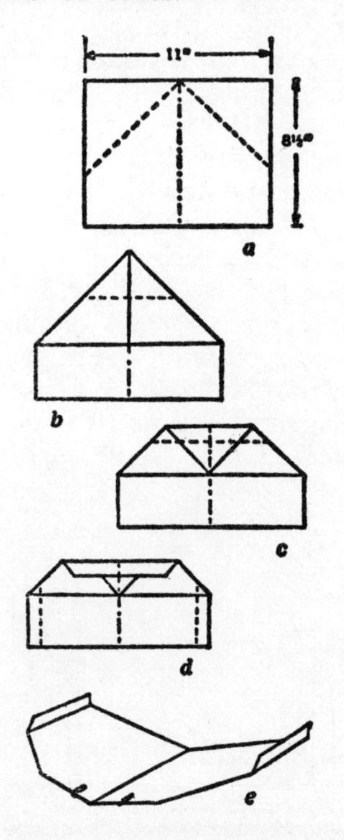

SCIENCE

How It Works

Do you ever wonder about how things work—refrigerators, automobiles, radios, TV sets, and tape recorders? They were all invented and built by human beings, so there is nothing mysterious about them. It is possible to understand even the most complicated machine, once you are told how it works. Here are two books, by Martin L. Keen, that describe some things you can see working every day and some, such as rockets, that you may not have a chance to see, but that you've heard a lot about.

How It Works, volume 1, tells you about electricity and electronics: how a light bulb works, a toaster, an electric iron, an electric fan, a vacuum cleaner. It describes engines that change energy into work: you can even find out how a ballpoint pen works.

Volume 2 has an extensive section on computers and other data processing machines. It explains the punched card, the sorter, the codes used, and how the input unit works. It also tells you about familiar things such as the aerosol container, sewing machine, barometer, speedometer, and odometer. There is a section on aircraft.

In both volumes there are excellent diagrams. Age 10 and up. Grosset & Dunlap. $4.95 each.

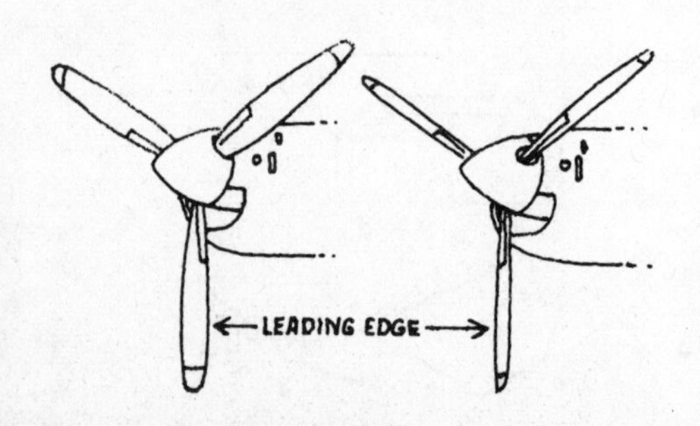

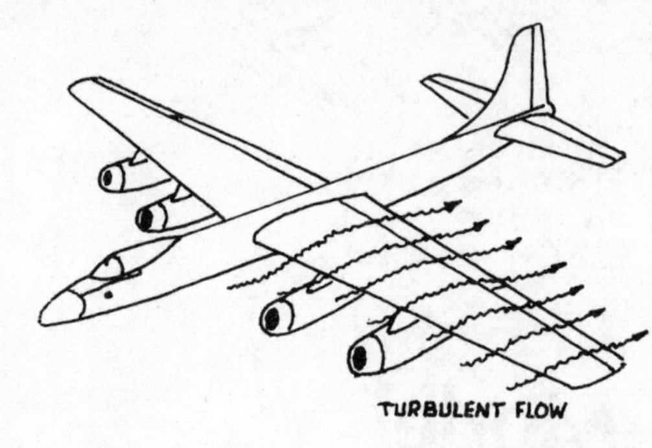

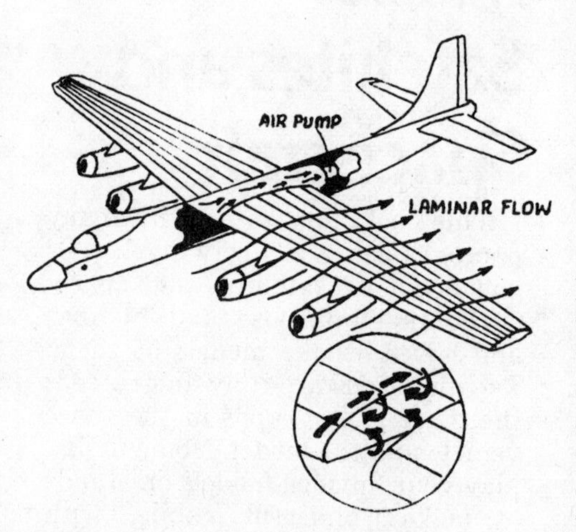

FREE

Your Body and How It Works

Especially created for children in their early elementary years, *Your Body and How It Works* has colorful pictures and simple text. It graphically shows you that your knee joint works like the hinge joint you see on a door and that your hip joint is similar to the ball-and-socket joint in a desk-pen set. One picture gives you X-ray eyes so you can see how the many bones in your body make up the skeleton. A good science primer book for the youngster who wants to learn the hows and whys of his body. For your free copy WRITE TO American Medical Association, Dept. of Health Education, 535 N. Dearborn St., Chicago, Ill. 60610.

Things of Science

Here's a fun way to discover the wonders of science. As a member of Things of Science, each month you'll receive a surpirse package filled with materials for conducting exciting scientific experiments. A booklet containing background information and detailed, easy-to-follow instructions for as many as 25 experiments is included. Each kit is completely self-contained. In the course of a year you'll gain firsthand knowledge of many fields of science. Among recent kits, for example, one demonstrated the principles of simple machines and contained all the materials needed to build them; another presented 25 experiments involving pendulums; a third explored the earth's past through fossils; and a fourth contained materials for building a model sextant and showed how to use it. Others have dealt with the principles of magnetism; the sense of touch; properties of glass; chemical models; weather; soilless gardening; liquid crystals; recycling; and electrostatics.

As you conduct the experiments you'll be following in the footsteps of scientists, learning about the principles and laws of science and how they were used for discovering new facts. Things of Science kits are created by and produced by Science Service. SEND $9.50 to Things of Science, 1719 N St., NW, Washington, D.C. 20036.

Light up a light bulb with 12 potatoes. Set up your own home lab and become a scientist. 1001 experiments and amusements. Why a color TV works. How a telephone works. What is a laser beam. Barometer. Odometer. Solar projects. The human body!

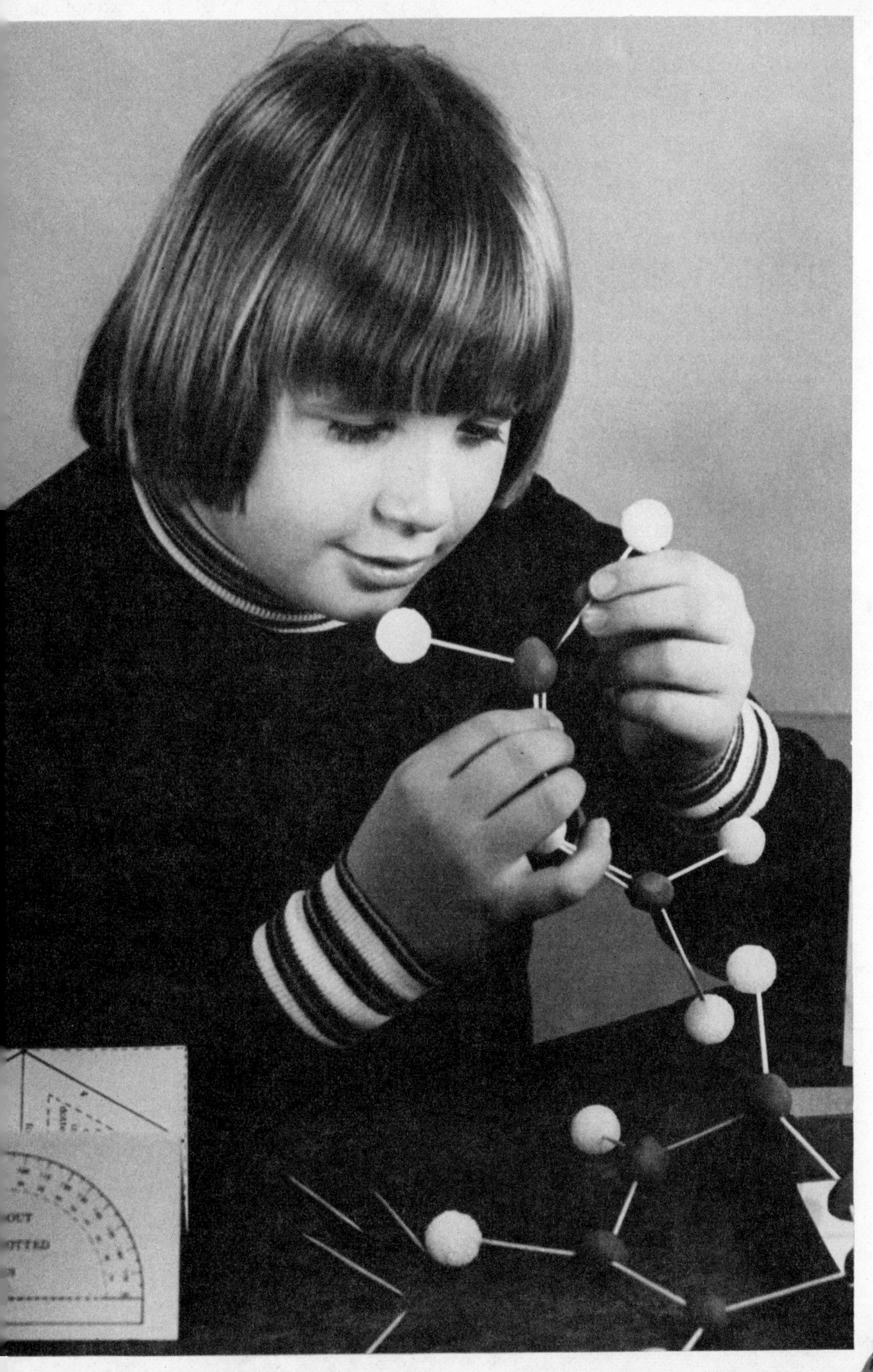

FREE Edmund Science Catalog

The big new Edmund catalog (over 160 pages) is crammed full of new and exciting scientific items; there are over 4500 listings, including science project kits, telescopes, windmills, plant-growing hothouses, a complete line of lab equipment, weather and electronic sets, solar furnace kits, science toys and games, science treasure chests, magnets, books, and puzzles. Topics include astronomy and telescopes, binoculars, lenses, microscopes, photography, fiber optics, gardening, ecology, and many others. Whether you are a hobbyist, a student, an experimenter, or a gadgeteer, Edmund (after 33 years) is recognized as America's marketplace for the science-minded. You'll be intrigued with their unusual items—but before you order, get the free catalog. WRITE TO Edmund Scientific Co., 555 Edscorp Building, Barrington, N.J. 08007.

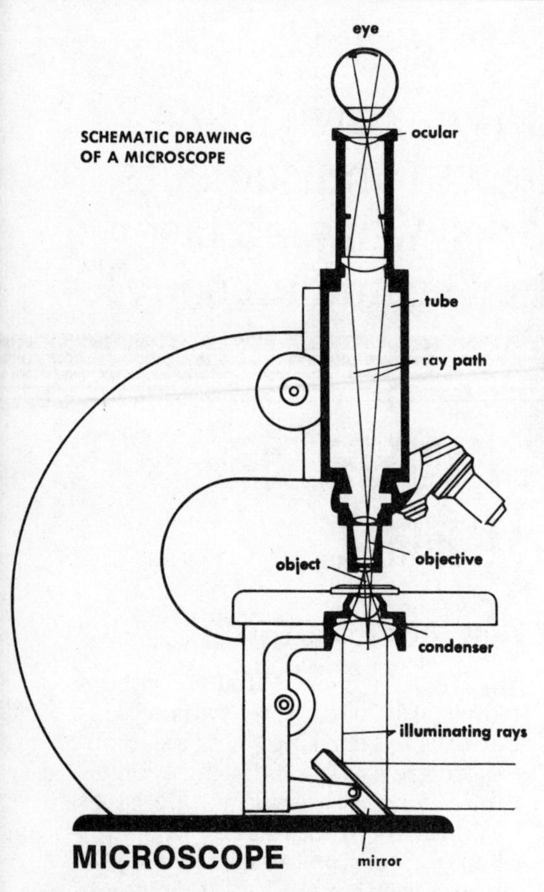

MICROSCOPE

The Way Things Work (Special Edition for Young People)

This edition of the international bestselling illustrated encyclopedia of technology is especially adapted for young readers. *The Way Things Work,* by T. Lodewijk and others, is designed to answer every question about batteries, escalators, color television—and gas turbines, jet engines, and nuclear reactors. There are 91 articles that provide clear and concise explanations phrased in simple, nontechnical language, with over 325 two-color drawings and easy-to-follow diagrams. When you look at a machine and wonder how it works you can probably find the answer in this book. Age 8 and up. Simon and Schuster. $9.95.

TELEPHONE

The telephone is a relatively simple but remarkable instrument consisting of a *transmitter* and a *receiver.* Its two functions are (1) to accept sound waves in the transmitter and convert them into electrical impulses and (2) to receive electrical impulses and change them back into sound waves.

When a person talks into a telephone, the sound waves from his voice are carried through the air. They enter the transmitter and cause a thin metal *diaphragm* to vibrate. Behind the diaphragm is a small metal box of *carbon granules* through which an *electric current* flows. The vibrating diaphragm exerts a pressure on the granules, compressing and relaxing them in direct relation to the vibrations from the person's voice. When the granules are compressed, or close together, they conduct more current; when relaxed, or loosely spaced, they conduct less. And so we now have a pulsating electric current that imitates the sound waves from the person's voice.

The current, strengthened by amplifiers called *repeaters,* can be carried great or short distances to its destination—the receiver at the other end of the phone call.

Now the electrical impulses have to be changed into sound waves again. This is done with the aid of an *electromagnet,* one of science's more outstanding inventions. Basically, an electromagnet is a piece of iron wound with coils of insulated wire. When an electric current passes through the wire, it magnetizes the iron—the more current, the more magnetism; the less current, the less magnetism.

Inside the telephone receiver is another highly sensitive metal diaphragm backed by an electromagnet. As the incoming electric current fluctuates, the strength of the magnet fluctuates, thus attracting and relaxing the diaphragm in varying degrees. The diaphragm, then, is converting the current back into the same sound waves that were transmitted. The sound waves re-create the caller's voice and, once again, travel through the air to the receiver's ear. The whole procedure outlined here is instantaneous.

OIL-BREAK SWITCH

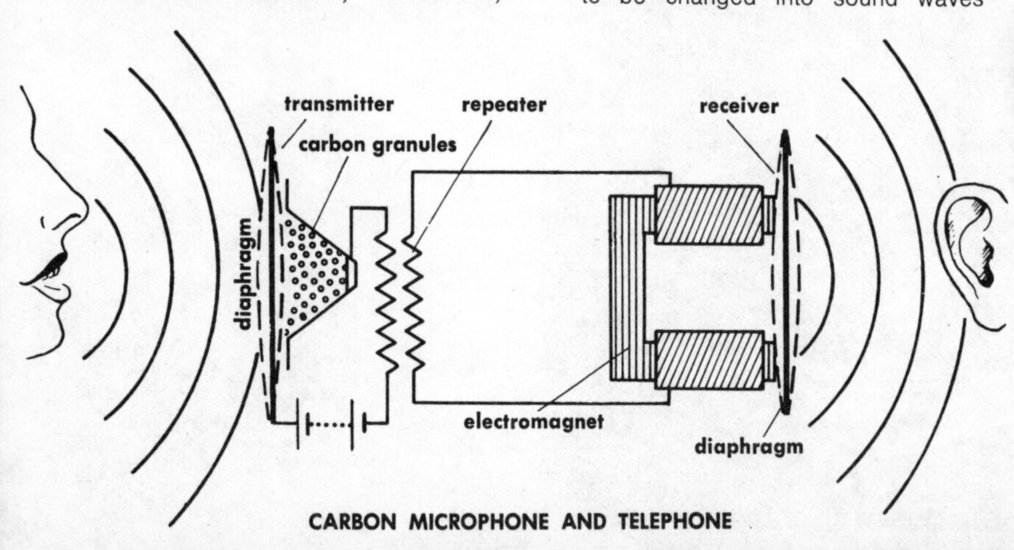

CARBON MICROPHONE AND TELEPHONE

Motors and Engines and How They Work

Young minds will ask "How does it work?" long before an answer is given in school. This book on motors will help fill the gap. Harvey Weiss explains in simple but graphic terms about engines powered by water, wind, gravity, springs, and steam. There are many experiments to help you discover for yourself how power works.

Motors and Engines and How They Work explains the workings of electric motors, gasoline engines, and rockets. The book points to the future and the importance of creating new types of engines to harness yet undiscovered sources of energy. Age 7 and up. Thomas Y. Crowell Co. $5.50.

Want to Know How and Why?

Weather rules our lives; *The How and Why Wonder Book of Weather,* especially written for the junior scientist, explains how air, sun, and water interact in a thousand different ways to produce an ever-changing yet predictable weather pattern. Dozens of basic questions are answered: What causes changes in the seasons? What happens to air when it is heated? Why does the wind blow? What makes your window cloud up? Why does it rain? What causes thunder?

This book, like all the How and Why Wonder Books, emphasizes experiments. Produced and approved by noted authorities, the series answers the questions most often asked about science, nature, and history. Each book contains many colorful and instructive illustrations. Young readers will want to collect the 48-page paperback books for their basic library. Some of the other titles available are:

Dinosaurs
Electricity
Rocks and Minerals
Rockets and Missiles
Insects
Reptiles and Amphibians
Birds
Our Earth
Beginning Science
Machines
The Human Body
Sea Shells
Atomic Energy
The Microscope
The Civil War
Mathematics
Ballet
Chemistry
Horses
Explorations and Discoveries
Primitive Man
North America
Planets and Interplanetary Travel
Wild Animals
Sound
Lost Cities
Ants and Bees
Wild Flowers
Dogs
Prehistoric Mammals
Science Experiments
World War II
Florence Nightingale
Butterflies and Moths
Robots and Electronic Brains
Light and Color
Winning of the West
The American Revolution
Caves to Skyscrapers
Time
Magnets and Magnetism
Guns
Famous Scientists
Old Testament
Building
Trees
Oceanography
North American Indians
Mushrooms, Ferns, and Mosses
The Polar Regions
Coins and Currency
Basic Inventions
The First World War
Electronics
Deserts
Air and Water
Stars
Airplanes and the Story of Flight
Fish
Boats and Ships
The Moon
Trains and Railroads

Age 7 and up. Grosset & Dunlap. 69 cents each.

Science Lab Kits

Flapping Airplane Kit. You can relive one of man's earliest dreams—to fly like a bird. This working ornithopter really flies, encouraging an exploration of flight. Can easily be rebuilt if wings are accidentally damaged. Age 8 and up.

Submarine Kit. In making and sailing this diving submarine you will discover what makes things float. You will learn from physical experience the principles of propulsion, displacement, and air and water pressure. Age 6 and up.

Fun with Magnets. All the mysteries of invisible magnetic force can be enjoyably explored with this pocket-sized kit. Attraction and repulsion, force fields, and earth magnetism are explained in play activities which range from building and using a floating compass to making magnetic pictures. Age 8 and up.

Solar Heater. This accurate solar heater dramatically demonstrates the power of the sun. It will boil water or cook a bit of food; using a light source it can even simulate a searchlight. Demonstrates many principles of reflection. Age 10 and up.

Astro Telescope Kit. Everything needed to build a working 30X telescope with tabletop tripod. You can see moon craters, explore other distant objects, and learn how a telescope works. Age 10 and up.

These are all Action Labs, learn-by-doing projects. There are 18 in all, including Test Car and Monorail Kit, Turning Mirrors, Printing Duplicator Kit, and more. Each Action Lab sells for $3.50; they're available at your toy or hobby shop and produced by Edcom Systems, 745 Alexander Road, Princeton, N.J. 08540.

More Science Lab Kits

Each Play Lab and kit contains simple assembly instructions and experiments to be performed.

Diode Radio. You can build a radio that works without batteries or outside electricity, and listen to radio programs on your private earphone. A radio is fun to listen to and also fun to build, and you will learn how a radio works. Age 8 and up. $4.00.

Electromagnetic Telegraph. The telegraph was invented in 1835 by Samuel F. B. Morse, who also invented the Morse code of dots and dashes. You will build a working telegraph, electromagnet, and buzzer, and learn the international Morse code and be able to send messages just as radio operators do. Age 10 and up. $3.00.

Experimental Three-Lens Camera. You build a real working camera, take pictures, and learn how cameras and film work. Age 10 and up. $4.00.

Electric Guitar. When you construct this guitar kit you will have both an acoustic and electric guitar. You will be able to play simple tunes and songs. You will learn how electronic musical instruments work and explore the nature of sound mechanically and electronically. Age 8 and up. $3.00.

These Play Labs and others are at your toy or hobby shop; they are educational kits made by Alabe Products, 184–10 Jamaica Ave., Hollis, N.Y. 11423.

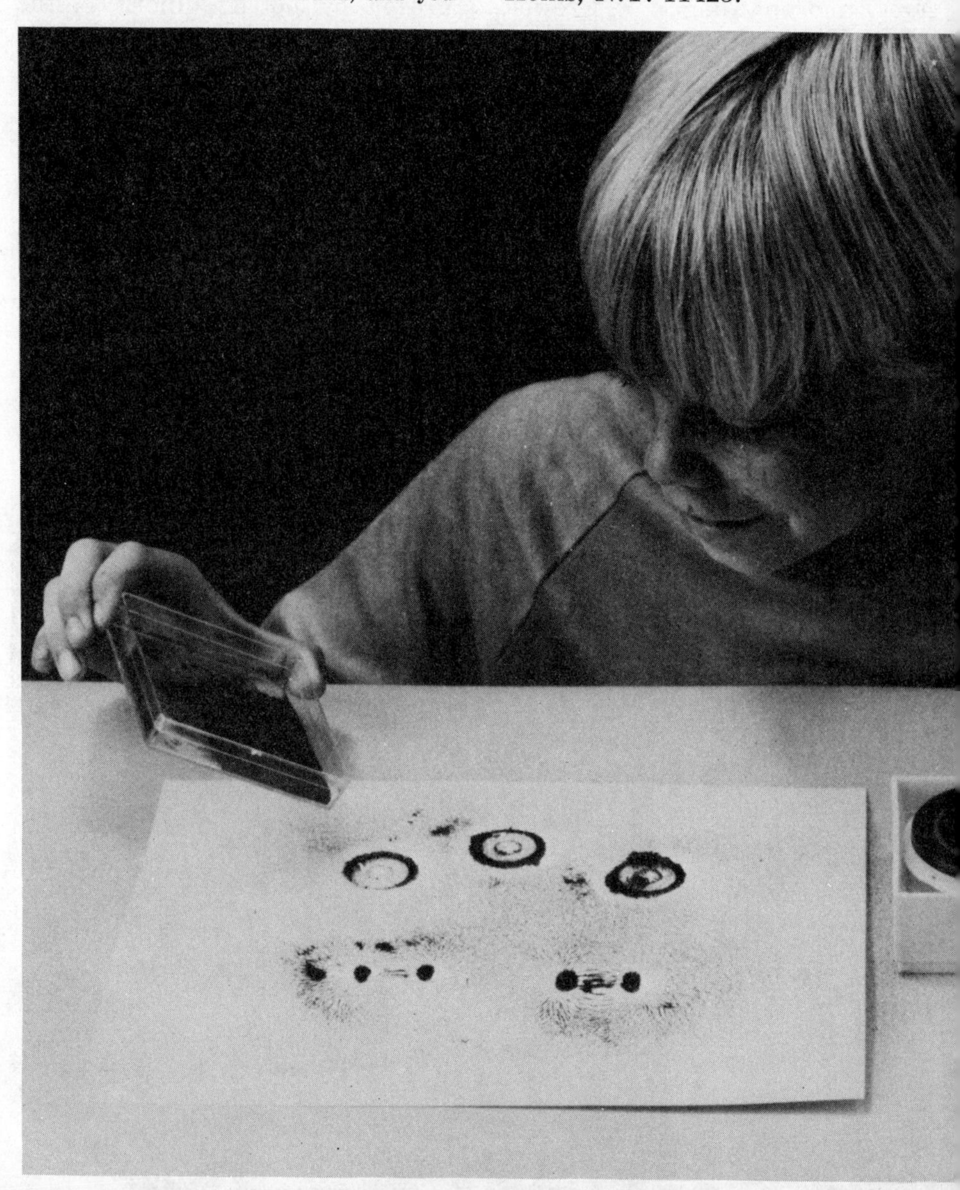

Still More Science Lab Kits

These are called MiniLabs; each kit contains building instructions plus activities and experiments.

Fingerprint Kit. With this kit you can take fingerprints, learn police and FBI methods. You'll bring out latent (hidden) prints when you sprinkle them with print powder, and preserve prints on special instant access cards. The kit contains black lifting powder, white lifting powder, 36 tape squares, 30 ID cards, dusting brush, five-power lens, and a carrying bag. Age 8 and up. $3.79.

Plant and Animal Life Kit. In these experiments you will hatch and raise brine shrimp from eggs. Brine shrimp are small cousins of crabs, lobsters, and shrimp: they don't live in the ocean, but in salty lakes in the western U.S. You will grow five kinds of plants (mung bean, sugar cane, radish, cress, and leek); no soil is needed. The kit contains growing cups, hatching tank, transplant pot, hand lens, plant and animal food, and complete instructions. Age 8 and up. $3.79.

Metric Balance Scale Kit. With this kit you can learn about the metric system of weights and measures. Everything you need to build your own accurate double-platform balance is included. You get a complete set of gram weights and full instructions with lots of experiments. The kit contains a metric converter which will allow you to change weights and measures in the metric system to the English system or vice versa without having to multiply or divide. Age 8 and up. $3.79.

Crystal Radio Kit. With this kit you build a working crystal radio. Wind your own coil. You need no soldering or batteries. Everything to make a radio of your own is included. The kit comes with an earpiece speaker so that only you can tune in. This radio works so well it doesn't need an antenna. There is a complete explanation of radios, and building instructions. Age 8 and up. $5.00.

Electric Motor Kit (with Boat). You can build your own motor with this kit. Learn about how boats work. Then outfit the boat with your motor and have lots of fun in the bathtub or pond with your own motorboat. You don't need any tools. The kit contains complete instructions and lots of activities and experiments. Age 8 and up. $5.00.

Bell/Burglar Alarm Kit. The kit contains all the parts needed to make a ringing bell. You build a burglar alarm system that can protect your room. The kit will help you understand electrical circuitry; it doesn't need tools or electricity to operate. Age 8 and up. $5.00.

You will find these MiniLabs and more at your local toy and hobby shop. MiniLabs are produced by the educational resources division of Educational Design, 47 W. 13th St., New York, N.Y. 10011.

Making Scientific Toys

You can make and enjoy all the toys in this book; you'll learn while playing, because each toy that you make demonstrates a scientific principle. Carson I. Ritchie, the author of *Making Scientific Toys,* describes and pictures the how-to of such optical toys as the kaleidoscope and the theatrical mirror illusion, such balancing toys as the magic stair walker, and such flying toys as the boomerang. There are acoustic toys, gravity toys, heat toys, weather and climate toys, chemical toys, even electric toys—and you can make them all. Age 7 and up. For your copy SEND $6.50 plus 50 cents for postage and handling to Thomas Nelson, 30 E. 42d St., New York, N.Y. 10017.

Exploring Science in Your Home Laboratory

To experiment is to begin to learn how today's scientists work, think, and talk. There is almost no field of science that cannot be studied in a home laboratory; *Exploring Science in Your Home Laboratory,* by Richard Harbeck, shows how easy it is to turn a corner of your home into a laboratory for scientific experiments. You will find detailed instructions for selecting the best location for your lab, building a suitable workbench, and assembling and using the necessary equipment.

Satisfy your curiosity as scientists have for centuries: become an explorer and range through the fields of biology, chemistry, physics, earth science, and space science. Age 10 and up. Scholastic Book Services. 75 cents.

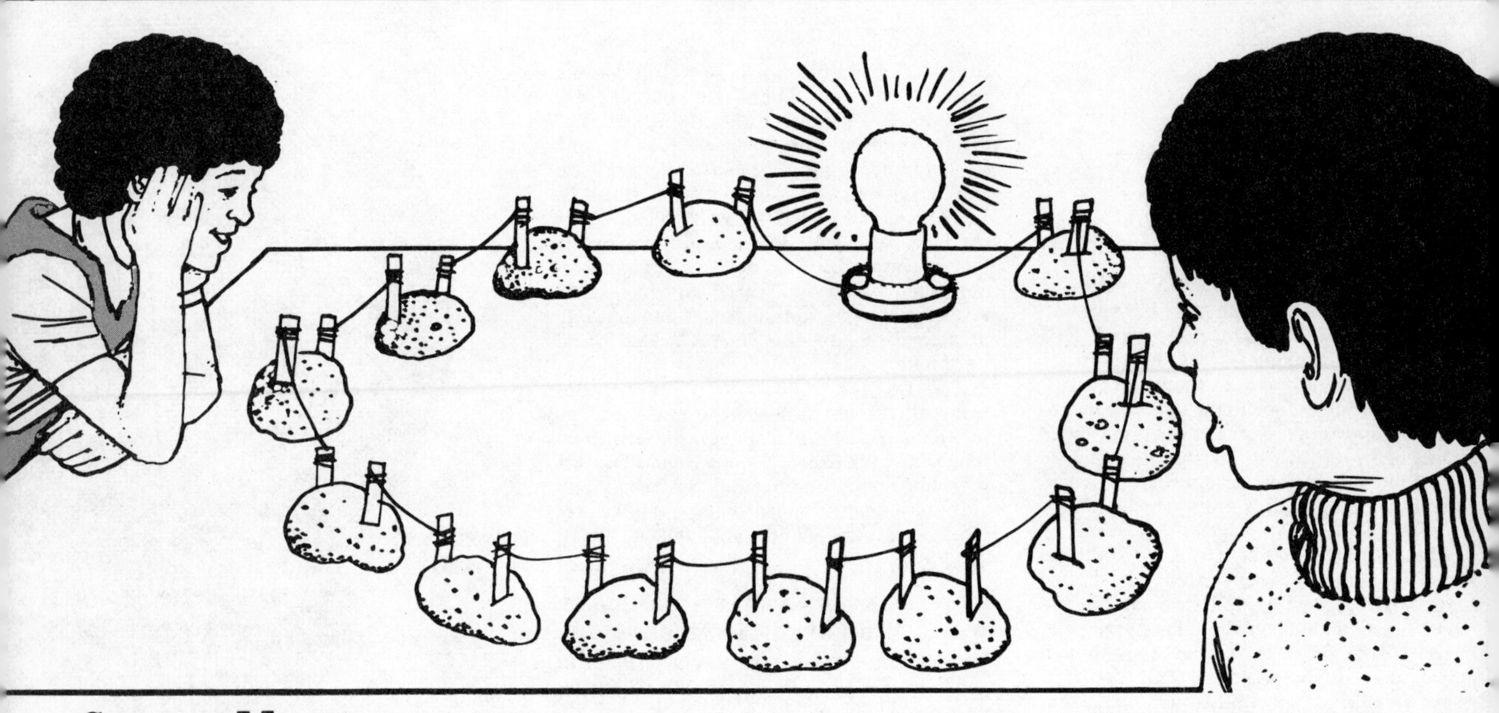

Science Magic with Physics

Can you light an electric light bulb with 12 potatoes? Can you see around corners? Can water flow uphill? The answer is yes to all three questions. You will discover the answers to many more questions as you try out the amazing experiments explained in *Science Magic with Physics*, by Kay Richards, which contains more than 100 "magic" experiments that are simple and safe, and use only household or inexpensive materials. The experiments will astound and mystify reader and onlookers; they are colorfully illustrated and the scientific theory behind each exercise is carefully explained in a lighthearted, easy-to-follow style. A great book for ages 8 to 14. Arco Publishing Co. $3.50.

You do not always need to use a battery to light a bulb. There are other kinds of battery, or cell, too. You can even make a short-lived one from potatoes.

The potato cell has two electrodes, made from strips of brass and zinc. Each strip is about 2 ins. long and ⅓ in. wide. They are stuck into the potato facing each other, about ⅓ in. apart. *Figure 1.*

To light a small bulb you will need twelve of these cells, connected in series. This means that the brass strip in one potato should be connected with a wire to the zinc strip in the next cell, and so on, in a long chain. *Figure 2.*

Connect the two ends of the chain to the bulb holder and the bulb should light! *Figure 3.*

To make sure the wires are in good contact with the strips, press them on with paperclips after winding several turns of wire around each strip.

If you do not know where to find strips of zinc, pull an old battery apart, as the case is made of soft zinc that could be cut up. If you cannot get brass strips, copper will do. There may be some at home, left over from door fittings.

Potato Power Station

YOU WILL NEED:
1·5-volt bulb in holder
13 pieces of copper wire
24 paperclips
12 potatoes
12 strips of zinc
12 strips of brass

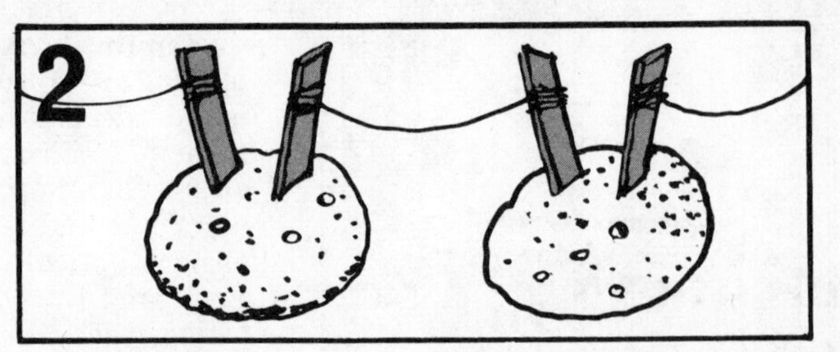

Color

YOU WILL NEED:
a white carnation
a dandelion root with leaves
2 jam jars
a pencil
red ink
black ink
a knife
a rubber band, or thread

Plants need water. You have probably helped your parents to water plants. If you have indoor plants, you may simply have added water to the containers in which the plant pots stand. Does water travel upwards through to the top of the plant? See if you can find out, using these experiments.

Half-fill each of the jars with water. To one add red ink to color the water red, and to the other add black ink to make black water. **Figure 1.**

Slit the stem of the carnation from the bottom for about 4 in. Arrange the split stem so that one side is in the black solution and the other in the red solution. Support the carnation by fixing a pencil to the stem using a rubber band or some thread. **Figure 2.**

Leave the carnation standing for several hours. Later you will find that one half of the flower is red and the other is black. What has happened?

All flowering plants have stems. Inside the stems are vessels, or tubes, through which water taken up by the roots travels upwards. The colored water in this experiment lets you see the movement of water.

Top up the jar containing red-colored water, so that it is half-full. Wash the dandelion root. **Figure 3.**

Put it into the red ink solution. Leave the root standing in the jar overnight. Remove the root and wash away the red ink. Using your knife, cut off the end of the root. What do you see? Carefully slice through the root. **Figure 4.**

The red ink has stained the tubes (vessels) which carry water through the root. Water travels through roots and through stems by way of water vessels in the roots and in the stems. It can move upwards through a plant.

Science Magic with Chemistry and Biology

Would you like to know how to make a flower change its color? How to turn a small egg into a big one? How to make your writing disappear right before your eyes? You can do these things and many more when you start working on the fascinating projects explained in this book. *Science Magic with Chemistry and Biology,* by Ted Johnston, shows you graphically and simply more than 100 "magic" experiments you can perform. You'll also learn the scientific theory behind each experiment. Ages 8 to 14. Arco Publishing Co. $3.50.

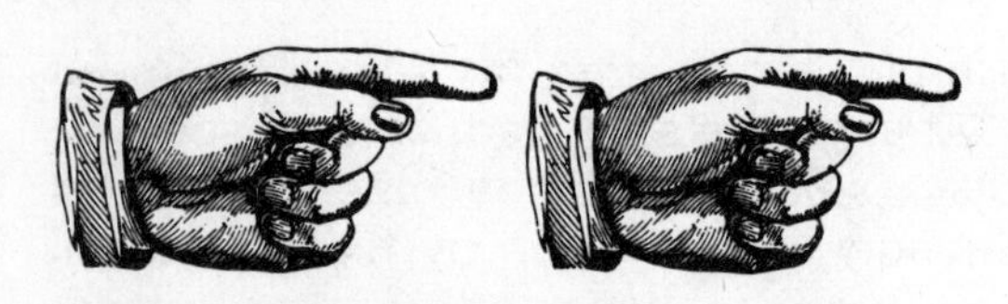

Safe and Simple Electrical Experiments

Here, in 101 entertaining experiments and projects, is a fast and reliable way of learning the basic principles of electricity. There is no possibility of injury because there is no call for high voltages, dangerous acids—even for connection to the house current.

Each experiment is presented by Rudolf F. Graf with instructions, illustrations, and a brief discussion of the often astonishing results to be expected. From cleaning phonograph records with saran wrap, telling time with a compass, and making a transformer or a telegraph sounder the young experimenter learns not only where to get electricity (from such unlikely sources as lemons or candle flames), but how to store and use it in a number of creative ways. For your copy of *Safe and Simple Electrical Experiments* SEND $2.50 plus 35 cents for postage and handling to Dover Publications.

A Hair-Raising Experiment

Materials you will need:

1. Comb
2. Head with dry hair

Have you ever noticed some crackling when you combed your hair on a cool, dry winter day and also that your hair seemed to "stand up"? You were actually generating very high electrostatic charges. What happens is that the comb removes electrons from your hair and aquires a high negative charge, whereas the hair, having lost electrons, becomes positively charged. If you comb your dry hair in front of a mirror on a dry day in a dark room, you will actually see many tiny sparks jumping from the comb to the hair.

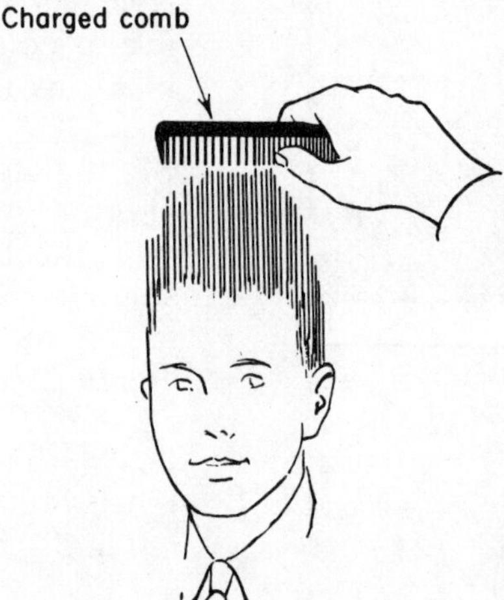

After you have combed your hair briskly, hold the comb about ¼ inch from a water tap. You will see a tiny charge jump from the comb to the tap, and you will also be able to hear it. Believe it or not, if the spark is about ⅛ inch long, you will have generated a charge of about 5,000 volts!

Knowledge Through Color

Here are books about science from the Knowledge Through Color series. Each book is magnificently illustrated in full color and written by a noted authority. Collect your favorite subjects from this series and build your own encyclopedia! Knowledge Through Color is published by Bantam Books.

The Human Body. Meticulously detailed studies of interactions between tissues and organs, with carefully etched drawings of the brain, nervous system, liver and biliary systems, sex and sense organs, and so on. By Paul Lewis and David Rubenstein. $1.95.

Microscopes and Microscopic Life explores plant leaves, insects, pond life, animal and human hair, cloth fibers, how microscopes work, and how to choose equipment and prepare specimens. By Peter Healey. $1.45.

Fossil Man describes the physical evolution of man from primates through apemen to *Homo sapiens,* including data on methods to determine the exact age of man's ancestors. By Michael H. Day. $1.45.

Archaeology begins with the diversified stone cultures in the Old and the New World, and advances to the great civilizations of the Near East. Includes 450 full-color illustrations. By Francis Celoria. $1.95.

Weather and Weather Forecasting gives explicit information on weather observations, forecasts, climates of the past and present, ancient weather lore, and the links between weather and air pollution. By A. G. Forsdyke. $1.45.

Electronics contains clear, nontechnical analyses of audio amplifiers, radio and TV receivers, tape and stereo equipment, counters, radar, solar cells—and the basic principles they operate on. By Roland Worcester. $1.95.

Atomic Energy contains lucid expositions of the nature of matter, thermonuclear power, atomic energy principles, and how energy is produced as heat in a reactor. By Matthew Gaines. $1.45.

Exploring the Planets relates the development of astronomy from its primitive beginnings to the intricately designed manned spacecraft of today. A comprehensive planet-by-planet account with the latest information on size, structure, surface appearance, moons, and speculations on possible life forms, with a chapter on guidelines for amateur home observation. By Iain Nicolson. $1.95.

Astronomy begins with Sumerian, Egyptian, and early Greek contributions to astronomy, and goes on to the development of new spatial concepts, with the latest observations and research. By Iain Nicolson. $1.95.

Five-Book Science Library

Now you can obtain five unusual books of experiments designed to teach you the fundamental laws and principles of modern science. What better way is there to learn than by doing? All the experiments are easy to perform and fun to do.

Physics Experiments for Children, by Muriel Mandell, with illustrations by S. Matsuda, contains more than 100 different things to do that illustrate important ideas and laws of physics. The projects are about what things are made of and how substances are affected by different forms of energy. Experiments show that air has weight, demonstrate air and water pressure, specific gravity, surface tension of liquids (making metal float on water), thread-spool pulleys, amplifying sound, magic color disks, and so on. The materials you need can be found around the house.

Electricity Experiments for Children, by Gabriel Reuben. It is surprising how much you can do with prosaic "found" objects, if you know how. Did you know that you can make a microphone out of dead flashlight batteries and miscellaneous junk, or a rheostat out of a pencil, or a generator out of old coat hangers? Or that you can tell time with a compass?

More than 55 experiments and demonstrations are devoted to aspects of magnetism, electricity, electronics, and atomic energy.

Chemistry Experiments for Children, by Virginia L. Mullin. This book describes in full detail more than 40 experiments and demonstrations involving chemistry. You can create light-sensitive paper and make lensless photographs, measure the amount of oxygen in the air, show that your breath contains carbon dioxide, and make a simple chemical fire extinguisher. You'll learn how to set up your own laboratory, bend glass, and handle chemicals.

Science Experiments and Amusements for Children, by Charles Vivian. If you have ever sucked liquid through a straw, you know that a finger held over the top of the straw will keep the liquid from flowing out the bottom. And you know that the metal screw top of a jar will come unstuck if held for a few moments under the hot-water tap. This book contains over 70 experiments.

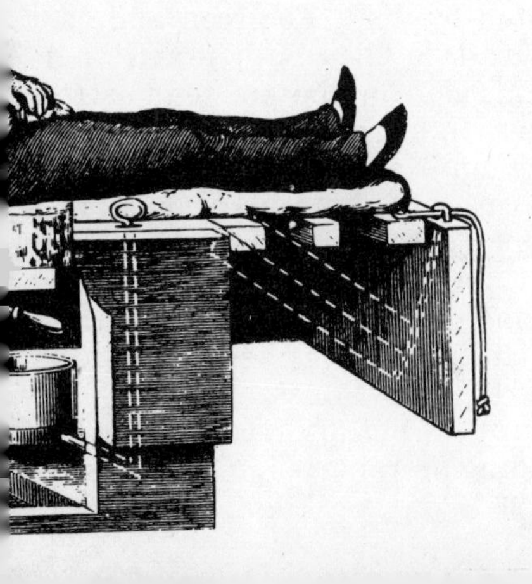

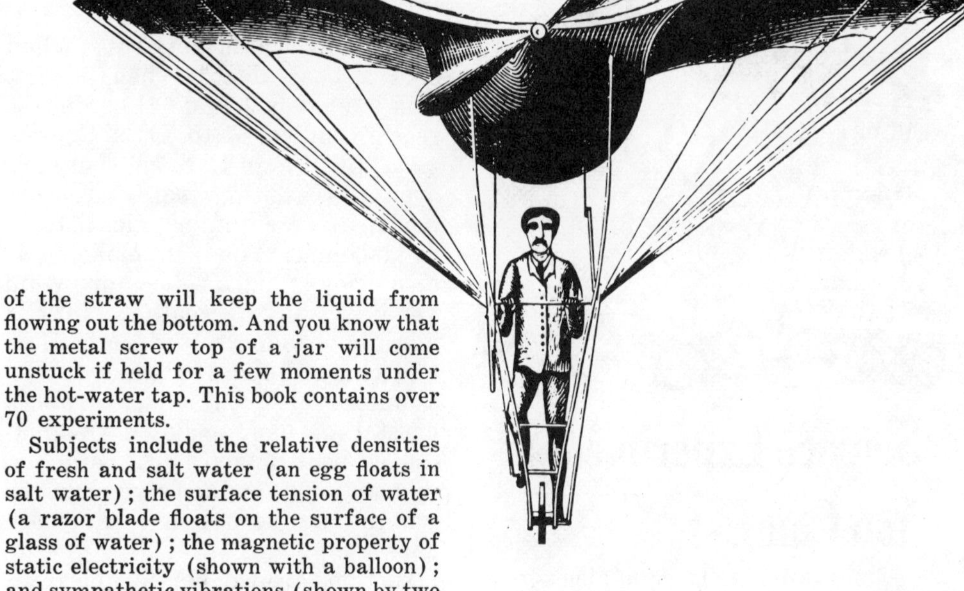

Subjects include the relative densities of fresh and salt water (an egg floats in salt water); the surface tension of water (a razor blade floats on the surface of a glass of water); the magnetic property of static electricity (shown with a balloon); and sympathetic vibrations (shown by two wine glasses and a string). There are full instructions for making a periscope, mariner's compass, walkie-talkie, boomerang, siphon, simple electric motor, pinhole camera, thermometer, and 3-D viewer. There are 102 photographs by S. A. Watts and numerous drawings to help explain the experiments.

Biology Experiments for Children, by Ethel Hanauer. For a broad, useful range of data, this book presents experiments on characteristics of all life, and specifically of plants, of animals, and of man. Topics include magnification, cells and the bacteria of decay, mushroom structure, and growing mosses in a terrarium. The book tells about the parts of the whole plant and of the flower, the reproductive system of plants, testing for starch, isolating chlorophyll, carnivorous plants, the frog life cycle, insect metamorphosis, circulation in a goldfish's tail, and the taste areas of the tongue.

SEND $1.25 for each book to Dover Publications. Be sure to state the title and author of the book you want, and add 35 cents for mailing and handling. If you order all five books, add only 70 cents for mailing and handling.

Polymer Science Rubber Kit

As an instructional aid, Uniroyal offers a kit of materials and experiments with rubber, one of the most amazing polymers, including samples of cured black rubber, smoked sheet, cured white rubber, foam rubber, rubber thread, synthetic rubber, and a jar of liquid latex from Far Eastern plantations. There is an informative booklet which tells about natural and synthetic rubber and some of its uses. A good kit for show-and-tell and science projects. To order the Polymer Science Rubber Kit SEND $1.00 to Public Relations Dept., Uniroyal, 1230 Ave. of the Americas, New York, N.Y. 10020.

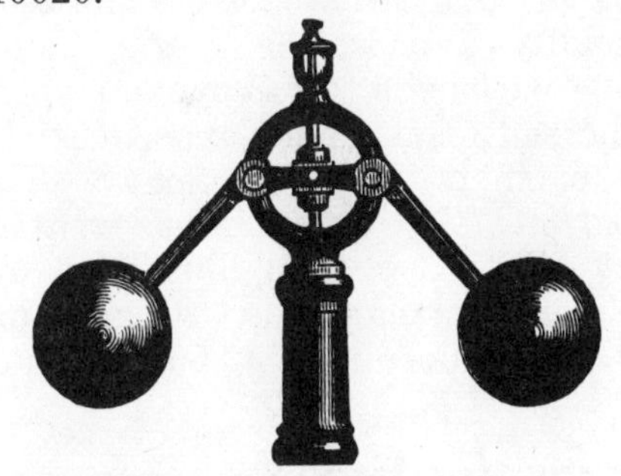

Science Project Suggestions

Are you looking for something to do in science? Do you want suggestions for a project report? Do you need inspiration for a science exhibit? Students can find answers to these questions in this 64-page booklet listing classified titles of exhibits shown at science fairs and/or produced as projects for the annual Science Talent Search. For a copy of *Thousands of Science Projects* SEND $1.00 to Science Service, 1719 N St., NW, Washington, D.C. 20036.

Science Experiments You Can Eat

Your kitchen is equivalent to a well-equipped laboratory, and there are many things in it worthy of scientific inquiry. In this book 39 experiments suggest ways to investigate the properties of solutions, suspensions, colloids, and emulsions; the cooking action of microbes and enzymes; and phenomena concerning carbohydrates, fats, and proteins.

You can make your own "litmus paper" from red cabbage and test acidity of foods. The freezing point of solutions is fun to explore when the end result of the experiment is a batch of popsicles. *Science Experiments You Can Eat* is written by Vicki Cobb and illustrated by Peter Lippman. Age 10 and up. J. B. Lippincott Co. $2.50.

Egg-Ventures

Here is an invitation to explore the fascinating world of the ordinary chicken egg. Find out how to tell a hard-boiled egg from a raw one—or a fresh egg from a rotten one—without cracking open the shell.

Egg-Ventures, by Harry Milgrom, shows young children how to measure an egg and how to make it stand on end. You learn about its different grades, sizes, and parts. With just a few eggs and the simplest kitchen equipment, you can launch yourself on an egg-venture —and have something to eat for lunch too. Age 7 and up. E. P. Dutton & Co. $4.95.

Let's Experiment

When is water not wet? When does a body float? When does it sink? What is sound? How does it reach your ears? In *Let's Experiment,* by Martin L. Keen, there are over 150 exciting, safe, easy experiments for junior scientists to do at home. You can make your own cloud, barometer, or wind vane; set up a weather station; learn how to locate the North Star and chart the stars; make a telephone, real or tin-can; grow a crystal garden; make a magnet-driven boat or your own rainbow; or map your tongue so you can tell where the sensations of sweet, salty, sour, and bitter are. The instructions are wonderfully clear, as are the illustrations. Age 7 and up. Grosset & Dunlap. $4.95.

Science Projects You Can Do

The purpose of this book is to suggest ideas for science projects that are not too difficult, even for beginners, and to provide just enough help to get you started on one that interests you. The main values of a science project come from doing the planning yourself, conducting the necessary research, and carrying the project forward to completion. To be eligible for science fair competition, an exhibitor is at liberty to draw ideas from any source but must work out the details for himself—otherwise his undertaking is not a bona fide science project.

Science Projects You Can Do, by George K. Stone, includes 101 projects, with diagrams for each, ranging from why the stars twinkle to stroboscopic motion, an earth-moon rotation model, and an electric air car. You are almost certain to find something you will want to do at home, demonstrate to the class at school, or enter in the next science fair. Prentice-Hall (Treehouse Paperback). 95 cents.

Something New Under the Sun

Seven solar energy projects you can build and use are included in *Solar Science Projects,* by D. S. Halacy, Jr. Make your own cooker, furnace, oven, water heater, motor, and radio—all powered by the only free fuel in the world: the sun. It's easy; no special skill is needed. It's inexpensive; just follow the directions for economical construction. It's fascinating; learn the principles behind the use of solar energy. It's practical; you'll use these projects again and again for fun, for experiments, and for practical purposes. Age 10 and up. Scholastic Book Services. 75 cents.

Experimenting with the Microscope

Many of the books that instruct the beginner in operation of the microscope are surprisingly poor, ranging from the often inadequate instruction manual that arrives with the child's instrument to highly technical material of little use to the amateur. This book fills the gap by providing a solid foundation for learning established techniques and exploring new ones. For your copy of *Experimenting with the Microscope,* by Dieter Krauter, SEND $2.00 plus 35 cents for postage and handling to Dover Publications.

The Sky Observer's Guide

This pocket book, colorfully illustrated by John Polgreen and easy to read, is filled with practical information for the amateur astronomer. In clear and simple terms R. Newton Mayall and others explain the use of binoculars and telescopes, what objects in the sky to look for, and when and how to see them best. Profusely illustrated with photographs, diagrams, charts, and tables, *The Sky Observer's Guide* is recommended by leading astronomers. Golden Press. $1.95.

Stars

This is a guide to the constellations, the sun, the moon, the planets, and other features of the heavens. *Stars,* by Herbert S. Zim and Robert H. Baker, is a Golden Nature Guide designed for everyone who wants to enjoy the wonders of the heavens and to understand more fully what he sees. Its 150 full-color paintings by James Gordon Irving include charts showing the location of the generally recognized constellations and tables that help to locate the planets. In easy-to-read language, this enjoyable, informative little handbook is ideal to take along on vacations as well as for use at home. Age 9 and up. Golden Press. $1.95.

Star Maps for Beginners

Amateur astronomers, especially beginners, have long been frustrated and discouraged by conventional star charts (planispheres), which depict the constellations arranged in a circle. The scheme is scientifically correct, but beginners have complained about the difficulty of matching up what they see in the sky with the patterns of the planispheres.

Drs. I. M. Levitt and Roy K. Marshall have solved the problem brilliantly by dividing the sky into quadrants. Their charts make it possible for you to orient yourself quickly and easily in any compass direction by turning the chart to correspond with the section of sky you are studying. Whether you use direct vision, binoculars, or a telescope, the charts enable you to locate yourself almost instantly anywhere in the northern hemisphere, at any hour of any night of the year.

Besides the sky maps, *Star Maps for Beginners* has a section on the history and development of the constellations and a discussion of the planets as bright stars. This is an excellent book for anyone who wishes to take the first steps toward knowing the sky and enjoying the delights of amateur astronomy. Simon and Schuster. $1.95.

FREE

Astronomy and You

Did you know that you use astronomy every day? Every time you look at a clock, watch, or calendar you are using astronomy. A day has 24 hours because it takes the earth that long to make a complete rotation on its axis; the year of 12 months is the time it takes the earth to orbit the sun.

Astronomy, the oldest science, is fascinating to read and learn about. *Astronomy and You* is a fully illustrated full-color, comic-type book which tells you about the history of astronomy, what astronomy is, how it is used, and how you can build your own telescope to look at the wonders in the sky. For your free copy WRITE TO Edmund Scientific Co., 555 Edscorp Building, Barrington, N.J. 08007.

Outer Space Wall Maps

"The Heavens." The map shows stars, the Milky Way, constellations visible in each hemisphere, zodiac signs, and star positions for each month. 34½ x 23 inches.

"Mars." Cratered topography combined with natural surface color; dust storm painting on reverse. Indexed with descriptive notes and diagrams. 38 x 23 inches.

"The Earth's Moon." Familiar near side and hidden far side; indexed with descriptive notes and lunar flight data. 42 x 28 inches.

To order WRITE TO National Geographic Society, 17th and M Sts., NW, Washington, D.C. 20036, and tell them which maps you want. The Society will bill you $2.00 for each map at the time of shipment, plus postage and handling.

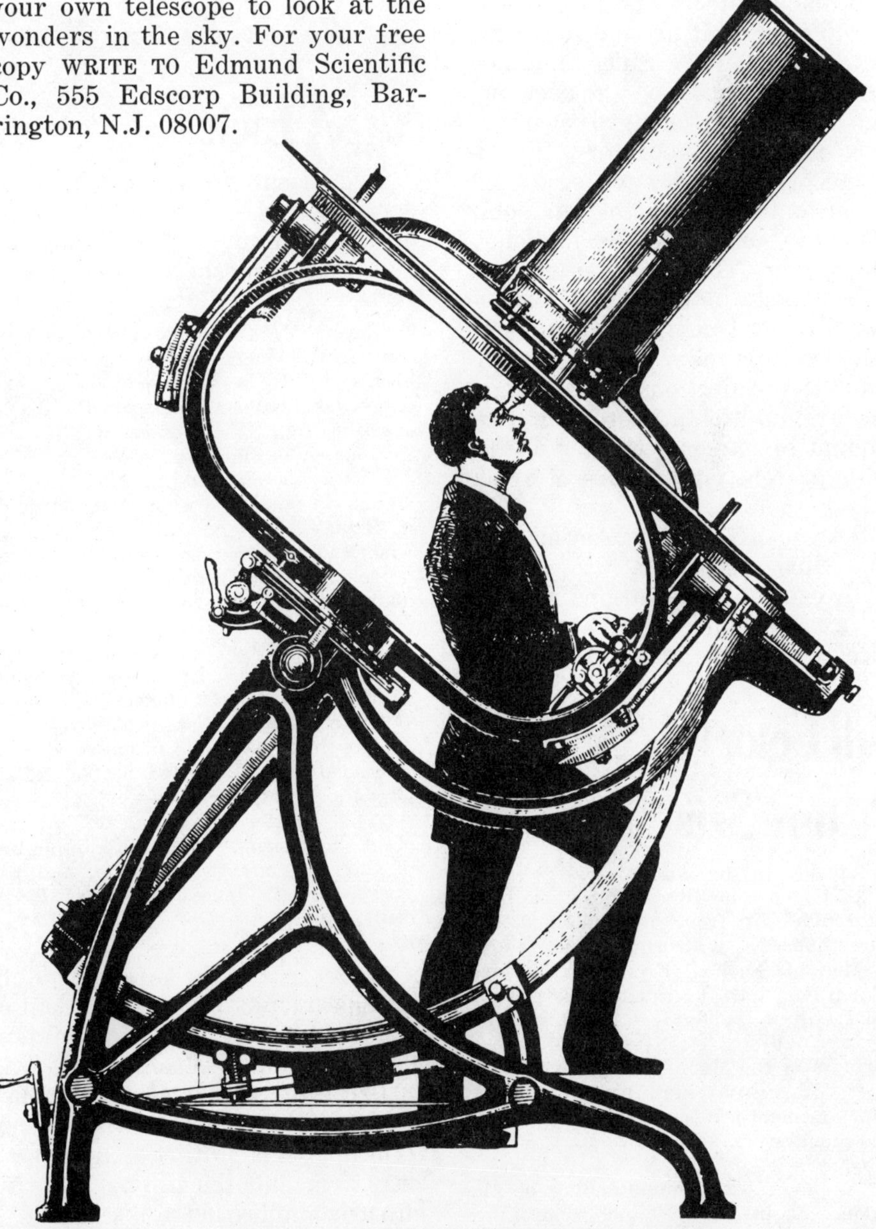

From Sputnik I to Apollo II, the incredible story of the fantastic voyages beyond earth. Rockets in space. Space shuttles. Space stations. Careers in space for you. Skylab. Probes of Venus and Mars. NASA pictures of man on the moon.

The Age of Space

The space age began when a man-made object, Sputnik I, was hurled into orbit around the earth on October 4, 1957. Since then men have gone to the moon and back and men have spent prolonged periods of time in a space laboratory. Instrumented spacecraft have been sent to the outer reaches of our solar system to send back data for scientific study. The concentration of development in the field of space has produced countless new technologies for the benefit of mankind; the projections for the space age are limitless. There is a vast amount of material written on the past, present, and future of space travel.

NASA (National Aeronautics and Space Administration) has many exciting publications, written by experts and illustrated with spectacular color photographs.

Full Color NASA Picture Sets

"Apollo—'In the Beginning...'" Seven 11 x 14 inch color lithographs that illustrate highlights from the Apollo 8, 9, and 10 missions. No. 1. 85 cents per set.

"Men of Apollo." Five 11 x 14 inch color lithographs that include portraits of the crews of Apollo 7, 8, 9, 10, and 11. No. 2. $1.00 per set.

"Man on the Moon." One 16 x 20 inch color lithograph that best illustrates man's moment of success, the first step in his conquest of space. No. 5. $1.85 per copy.

"Apollo 12—Pinpoint Landing on the Moon." Eight 11 x 14 inch color lithographs and two 11 x 14 inch black-and-white lithographs illustrating man's return to the moon. No. 6. $2.15 per set.

"Apollo 15." Nine 11 x 14 inch color lithographs illustrating the journey to the moon of Endeavour and Falcon. No. 7. $2.10 per set.

NASA Facts

Each issue of *NASA Facts* describes a NASA project or discusses techniques. The issues are full-color wall sheets, 4 feet x 21 inches.

"Saturn V." Color wall sheet depicting Saturn V, America's largest rocket vehicle, used to launch Apollo. Its upper stage was modified to serve in the Skylab space station. NF-33. 25 cents.

"Journey to the Moon." Color wall sheet depicts Apollo manned lunar landing from launch to splashdown. NF-40. 60 cents.

"Skylab." Color wall sheet features America's first manned space station, the launch sequence, living quarters, and many of the scientific experiments. NF-43. 60 cents.

"Space Shuttle." Color wall sheet depicts space launch and recovery system of the future. NF-44. 60 cents.

"NASA Aeronautics." Color wall sheet depicts the present and proposed air- and spacecraft of the United States. NF-46. 50 cents.

"The Jupiter Pioneers." Color wall sheet showing man's first closeup pictures of Jupiter and other planets, and proposed Pioneer space travel. NF-50. 60 cents.

To get any of these NASA publications WRITE TO Superintendent of Documents, U.S. Government Printing Office, Washington, D.C. 20402. Payment by check or money order must accompany your order. (Do not send postage stamps.) Be sure to include the number and title of each publication you want.

Man's Conquest of Space

This book retraces man's first steps in the exploration of space. Experience the drama of the pioneer flights of lunar astronauts and cosmonauts. Orbit earth for 84 days in Skylab, scanning our planet and far reaches of the universe. 200 pages. To order *Man's Conquest of Space,* by William R. Shelton, WRITE TO National Geographic Society, 17th and M Sts., NW, Washington, D.C. 20036. The Society will bill you $4.25 at time of shipment, plus postage and handling.

Ask Me a Question

Here is a book that will help you to understand and enjoy the exciting new language of the space age. It is in a clear question-and-answer form that will enable you to attain a knowledge of the important scientific principles involved in rockets and space travel.

Many of the most puzzling and intriguing questions asked are included in *Ask Me a Question About Rockets, Satellites and Space Stations.* "What keeps a satellite from falling to earth?" "What causes weightlessness?" "How can a satellite be made to appear motionless in the sky?" These and many other questions are answered in an entertaining and interesting manner by Sam Rosenfeld with the aid of many photographs and drawings. Harvey House. $5.35.

MONTAGE FROM NASA PHOTOS

HISTORY

History Can Be Colorful!

Yes, history can be colorful with these very special historical coloring books. The titles range from *A Coloring Book of Incas, Aztecs & Mayas* to *A Chaucer Coloring Book,* from *A Coloring Book of Ancient Greece* to *A Coloring Book of Pirates.* A far cry from the usual simple-outline coloring books that children are used to, these fabulous books are filled with detailed pictures from the artists of the times, each one from a historical source. They are, indeed, a joy to behold . . . and you have all the fun, because you do the coloring! You can use crayons, colored pencils, felt-tip pens, even watercolors, and your finished pictures may be beautiful enough to have framed and hang in your room. There are historical notes about each picture and instructions about the best colors to use. The books are available at your bookstore or SEND $2.50 plus 25 cents for postage and handling for each one to Bellerophon Books.

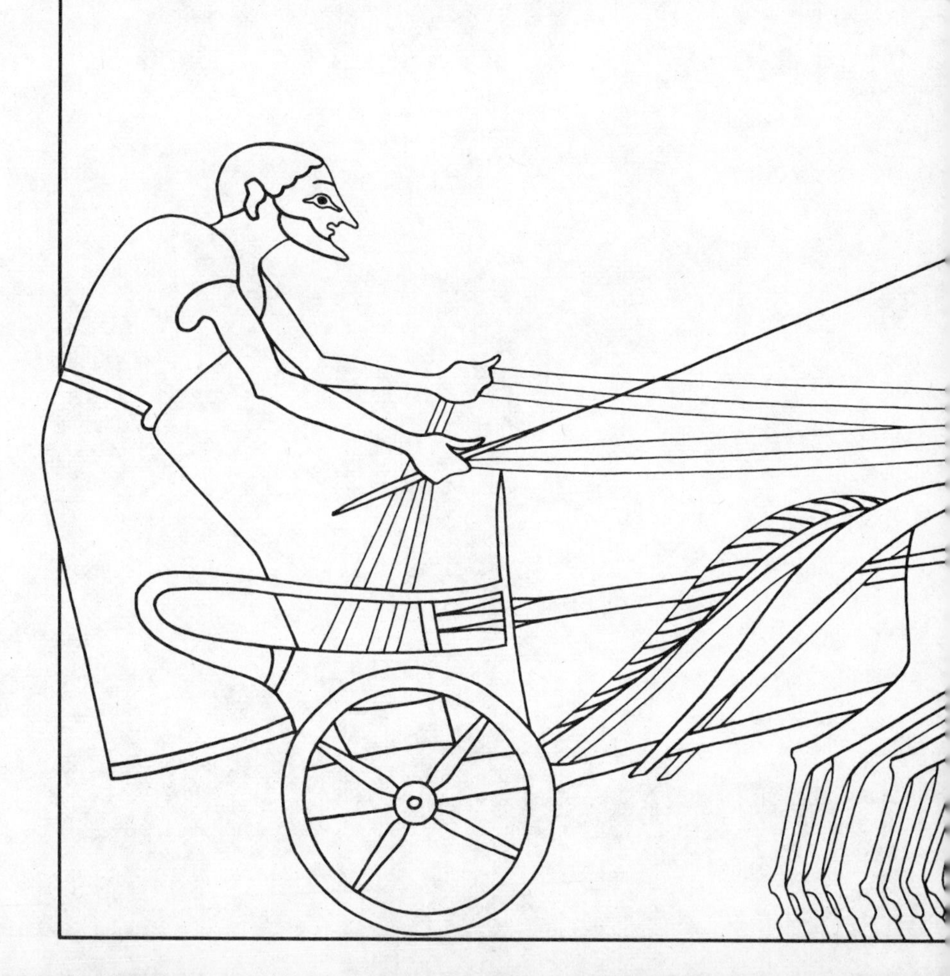

A Map of the United States, Today!

We've come a long way from those first 13 states! And how often do you say: "I wish I had a good map of the United States, today." Here is a beautiful National Geographic map, printed in full color with fadeproof inks on heavy chart paper. It's 68 x 47 inches and just right for the wall in your room. To order WRITE TO National Geographic Society, 17th and M Sts., NW, Washington, D.C. 20036. The Society will bill you $5.00 at time of shipment, plus postage and handling.

Cut and color Henry VIII and his wives. Create an early American village. Make a Roman helmet emblazoned with gold. Enjoy panoramas, posters, maps, models. Travel from ancient Greece to the 20th century between lunch and dinner.

OTHER BELLEROPHON HISTORICAL COLORING BOOKS

Coloring Book of Ships, $1.95; *A Coloring Book of Japan*, $1.95; *Coloring Books of Animals, Birds, Horses* (from 5,000 years of art), 59 cents each; *A Coloring Book of the Renaissance*, $1.95; *The Miller's Tale* (Chaucer), $1.95; *The Wife of Bath* (Chaucer), $1.95; *A Shakespeare Coloring Book*, $1.95; *A Coloring Book of the Ancient Near East*, $1.95.

History Is What You Had for Breakfast This Morning

This book says there is more to history than memorizing the names of presidents and war dates. It's what you had for breakfast this morning, and finding your grandfather's top hat in a box in the attic, and all the mysterious-looking things you find in old trunks that haven't been opened for years. Where else do you find it? In cemeteries and cornerstones, in diaries and the Sears catalogs of yesteryear. *My Backyard History Book*, a Brown Paper School book written by David Weitzman and illustrated by James Roberston, is the first local history book published for kids. There are projects and activities that have to do with making your own family tree, old-time music, slang, old recipes, license plates. You'll learn how to make a time capsule, and find out how your town came to be called Henry's Hunch. Whatever your age and wherever you live, you can leave behind your old ideas about history and dive into more than 101 projects, puzzles, games, and mysteries, and discover what the past has in store for you. Little, Brown and Co. $5.95 cloth, $2.95 paper.

A RACING FOUR-HORSE CHARIOT

From a black-figure lip cup in the Museo Nazionale, Tarquinia.

Historical American Documents

The six most important American documents are now available on tan parchment: the Declaration of Independence, the Constitution, the Bill of Rights, the Monroe Doctrine, the Gettysburg Address, and "The Star-Spangled Banner." Large size—about 10 x 14 inches. Easily handled: each set of six is bound into a booklet; each document is readily removed for individual use or framing. WRITE TO Buck Hill Associates, Garnet Lake Road, Johnsburg, N.Y. 12843. $1.50 per set.

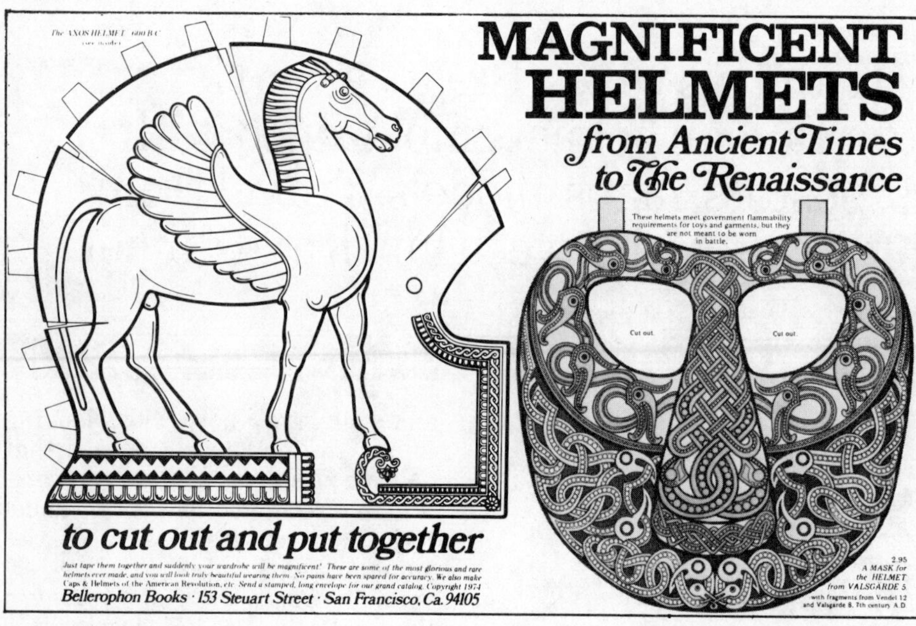

Magnificent Helmets from Ancient Times to the Renaissance

Here are four of the most beautiful, dazzling helmets you have ever seen. They are printed life-size in gold and silver and green and pink, and they're all in a giant book, waiting for you to cut them out and put them together and put them on your head. No pains have been spared for accuracy. Historical notes accompany each helmet.

There is an Axios helmet (600 B.C.); a helmet from Valsgarde (7th century A.D.); a Magnificent Morion (made in 1600 for the state guard of Christian II); and a silver Scythian helmet (300 B.C.). You will feel like a prince or princess with such noble headgear, which no household armory should be without. (A note from the publisher says: "These helmets meet government flammability requirements for toys and garments, but they are not meant to be worn in battle.")

To make your study of history more exciting (and more fun) get a copy of *Magnificent Helmets from Ancient Times to the Renaissance* at your bookstore or SEND $2.95 plus 25 cents for postage and handling to Bellerophon Books.

Queen Elizabeth I

We are fortunate that Queen Elizabeth I liked having her picture painted, for there are many splendid portraits of her in gorgeous attire. *Queen Elizabeth I: Paper Dolls to Cut Out and Color* reproduces her most beautiful dresses in all their splendor. Many have come down to us only in engravings, to be colored now (by you) for the first time in hundreds of years. There is much detail in all the historical costumes, and you can color them, cut them out, and change Queen Elizabeth's wardrobe whenever you want. Sir Walter Raleigh, Mary Queen of Scots, King James I, Sir Francis Bacon, and other notable folk are included, with fascinating text material. They're sophisticated paper dolls: it's a better way to learn history than going to the movies, and cheaper too. *Queen Elizabeth I* is available at your bookstore or SEND $2.50 plus 25 cents for postage and handling to Bellerophon Books.

Eliza, rich and royal, fair and just—

Giant Historical Posters to Color

If you have never seen (or owned) a Bellerophon poster, you've missed a lot of fun, and a colorful way to learn history. Each one is giant-sized, almost 2 x 3 feet, and you have the pleasure of coloring the historical drawings before you tack them up on the walls of your room.

The subjects range from Blackbeard the Pirate to Indian Chief Four Bears, from racing chariots to four-masted ships. Each drawing has great detail; all of the costumes are historically authentic—a visual treat that will give you hours of pleasure and fabulous decoration for your room. Choose from this list: "African King"; "Bottom (from *A Midsummer Night's Dream*)"; "Dinosaurs More Dinosaurs"; "Fiddling Lion"; "Gilgamesh"; "Globe Theatre"; "Great Mayan"; "Henry VIII"; "Henry & Hotspur"; "Horses"; "Humpty Dumpty & Mother Goose"; "Isis"; "Japanese Dancer"; "Japanese Dragon"; "Japanese Goddess"; "Lorenzo the Magnificent"; "Lucrezia Borgia"; "Mayan Princess"; "Pirate Ladies"; "Plains Indian"; "Power Howler"; "Queen Elizabeth I"; "Racing Chariots"; "Samurai"; "Serpent King"; "Ships"; "Titania"; "Trojan Horse"; and "Tutankhamen." There are portrait posters, too: Shakespeare, Beethoven, George Washington, Mozart, Bach, and Handel.

Giant Bellerophon posters are $1.00 each, or 7 for $5.00, plus 25 cents for postage and handling no matter how many you order. WRITE TO Bellerophon Books.

From the engraving by William Rogers after Issac Oliver, c. 1595. British Museum.

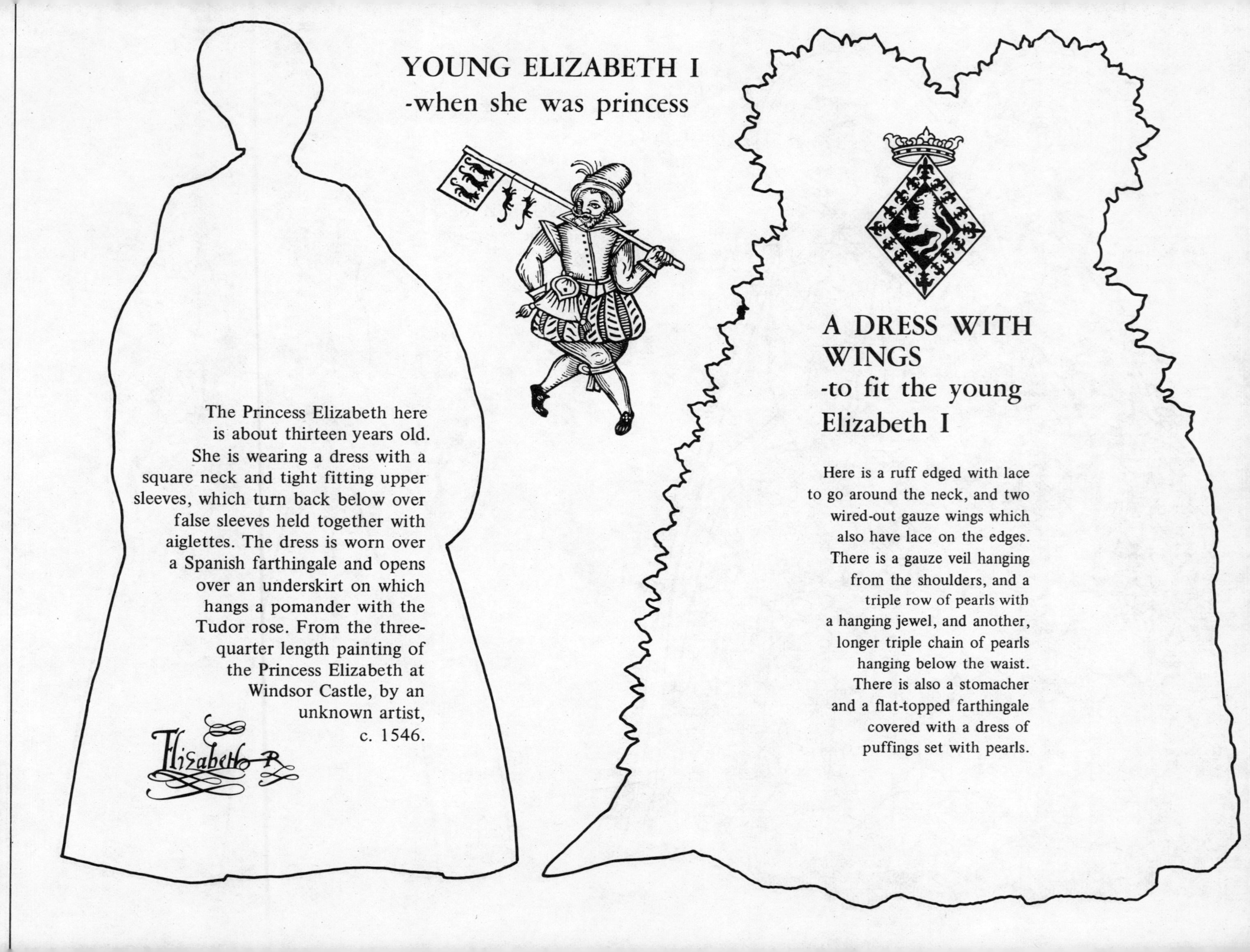

YOUNG ELIZABETH I
-when she was princess

The Princess Elizabeth here is about thirteen years old. She is wearing a dress with a square neck and tight fitting upper sleeves, which turn back below over false sleeves held together with aiglettes. The dress is worn over a Spanish farthingale and opens over an underskirt on which hangs a pomander with the Tudor rose. From the three-quarter length painting of the Princess Elizabeth at Windsor Castle, by an unknown artist, c. 1546.

A DRESS WITH WINGS
-to fit the young Elizabeth I

Here is a ruff edged with lace to go around the neck, and two wired-out gauze wings which also have lace on the edges. There is a gauze veil hanging from the shoulders, and a triple row of pearls with a hanging jewel, and another, longer triple chain of pearls hanging below the waist. There is also a stomacher and a flat-topped farthingale covered with a dress of puffings set with pearls.

Animals in History

Animals, birds, and horses have their place in history too; and to prove it Bellerophon has produced *A Coloring Book of Animals, A Coloring Book of Birds,* and *A Coloring Book of Horses.*

Get out your crayons and colored pencils, for here are:

HISTORICAL ANIMALS

"Elegant Lions in Medieval Japan" (from a folding screen printed in 1543–90, Imperial Palace, Kyoto).

"Baboons in a Fig Tree in Ancient Greece" (from a wall painting in the tomb of Khnemhotpe, Beni Hassan, 1900 B.C.).

"A Frolicking Water Buffalo in Old India" (from a 17th-century drawing, India. British Museum).

HISTORICAL HORSES

"Three-Horse Chariots in Ancient Assyria" (from a lost relief, Nimrud, 8th century B.C.).

"Queen Elizabeth I Riding over Her Enemies" (from an engraving by Thomas Cecill, c. 1625–40. British Museum).

"The Trojan Horse" (from an amphora, Mykonos, 670 B.C.).

HISTORICAL BIRDS

"Birds in an Ancient Egyptian Pont" (from a relief, Saqqara, c. 2450 B.C.).

"Riding on a Swan in Ancient Greece" (from a red-figure vase by the Zephyros Painter: Kunsthistorisches Museum, Vienna).

"A Quail in America About a Thousand Years Ago" (from a Mimbres bowl, New Mexico, 950–1100 A.D.).

And there are lots more! The coloring books are available at your bookstore or SEND $2.00 plus 25 cents for postage and handling to Bellerophon Books for all three.

VICTORIAN DICKEY HOUSE

Historic Buildings Model Kits

Now you can create your own restoration of eight three-dimensional models of historic American buildings, and with the colorful notes that are included you will be able to relive a part of our history. These authentically rendered paper replicas are executed to scale (they vary from ¼ inch to 1⁄16 inch) by San Francisco architect Roy F. Killeen; each model has been chosen for its historical significance as well as its architectural charm. The models are printed in color on heavy paper and are ready for you to cut out and construct. Each kit (15½ x 10½ inches) includes a heavy cardboard base, an instruction sheet, and a history of the building. Among the Mini-Mansion Model Kits available are:

"Victorian Dickey House." The house is located on a wildlife sanctuary and operated by the National Audubon Society in Tiburon, California.

"Old Lighthouse at Point Loma." The lighthouse is now part of the Cabrillo National Monument near San Diego and a major tourist attraction.

"Old Plaza Firehouse No. 1." The firehouse is a state historical landmark in the restored old pueblo section of downtown Los Angeles.

"St. Paul's Chapel of Trinity Church." This is the oldest public building in New York, located at the tip of Manhattan Island.

"The Cabildo, New Orleans." This was originally the seat of city government and now houses the Louisiana State Museum.

"Captain Penniman House, New England." This is located in Eastham, Massachusetts, part of the Cape Cod National Seashore.

"Maxwell House, Colorado." Located in the colorful mining town of Georgetown, this is considered one of the finest Victorian houses in America.

"John Muir House." The famed naturalist's home, in Martinez, California, is now a national historic site administered by the National Park Service.

To order Mini-Mansion Model Kits SEND a check for $4.95 (postage and tax included) per kit to 101 Productions. Be sure to specify the name(s) of the kit(s) you want.

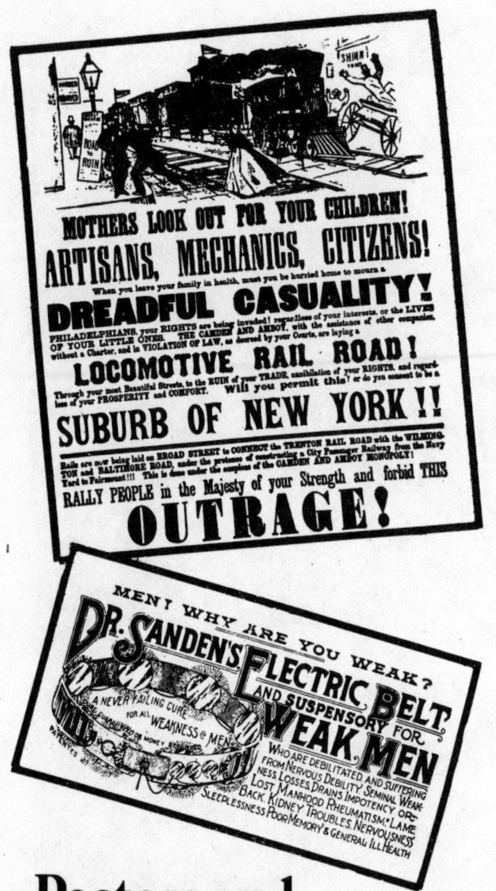

Posters and Handbills from America's Past

Here's a fabulous catalog that lists over 1000 historical posters, handbills, broadsides, prints, and advertisements. Authentic reproductions trace the political and social history of America from earliest times to the recent past. These on-the-spot records of America's heritage will make history come alive for you. You'll find such records of antiquities as:

"Call to Arms," by George Washington. Illustrated recruiting poster offers bounty of $12; 13 x 11 inches. No. R10. 75 cents.

"British Recruiting." Poster offers "50 acres where gallant hero may retire and enjoy his bottle and lass"; 13 x 18½ inches. No. R35. 95 cents.

"Enemy Alert" (Philadelphia, 1776). Warns of General Howe's approach: delay "may be fatal"; 11 x 15½ inches. No. R37. 75 cents.

"Cornwallis Surrender, 1781." To be celebrated in an orderly way with no rioting; 10 x 15 inches (broadside). No. R38. 75 cents.

"Gentlemen Volunteers." Wanted for His Majesty's ships (British recruiting poster, Philadelphia, 1777); 8½ x 11 inches. No. 6. 65 cents.

"Cornwallis Retreating." "His horse shot under him," report from battlefield, 1781; 8½ x 11 inches. No. 34. 50 cents.

"Signing of Declaration of Independence." Picture of 48 signers, their signatures; text of Declaration; fancy border with state seals and early presidents. Broadside engraved 1841. Nice for framing; 12 x 16 inches. No. 60. 95 cents.

"Washington's Pledge" (to defend the United States). Signed at Valley Forge, 1778; 8½ x 11 inches. No. 151. 50 cents.

"Washington's Appointment as Commander in Chief" (1775). Signed by John Hancock; 14 x 7½ inches. No. 158. 65 cents.

If you're a history buff and want some fascinating wall decorations, write for a copy of this exciting catalog. SEND 25 cents to Buck Hill Associates, Garnet Lake Road, Johnsburg, N.Y. 12843. There is a minimum order of $3.95, so please write for the catalog before you order any posters.

Historical Wall Maps

They're large. They're informative. They're beautiful. They're produced by the National Geographic Society. And they'll make handsome wall decorations for your room.

"Historical Map of the U.S." From early explorations to the present. 41 x 26½ inches.

"Battlefields of Civil War." Historical details, campaigns, battles, place-names, roads, and railroads. 405 historical notes. 30 x 23 inches. Printed on both sides.

"Indians of North America." Archaeological/ethnological. 61 illustrations, 30 descriptive notes. 32½ x 37½ inches. Printed on both sides.

"Africa. Its People and Its Past." 39 illustrations, 56 descriptive notes, language key of African past. 23 x 28½ inches. Printed on both sides.

"Classical Lands of the Mediterranean." Insets of ancient civilizations, descriptive notes. 39 x 26½ inches.

"Lands of the Bible Today." 342 descriptive notes, 4504 place-names.

"Holy Land Today." Biblical events, archaeological digs, Crusades, walled City of Jerusalem. 32½ x 42 inches.

To order WRITE TO National Geographic Society, 17th and M Sts., NW, Washington, D.C. 20036, and tell them which maps you want. The Society will bill you $2.00 for each map at time of shipment, plus postage and handling.

Great Women Paper Dolls

20 of the great women in history, from Cleopatra to Golda Meir, are pictured with splendid costumes for you to color, cut out, and use to dress the figures. Included are Queen Boudicca, Theodora, Lady Murasaki, Eleanor of Aquitaine, Joan of Arc, Pocahontas, Madame de Pompadour, Queen Victoria, Florence Nightingale, Sarah Bernhardt, Madame Curie, Anna Pavlova, and Bessie Smith. There's even a picture of Amelia Earhart and her airplane. There are two historically accurate costumes for each woman: gowns, uniforms, headdresses, helmets and shields, even the royal scepter. *Great Women Paper Dolls* is informative and enjoyable. It is available at your bookstore or SEND $2.50 plus 25 cents for postage and handling to Bellerophon Books.

Panorama Coloring Book

Here for you to color is a panorama of early American scenes that is over 14 feet long! The scenes show houses, churches, and public buildings, people, animals, and ships. When you finish coloring the drawings, you can make a panorama around your room, just as they did in the 19th century. In those days enormously long drawings of a great stretch of a city or a huge naval battle would be hung around the wall of a circular room, and if you stood in the center you could easily feel that you were part of the scene. The panorama of Massachusetts from which Suzanne Chapman made her drawings actually hangs in the Museum of Fine Arts in Boston and is 42 feet long. It's fun, it's history, and it's decorative. To order your *Coloring Book of Early American Scenes* SEND $2.00 plus 50 cents for postage and handling to Museum Shop, Museum of Fine Arts, Boston, Mass. 02115.

JOAN of ARC

Military leader & saint. 1412 - 1431

From the statue by Princess Marie d'Orléans at Versailles, the skirt being removed since Joan didn't wear one with her armor. The pattern on the dress is from the painting by Henri Bellechose of Saint Denis, from about the time of Joan, in the Louvre.

JOAN of ARC
Military leader & saint. 1412 - 1431
I will raise such a battle-cry that you will remember it forever! . . . We will strike with great thunder, and we will see who has their rights . . . . And better now than tomorrow, and tomorrow than the day after.
Je vous ferai un tel hahay, dont vous aurez éternelle mémoir . . . Nous frapperons dedans à grands horions, et verrons lesquels meilleur droit auront . . . Plus vite maintenant que demain, et demain qu'àpres . . .

KING HENRY VIII

Henry VIII spent 16,000 ducats yearly on his wardrobe, over $1,000,000.00 in money today. "His robes are the richest and most superb that can be imagined, and he puts on new clothes every Holy Day."

—*The Venetian Ambassador*

Henry VIII & His Wives

Henry VIII & His Wives has page after page of detailed drawings of the authentic dress: suits of armor for the king and regal robes for his wives. You have the fun of coloring them, and then, because they are paper dolls, you can cut them out and dress the figures with historical accuracy. Jewels, furs, damask, and elaborate headdresses are reproduced with the greatest care, and still the pictures retain simplicity. You will have hours of fun creating your own color schemes. The story of each queen is told, and it's a great way to learn history. The book is available at your bookstore or SEND $2.50 plus 25 cents for postage and handling to Bellerophon Books.

MATH CAN BE FUN

FREE Creative Publications Catalog

An exciting, full-color, 84-page catalog brimming over with mathematics curriculum materials that can be used in the school or at home. The careful selection of products represents the combined efforts of authors, editors, classroom teachers, and consultants. Their purpose: to make learning mathematics exciting.

Here you will find laboratory materials, games and puzzles, colorful posters, tangrams, polyhedra kits, SuperStructures, and attribute games. As Creative Publications says: "Leonardo da Vinci said that mathematics is the language of the universe. We bring people together to increase their awareness and understanding of the mathematics in our daily lives. Our goal is to provide materials and inspiration so students can be helped to understand the universe around them."

For your free copy of the catalog WRITE TO Creative Publications.

Colorful Design Posters

These beautiful, colorful designs will add handsome decor to your room. Both sets of four posters (25 x 25 inches) are full-color reproductions of beautiful mathematical arrays. The Math Art Posters, by Sonia Forseth, are based on and illustrate multiplication and addition tables in modular systems, as well as the design possibilities in mathematical patterns. A 16-page booklet, by Andria Troutman and Sonia Forseth, explains the many fascinating aspects of mathematics in art; the booklet also includes color reproductions of more Math Art posters. The posters are packaged in a heavy-duty mailing tube that is also suitable for storage. Each set of four posters costs $5.50. WRITE TO Creative Publications. Specify No. 60100 (set A) or No. 60105 (set B).

Think Metric!

Say good-bye forever to miles, quarts, bushels, pounds, and tons. Soon you will measure things the way the rest of the world does—in meters, liters, and grams. We now have to cope with units of measure divisible by 12, 16, 36, or 5280. In the metric system—officially known as the International System of Units—everything is more conveniently divisible by 10.

Author Franklyn M. Branley traces the evolution of weights and measurements from the days when a yard was arbitrarily determined by the distance from the tip of King Edgar's nose to the tip of his middle finger when his arm was stretched out. There is a set of entertaining measurement projects and a set of English-metric conversion charts to get you ready to *Think Metric!* Illustrated by Graham Booth. Thomas Y. Crowell Co. $4.50.

little bitty area —
just too small to fuss with
maybe a sesame seed? —
square millimetre

LOGICAL THINKING
(or, there is egg on your face!)

HERE IS ANOTHER ONE OF THOSE QUESTIONS THAT SOUNDS LIKE A TRICK UNTIL YOU REALIZE IT'S REALLY PRACTICE IN STRAIGHT THINKING.

How to stop hating math! Arithmetricks, mathematical magic, graph games, space forms. Learn about mathematics in art and geometric playthings. Construct mysterious Mobius bands. Make your own computer. Enter the new world of metrics!

The I Hate Mathematics Book

Here's what the book is all about:

This book is for nonbelievers of all ages. It was written especially for kids who have been convinced (by the attitudes of adults) that mathematics is (1) impossible (2) for those smart kids who can't play stickball, and (3) no fun anyhow. But this book will also do wonders for parents, teachers, or any adult who likes kids. This book says that mathematics is nothing more (nor less) than a way of looking at the world and is not to be confused with arithmetic. The content of mathematics is the same as the content of any kid's life. Why not? In this book you'll find several hundred mathematical events, gags, magic tricks and experiments to prove it.

And here's what the editor of *The I Hate Mathematics Book* has to say about it:

How would you like to have the last laugh on your parents when they tell you to stop playing around with those magic tricks and get on with your homework? Are you a permutation ice cream eater? What offers you thirteen new riddles, sidewalk games, things to do when you have the flu, and a sneaky way of getting around drying the dishes?

If you've always thought mathematics was a diabolical plot on the part of your teacher to make you miserable, you need help. Try *The I Hate Mathematics Book.*

This most unusual (and fun) Brown Paper School book is written by Marilyn Burns and illustrated by Martha Hairston. Little, Brown and Co. $5.95 hardcover, $2.95 paper.

ONE MINUTE TO GET ACROSS

FIRST, TRY THIS YOURSELF. (IT WILL HELP KEEP YOU HUMBLE.)

LOOK AT THIS DRAWING. WHAT IS THE DIAMETER OF THE CIRCLE?

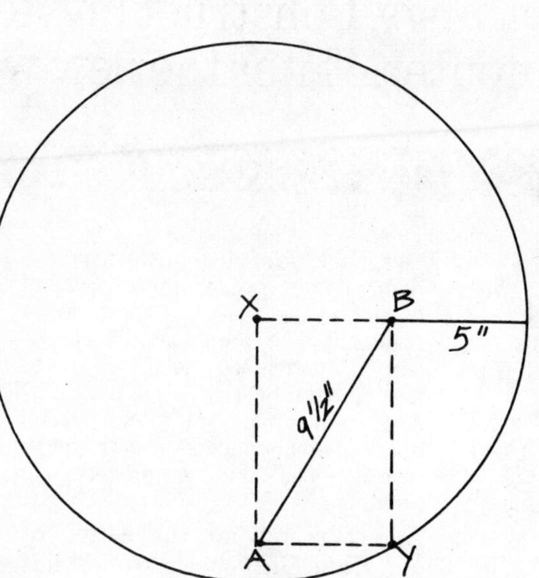

TAKE A MINUTE TO FIGURE IT OUT. YOU DON'T NEED TO DO ANY FANCY CALCULATIONS.

JUST LOOK.

WHEN YOU'VE SHOWN THIS TO YOUR GROWNUP, AND THEY HAVE GIVEN UP, EXPLAIN TO THEM:
"ISN'T THE DIAGONAL **AB** THE SAME IF I DRAW IT FROM **X** TO **Y**?"
(THE ANSWER TO THAT IS YES.)
"ISN'T **XY** A **RADIUS**?" (DISTANCE FROM THE CENTER TO ONE EDGE.)
(YES.) "NOW CAN YOU FIND THE DIAMETER?"

AND DON'T SMIRK. YOU DIDN'T GET IT EITHER.
(DID YOU?)

Measurements and How We Use Them

When you want to know—
how soon your birthday will be, how high you can reach, how tall you are, how much your dog weighs, who is heavier—you or your best friend, how warm your bathtub water is . . .
you *really* want to find out about—
the passing of time, how far or near, how tall or short, how heavy or light, how hot or cold things are.

You can guess, of course, and the answers might be soon, fast, slow, heavy, high, low, short, hot, cold, very far, or very near.

But when you use these words, you will not get precise answers. You need something to help you measure these things.

As you read *Measurements and How We Use Them*, by Tillie S. Pine and Joseph Levine, with its many pictures and drawings, you will find out how to measure things, what we use to measure, and why we need to measure things exactly. Age 6 and up. McGraw-Hill Book Co. Available at your local library.

Building Tables on Tables

This is a book about multiplication. It challenges the commonly held notion that there is just one right answer to every multiplication problem, and shows that addition, subtraction, and multiplication are happily related to one another. The multiplication tables are transformed from boring things you have to memorize to flexible constructive tools. *Building Tables on Tables*, by John V. Trivett, a book of relationships, shows you how to create a computer by stringing a few simple multiplication tables across the room. Good for the young mathematician, age 9 and up. Illustrated by Giulio Maestro. Thomas Y. Crowell Co. $4.50.

SuperStructures

These are unique new building pieces. The inherent geometry of SuperStructures is founded on principles of modular structure in nature. Construction possibilities are unlimited and the modular materials permit error-free, unfrustrating exploration. The set combines a fantastic connector, the Universal Node (it has 26 spokes), with color-, shape-, and length-coded branches so you can't go wrong. Brightly colored PolyPanels snap easily into position to further enhance the structures you create.

You can build a bridge, a house, a pyramid, futuristic cities, whimsical birds—or things scientific such as polygons, polyhedra, and periodic networks. This is a fascinating building toy for young children and older students. Kits come with durable plastic trays compartmented to store the pieces. Ages 5 to 15. Look for the sets at your toy or hobby shop. If you can't find them WRITE TO Creative Publications. SuperStructures (medium—128 pieces) is $9.95; (large—300 pieces) is $17.95. No charge for postage and handling.

Worlds Within Worlds

John Reitzel makes available for the first time an inexpensive model of the nesting relationships of the five regular polyhedra. Printed in brilliant colors on sturdy coated stock, they are die-cut and scored so you can easily remove them from the sheet and assemble them. No glue or fasteners are required. Detailed step-by-step illustrated instructions make the assembly easy enough for the very young. It's also a learning tool for older children.

When the models are completed they nest inside one another from the smallest, the icosahedron, through the tetrahedron, octahedron, and dodecahedron to the largest, the cube. Worlds Within Worlds (No. 34220) is fascinating to all ages. ORDER FROM Creative Publications. $4.50.

A First Book of Space Form Making

Space forms are sets of surfaces that enclose space. A pyramid is a space form; so is a cube. Can you take a flat sheet of paper and draw lines on it so that when you cut it out and fold it you will have a many-sided space form? George R. Fouke will show you how to do it. All you need is a compass, ruler, protractor, scissors, paper, and glue; and decorating materials for the added touch. *A First Book of Space Form Making* gives you a small lesson on how to use each tool. Then you will begin to make triangles, rectangles, trapezoids, hexagons and meanwhile you will also be learning about them. Using these, you will learn to make simple space forms. As you progress, the forms become more complicated and interesting. You'll find you can make hexahedrons, 13-sided space forms, 15-sided space forms, decahedrons, and many others. Then you can combine them to make trains, rockets, circus tents, abstract creatures from Mars—whatever suits your fancy. GeoBooks. $2.95

MAKE MONEY!

Good Cents: Every Kid's Guide to Making Money

Good Cents is just about "must reading" for every boy and girl, in between selling lemonade on a hot day and washing the neighbor's car. It says there is no reason why kids shouldn't earn money, and if they do, the experience will probably be a good one. It is a strong pat on the back for kids who would rather do it themselves, and a guide for adults who would like to help without getting in the way.

The people who wrote this book call themselves the Amazing Life Games Company (and friends): a free-form association of writers, artists, teachers, and kids who assemble to produce books, films, and learning environments. They believe that learning, which has gotten a bad reputation, shouldn't be a dirty word. They also believe that learning can't be enclosed in a school yard or put somewhere it isn't wanted, and that it needs all the friends it can get. Illustrated by Martha Hairston and James Robertson, Houghton Mifflin Co. $6.95 cloth, $3.95 paper.

About This Book

If you just want to have fun, read some other book. This one is about having fun (and making money) by doing work. Hard work. You can't have one without the other.

Remember this: you'll have to do some serious thinking on your own. No book can tell you what's right for you. Any book that says different is a liar.

This book is also about caring. You can't have much fun if you don't care about the things you do.

Everybody is good at something. *Everybody*. And usually (but not always) people like to do what they are good at. It makes them feel right inside. It's important to try to figure out what you're good at.

A lot of the ideas adults have about money and work are pretty crazy. Don't be fooled. Especially don't pay much attention to people who think that making a lot of money is equal to being happy. They are the craziest of all. The next craziest are those people who spend all day doing something they hate. The ones who say, "I'm not good at anything," aren't crazy, they're just sad.

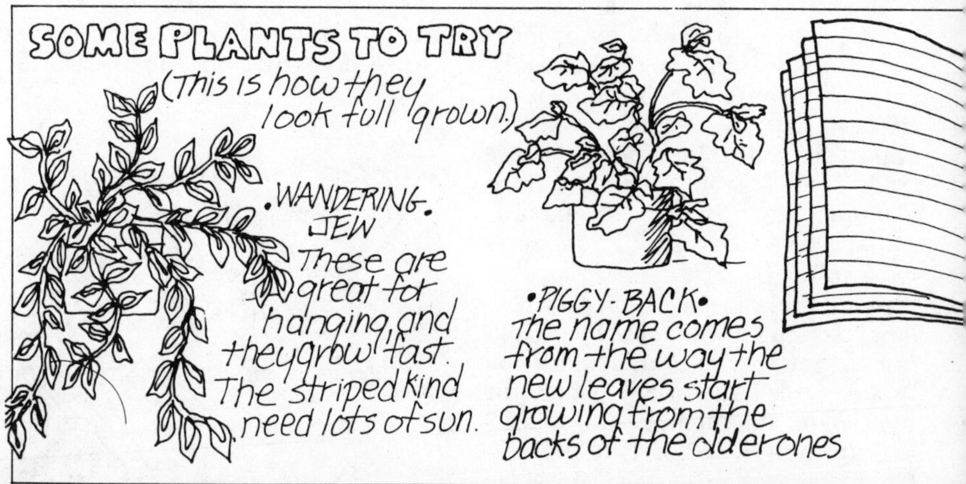

Walk a dog. Shovel snow. Deliver newspapers (only on Sunday). Raise worms. Open a lemonade stand. Shampoo a dog. Want a big money-maker (plus a lot of fun)? . . . dye colorful T-shirts for friends. Start a business and earn your own money!

The White Tornado Basement Cleaners and Garage Sale Company

THIS IS TWO IDEAS IN ONE. FIRST: CLEANING BASEMENTS OR GARAGES AND HAULING AWAY JUNK. SECOND: SELLING JUNK YOU HAUL AWAY FROM THE PLACES YOU CLEAN. TRY IT AS A WEEKEND BUSINESS. CLEAN BASEMENTS ON SATURDAYS UNTIL YOU HAVE GATHERED ENOUGH GOOD JUNK TO SELL. THEN HAVE A GIANT GARAGE OR BASEMENT SALE THE NEXT SATURDAY. YOU CAN MAKE MONEY BOTH WAYS.

THE HAND and FOOT PRINTER

HANDMADE GIFT CARDS, WRAPPING PAPER, STATIONERY ETC PRINTED HERE. REASONABLE RATES 15 CEDAR STREET.

FREE
Fun with Staples

Let's Have Fun Making Things with Staples is the name of a free booklet that will show you exactly how to do it. Among the many "things" you can make are a boat, a doll bed, a pom-pom, flags and pennants, bean bags, a puppet stage, an Indian headdress, and all sorts of party favors and presents. With odds and ends you can find around the house, a stapler, and these simple illustrated instructions, you'll be on your way to hours of fun. For the free booklet SEND 10 cents for postage and handling to Postitch, Retail Products Division, 11 Briggs Drive, East Greenwich, R.I. 02818.

FREE
Pointers for When You Baby-Sit

It's a big responsibility to be asked to baby-sit. Baby-sitting doesn't mean sitting and watching baby, it means actively participating in baby's day, perhaps in bathing and dressing, feeding, playing, entertaining—being in general an alert, attentive, useful aid. For some helpful hints and pointers on how to baby-sit and what to do under specific circumstances, get a free copy of *When You Baby-Sit*. WRITE TO Dr. Barbara Fite, Specialist in Human Development, Alabama Cooperative Extension Service, Auburn University, Auburn, Ala. 36830.

FREE
Sitting Pretty

Baby-sitting—and the safety of children and sitters—is a subject of major concern to the community, to parents, and to the sitters. *Sitting Pretty* is a free guide to baby-sitting; it contains advice for parents employing sitters, for sitters, and for parents of sitters. For your free copy of this important guide WRITE TO Greater New York Safety Council, 302 5th Ave., New York, N.Y. 10001. Enclose a stamped self-addressed envelope.

Baby Sitting: A Concise Guide

The author, Rubie Saunders, says,

> Baby sitting is a profitable enjoyable profession for the girl or boy who takes it seriously.... Furthermore, you may find that learning how to accept responsibility is not a drag at all; instead, it's rather a good feeling to know that people have learned that they can really rely on you.
>
> So, welcome to the ranks of the youngest profession; you're in excellent company!

Baby Sitting: A Concise Guide deals briefly with each step taken in preparation for the first job. Covered are how to find a job, how much to charge, the sitter's responsibilities, safety precautions to remember, and of course the care and feeding of the child. It also has an excellent check list of dos and don'ts. Pocket Books (Archway Paperback). 75 cents.

Child Care

WANTED: Baby-sitter for young children. Conscientious, reliable, good judgment, fun-loving. Will train.

If you answer this ad, are you aware of what you will need to know about baby-sitting? This 37-page book will give you the basics a good baby-sitter should know: helpful hints, charts, a baby-sitter's directory. It will give you confidence. The "Basic Care" chapter tells you about feeding and giving meals to infants and older children; diapering, naps, bedtime, temper tantrums. Other chapters include "Play"; "Emergencies"; "Accidents and Illness," with a section on "What to do if . . ."; "Group Baby-Sitting"; and "Baby-Sitting Is a Job." Part II is about how to learn and teach baby-sitting.

The book is not only an excellent baby-sitter's guide but contains useful information for parents, counselors, teachers, playground aides—anyone who is concerned about children. To get your copy of *Child Care* SEND $1.00 to Supply Division, Camp Fire Girls, 450 Ave. of the Americas, New York, N.Y. 10011.

Free samples! Free posters! Free pictures! Free books! Free recipes! Free coloring books! Free membership in Sparky's Fire Department. Free baby-sitting guides. Free story of the Texas Rangers. Learn how to play the harmonica, free!

FREE
The Pocket Guide to Babysitting

When as a young person you agree to take care of a child or children you are assuming a lot of responsibility. And here's a lot of help: a 48-page booklet which is chock-full of advice that may make you the most popular babysitter on the block. You'll learn what to put in the babysitter's service kit (a collection of toys, surprises, and emergency needs); how to keep babies safe and sound; games to play, hints on feeding, and so on. An important chapter is titled "What to Do If . . ." A babysitter sometimes has to be a trouble-shooter. You have to be ready for the unexpected. The book will fit in the pocket of your jeans. For your copy of *The Pocket Guide to Babysitting* (No. OHD 74–45) WRITE TO U.S. Department of Health, Education, and Welfare, Washington, D.C. 20201.

Africa Safari Poster

Here's a large (almost 3 x 4 feet), colorful poster that shows pictures of 10 African big-game animals (from leopards to antelopes, from rhinos to lions); a map of Africa; a picture and description of an African guide; and stories about man and his environment, and hunting and conservation. This informative and decorative safari poster of Africa will make a fine wall decoration for your room. To get it SEND 50 cents to Daisy, Rogers, Ark. 72756.

Books on Birds and Animals

Here are two attractive, worthwhile, low-priced booklets printed in full color to add to your collection.

Upland Game Birds. This is a real who's who of United States and Canadian upland game birds. 23 species are featured (21 in full outdoor color) on its 6 x 9 inch pages. All 32 pages are packed with information you'll enjoy reading. Information on each bird includes its living, eating, and breeding habits.

Big Game Animals. This features full-color photographs (6 x 9 inches) of United States and Canadian big game animals taken where they live. There are 24 pages filled with interesting facts on native animals from moose to wild boar. The text covers 19 species and pictures them with professional accuracy. A map shows where each animal lives.

For a brace of books for the outdoorsman WRITE TO Federal Book Offer, P.O. Box 7200, Maple Plain, Minn. 55359. 50 cents each.

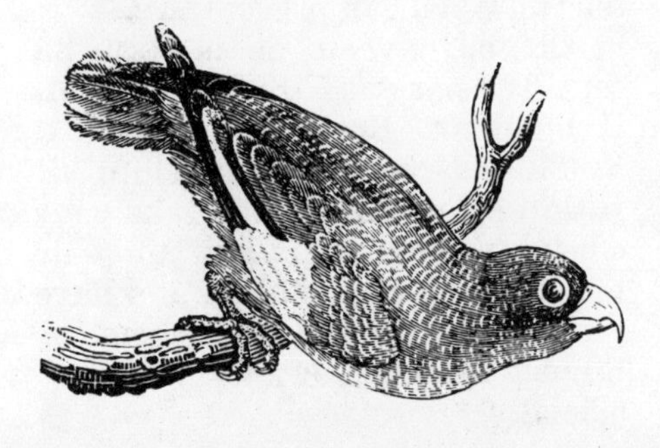

FREE
Shall We Dance?

Capezio, the dancers' cobbler since 1887, has three excellent free booklets to teach you about dancing. *How to Achieve a Dancer's Body Through Dance Exercise,* written by Olga Ley, a well-known teacher and dancer, will help give you the look of a dancer. The booklet provides 12 easy, fun-to-do exercises which can help you develop that lithe, graceful look. There are illustrations of various styles of dancers' shoes and leotards.

Did you know that hard-toe, "boxed" toe shoes have no rights or lefts? *The Care and Fit of Toe Shoes* will tell you about toe shoes and how to choose them. Equally important and interesting is the pamphlet *Why Can't I Go on My Toes?* which gives the physical requisites you should meet before you dance on toe shoes.

For free copies of these pamphlets WRITE TO Capezio Ballet Makers, Dept. M, 543 W. 43d St., New York, N.Y. 10036.

FREE
Let's Make Puzzles

Let's Make Puzzles tells an ingenious way for you to one-two-three whip up a puzzle to play with. If you're tired of the puzzles you have and want something really different to do, here's your chance to make whatever your heart desires. As a special aid this pamphlet teaches you how to make your own flour and water paste. To get your free copy WRITE TO Dr. Barbara Fite, Specialist in Human Development, Alabama Cooperative Extension Service, Auburn University, Auburn, Ala. 36830.

FREE
Fun in the Making

This illustrated 30-page booklet contains some excellent ideas for making toys and games. Almost everything is made from throw-away materials; there are recipes for homemade paste, play-dough clay, and a bubble solution. Toys are described that you can make from egg and milk cartons; gardens that you can make from onion and carrot tops; playthings that you can make from spools, boxes, and empty margarine tubs; and paper-bag puppets and homemade kazoos. Simple diagrams and simple instruction make this an excellent booklet for your house. To get your copy ask for *Fun in the Making* (No. OCD 73-31) and WRITE TO U.S. Department of Health, Education, and Welfare, Office of Child Development, Washington, D.C. 20201.

FREE
How to Play the Harmonica

One musical instrument you can carry in your pocket is the harmonica. You can learn to play it easily and quickly with this wonderful free booklet, *How to Play the Harmonica.* Written by the famous "tune detective" Sigmund Spaeth, this 24-page instruction booklet outlines a "new easy method for beginners." Simple diagrams show you how to play all your favorite tunes—everything from "Hot Cross Buns" to Brahms's "Lullaby." Christmas carols include "Jingle Bells," "Joy to the World," and "The First Noel." It's easy! It's fun! For your free copy WRITE TO M. Hohner, Dept. S, Andrews Road, Hicksville, N.Y. 11802.

FREE
And the Band Plays On

The band is the most popular form of large-group music making in the United States. *The Band: Past, Present and Future,* a completely illustrated history of the beginning and development of bands around the world and in America, includes a discussion of the present and future status of bands. If you're a budding musician, or are simply interested in this musical form, get your free copy; WRITE TO American Music Conference, 150 E. Huron St., Chicago, Ill. 60611. Enclose 25 cents for postage and handling.

Join Sparky's Fire Department

Children of ages 3 to 10 can join the famous Sparky the fire dog in his own fire department. Join right away and become an official inspector in Sparky's fire department.

As an inspector your first duty, like that of any real fireman, is to save others from being hurt by fire. As a member, you'll learn how to prevent fires, and how to help your family and friends stop fire before it starts.

And you can wear the official Sparky inspector's badge, shaped like a real fire department badge, in bright red plastic, and carry the official inspector's manual, which tells how you can prevent fire.

All you have to do is SEND 35 cents to Sparky's Fire Department, P.O. Box 32, Boston, Mass. 02101. You'll get your membership kit within a few days. Then take your official membership card to your neighborhood fire station and have it signed by a fireman! Remember, it must be signed by a fireman to be official.

FREE
Sparky's Junior Fire Prevention Packet

This kit is jam-packed with comic and coloring books, songs, Sparky's official home fire inspection blank, and a do-it-yourself home fire escape planning guide the whole family can work on together. Sparky's Junior Fire Prevention Packet is fun and educational. Look at all the things you'll get:

1. *Sparky's Coloring Book.* Fire prevention in song and story. 15 pages of drawings to color.
2. *Hi—Mr. Fire!* A Sparky fire prevention comic for young children, and for children old enough to read who have younger brothers and sisters who must be kept safe from fire.
3. *Sparky's Helpers Answer the Alarm.* Another comic book, showing where to look for fire dangers around the home, and what to do in case fire does start.
4. *Alfie Looks for Fire Trouble!* Sparky's friend Alfie and his magic helicopter do a great cleanup job and chase "mean Mr. Fire" out of the house in this eight-page comic book.
5. "Sparky's Firemen." A song and game sheet written by a class of kindergarten children for Sparky's friends!
6. "Sparky's Office Home Fire Inspection Blank." On one side is a 20-question checklist of home fire hazards with places for yes or no answers. Each no answer means a fire hazard. On the other side is Sparky's inspector practice page: an inside look at a house nearly bursting with fire hazards, to be located and listed at the bottom of the sheet. Find the ten fire hazard clues in the practice house, then make sure none of these dangers are in your real house.
7. "Family Night Fire Escape Planner." A handy folder on what to do in case of fire at night, when the worst home fires break out. How to make and rehearse a plan for everyone to get out safely; space to draw a floor plan for escape from the bedrooms; and a

chart for keeping tabs on the home fire-drill schedule (at least one drill every 6 months). This is life-saving information the whole family needs, in one convenient folder.

8. Information on joining Sparky's Fire Department.

To get your Sparky's Junior Fire Prevention Packet SEND 25 cents to cover postage and handling to Sparky Junior Packet, Public Affairs Division, National Fire Protection Association, 470 Atlantic Ave., Boston, Mass. 02210.

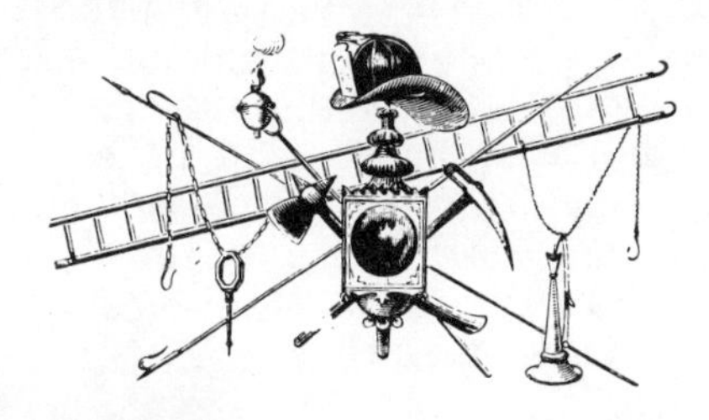

FREE A Fire Extinguisher

Baking soda is not only a cooking and cleaning aid, it can help save your life. Here from Church & Dwight is a special FIRE label for you to wrap around an empty coffee can which you have filled with baking soda. If you keep this fire extinguisher handy, you can use it to put out small cooking fires in your kitchen or at a campsite, as well as electrical and liquid fires. You can even use the baking soda to treat minor burns. For your free FIRE label and information on fire fighting WRITE TO Church & Dwight Co., 2 Pennsylvania Plaza, New York, N.Y. 10001.

FREE The Texas Rangers

No more colorful body of fighting men has graced the pages of American history than the Texas Rangers, our nation's oldest law enforcement body. Its members have become legendary as a result of their long-established traditions; today it is proud men who wear a Ranger badge. *The Texas Rangers,* by Robert W. Stephens, is a fully illustrated, 44-page history of the 150-year-old law enforcement agency. For your copy SEND 25 cents for postage and handling to Daisy, Rogers, Ark. 72756.

Western Paintings

Here are lithographs of four colorful Fred Harman western paintings. Harman was the artist-author of "Red Ryder" and "Little Beaver," and he is now recognized as an outstanding western illustrator. His paintings hang in a number of prominent galleries and he was recently commissioned by actor John Wayne to paint five western scenes. The pictures in these color lithographs are titled "Grub Time"; "Bear Scent"; "The Point Rider"; and "Wagon Meat." The pictures are 10 x 13 inches and are available at $1.00 per set of four. WRITE TO Daisy, Rogers, Ark. 72756.

Red Ryder Poster

The title of this full-color poster is "Red Ryder Rides Again"; it pictures the famous western hero and Little Beaver galloping their horses. Almost 2 x 3 feet, the poster will make a great wall decoration for your room. To get this exciting poster SEND 50 cents to Daisy, Rogers, Ark. 72756.

FREE All About the Secret Service

The Secret Service protects the president and his family—everyone knows that—but do you know that the Secret Service was founded in 1865 as a bureau of the Department of the Treasury with the main task of suppressing the widespread counterfeiting of currency in the United States? It wasn't until 1901, after the assassination of President William McKinley, that the Secret Service was made responsible for protecting the president. Not until 1906 was legislation enacted to officially authorize such protection. To learn more about the Secret Service, here are two informative pamphlets which can be yours free: *The Secret Service Story* and *The Secret Service and Its Protective Responsibilities.* WRITE TO U.S. Secret Service, Suite 805, 1800 G St., NW, Washington, D.C. 20223.

FREE
Learn About Money

Would you like to learn more about how to use your money wisely? Here's an excellent 36-page booklet called *Children's Spending* that will help you develop money responsibility. It gives you some good tips on how to handle your allowance, how to work for more money, how to save, and how to avoid spending money foolishly.

Amusingly illustrated, this is one of the best sources of information on money management. For your copy WRITE TO Money Management Institute, Dept. AWWC, Household Finance Corp., Prudential Plaza, Chicago, Ill. 60601. Enclose 25 cents for mailing and handling.

FREE
It's Fun to Write Letters

If letter writing is a problem, here's a 32-page booklet for you: *It's Fun to Write Letters.* After reading it you'll agree, and your letters will be more fun to read. The booklet is chock-full of helpful advice. One chapter is "Ten Tricks to Make Letters Sparkle"; another, "The Letters That Must Be Written." You'll learn about how to close a letter with a signature, and about what occasions you'll want to remember with a letter. Enclose 25 cents for postage and handling and WRITE TO Eaton Paper Co., Pittsfield, Mass. 01201.

FREE
Learn How Paper Is Made

You will learn from this picture-pamphlet how paper is made from trees. It has photographs that take you on a tour through the process at the large Hammermill paper plant. The captions explain about stripping the logs, chopping, cooking, refining, pressing, drying, and rolling—17 steps in all. For your free copy of *From Forest Tree to Fine Papers* WRITE TO Educational Services, Hammermill Papers Group, Erie, Pa. 16512.

FREE
How You Can Make Paper

Would you like to make your own paper? Here is a booklet, *How You Can Make Paper,* that explains in six easy-to-follow, step-by-step photographs how you can do it. You need only inexpensive supplies found around your home: facial tissues, laundry starch, an eggbeater, an old metal pan, and in no time at all you can make paper. It's a great learning experience. To get your free booklet WRITE TO Things to Do at Home, American Paper Institute, 260 Madison Ave., New York, N.Y. 10016.

FREE
How Paper Came to America

The story of how paper came to America spans over a thousand years and many thousands of miles. This map/chart (2 x 3 feet) will help you identify the important chronological and geographical points in the history of paper. It has pictures of paper being made in old China, of how the Mayans invented bark paper, and of early papermaking machines. To get your free map/chart WRITE TO American Paper Institute, 260 Madison Ave., New York, N.Y. 10016.

FREE
Our Redwood Heritage

Of course you know that redwood trees grow in California, but what else do you know about them? To read about the coast redwood forests—their history, recreation in the redwoods, the commercial redwood forest, managing and harvesting the redwood forest, the lumber mill in action, products of the forest, redwood industry and our economy, and forest enemies and friends—get the free booklet *Our Redwood Heritage, the Story of the Coast Redwood Forests,* which includes a list of educational activities to broaden your understanding of the redwood forests and industry. WRITE TO Service Library, Dept. ORH, California Redwood Association, San Francisco, Calif. 94111.

FREE
A Brief Story of Aluminum and Alcoa

A Brief Story of Aluminum and Alcoa is a must for your reference shelf. This colorful poster-type illustrated brochure reviews in nontechnical language the history of aluminum. You'll learn about the discovery of an inexpensive process for extracting aluminum by a 22-year-old inventor, Charles Martin Hall, in his homemade laboratory in the woodshed behind his parents' home, and about Alcoa's role as the world's pioneer aluminum producer. This educational brochure includes flowcharts that show refining, smelting, and fabricating processes; a map depicting the company's operating locations and sources of new materials; a product listing; and a summary of Alcoa activities in environmental protection, recycling, and research and development. It also talks about careers in the aluminum industry. For your free copy WRITE TO Aluminum Company of America, 1254 Alcoa Building, Pittsburgh, Pa. 15219.

FREE
Informational Aids on Aluminum

Science and chemistry enthusiasts will want a copy of *Alcoa Informational Aids,* a 42-page booklet filled with listings of all sorts of films and booklets that will help when you want to study and write about aluminum. For your free copy WRITE TO Aluminum Company of America, 1254 Alcoa Building, Pittsburgh, Pa. 15219.

FREE
Copper Sets the Stage for the '70's

What do these have in common: Stone Age weapons, the Statue of Liberty, Pharaoh Cheops's bath, home furnishings, roofing, spacecraft, and your TV set? They all depend on copper and copper alloys for construction. Why copper? Copper doesn't rust, it resists abrasion, it conducts heat and electricity, it's nonsparking and nonmagnetic, it's strong and easy to work with. To read more about this important metal, send for a free, colorfully illustrated *New York Times* supplement on copper, *Copper Sets the Stage for the '70's,* which tells the story of copper from its first use about 8000 years ago to today—and tomorrow. WRITE TO Copper Development Association, 405 Lexington Ave., New York, N.Y. 10017.

FREE
Wheat Sample

Here is a packet of soft red winter wheat seeds for you to grow. This is one of many kinds of wheat; it usually comes from Ohio or Indiana.

Plant the grains in a flowerpot or wooden box with at least 6 inches of rich garden soil. Plant the grains about ½ inch deep, each grain 2 to 3 inches from the next grain. Keep the soil moist, but not wet. (You can also try sprouting the grains in moist cotton.) Nabisco's Graham Crackers are made from this wheat and you can get your free wheat sample if you WRITE TO Nabisco, P.O. Box 29, New York, N.Y. 10017.

FREE
All About Rubber and Tires

Here are two free informative booklets about rubber and tires:

The Story of the Tire. This is a 16-page, two-color illustrated booklet that describes in nontechnical language the story of "how it all started." Some credit Christopher Columbus with having discovered rubber, but Central and South American Indians were using rubber long before Columbus landed. You'll also learn how tires are made today.

The Miracle of Rubber. This is a 24-page, two-color booklet with many photographs that describes the history of rubber and the myriad products made from rubber. It includes the story of the exciting and accidental discovery of vulcanization by Charles Goodyear in 1839.

For free copies of these educational booklets WRITE TO Goodyear Tire & Rubber Co., Akron, Ohio 44316.

FREE
Learn About Rubber!

Want to learn about rubber? *Rubber,* a 32-page, completely illustrated booklet written in answer to many requests from schools for authentic, up-to-date information on the subject, tells a comprehensive story. (There are over 65 pictures.) For your free copy WRITE TO Dept. of Public Relations, Firestone Tire & Rubber Co., Akron, Ohio 44317.

FREE
Very Salty Story

You have heard the expression "not worth his salt," but do you know its derivation? The answer: In ancient Greece salt was traded for slaves; some times a slave was said to be not worth his salt. This booklet *is* worth its salt. Fully illustrated, *Salt Today and Yesterday* relates the history of salt, explains different methods of salt production (solar evaporation, mining, evaporation from brine), and discusses some of the important uses of salt (for health, in the home, for soft water, on the farm, in industry, and in transportation). For your free copy WRITE TO Morton Salt Co., 110 N. Wacker Drive, Chicago, Ill. 60606.

FREE
Pretty Sweet Story

This sweet story is entitled *The Story of Chocolate and Cocoa at Hershey's Chocolate World:* it contains interesting facts about our favorite sweet. Did you know that Christopher Columbus brought back cacao beans (our source of chocolate) from his voyage to the New World? You'll find out how cocoa and chocolate are produced, and the history of chocolate and Hershey Foods. For your free copy WRITE TO Hershey's Chocolate World, Hershey, Pa. 17033.

FREE
All About Cookies and Crackers

Cookies, Crackers and Nabisco tells the story of the cracker from its early days in old-time storekeepers' cracker barrels to its present-day sealed packaging in cardboard boxes. Complete with footnoted definitions of technical terms, this leaflet describes how cookies and crackers are made. SEND FOR your free copy; address Nabisco, P.O. Box 29, New York, N.Y. 10017.

FREE
Sidney Sheep Coloring Book

This coloring book tells of the adventure of a young American sheep named Sidney. As you color the 28 pages you will learn a lot about Sidney's wool coat, why it's called nature's wonder fiber, and why it keeps you warm. You will learn about the shepherd and his flock, the sheep dog, and the trouble Sidney has when he runs away from the flock. *Sidney Sheep Coloring Book* is fun, it's in English and in Spanish, and you can get it free. WRITE TO Sidney Sheep Coloring Book, Wool Bureau, 200 Clayton St., Denver, Colo. 80206.

FREE
Sheep in America

Do you know that sheep were first brought to America by Christopher Columbus on his second voyage in 1493? Do you know that we get glue, strings for musical instruments, and lamb to eat from sheep? Do you know how the hair of the sheep is made into the wool clothing you wear?

You can find these facts and more in the free booklet *Sheep in America.* To get your copy WRITE TO Wool Bureau, 200 Clayton St., Denver, Colo. 80206.

FREE
Young Cooks Baking Books

"A piece of it in a baby's crib keeps the ghosts away."

"If you dream about it, you will make money."

"If you take the last piece of it, you will be an old maid."

Do you know what "it" is? The answer is "bread." You can learn all about bread and how to make it in a special cartoon-type booklet created especially for the younger set. *The Young Cook's Bake-a-Bread Book* is free for the writing, as is *The Young Cook's Bake-a-Bun Book.* You'll receive a colorfully illustrated history of buns with complete instructions on how to make them. For your baking books WRITE TO Fleischmann's Educational Services, P.O. Box 2695, Grand Central Sta., New York, N.Y. 10017.

FREE
The Story of Bread

How do bakers make bread? *Bread in the Making* is a colorfully illustrated story of a school class that goes to visit a bakery. As you read, you get a complete tour of the bakery, and the machines and processes are explained to you. For this interesting booklet SEND 25 cents for postage and handling to American Institute of Baking, Nutrition Education Dept., 400 E. Ontario St., Chicago, Ill. 60611.

FREE The Story of Golden Grahams

Everyone knows and loves Graham Crackers. But did you ever stop to think about how they were invented? *Golden Grahams* is the story of Dr. Sylvester Graham, who discovered in the 19th century that people who ate whole wheat flour were generally healthier than people who ate the unenriched white flour of that time. He urged people to eat unsifted whole wheat flour, which became known as Graham flour, the basic ingredient for Graham Crackers. To learn more about Dr. Graham and to read about how Graham Crackers are made today, send for this free, colorfully illustrated story. WRITE TO Nabisco, P.O. Box 29, New York, N.Y. 10017.

Cooking Is Fun

Gaily illustrated and full of fun recipes, this is a story-type cookbook for beginning-to-read youngsters. Children will enjoy the chance to do some of the things they read about in this book and will learn about foods as they work. Help from adults is emphasized because children need guidance in beginning to cook and in learning caution and safety. A spirit of working together will make this a happy experience for all. For your copy of *Cooking Is Fun* SEND 30 cents to National Dairy Council, 111 N. Canal St., Chicago, Ill. 60606.

FREE Think Spaghetti!

How's this for a different after-school snack? It's called spaghetti snack-a-roni.

Ingredients

8 ounces spaghetti, broken in half, cooked according to directions on box
hot salad oil for deep frying
onion salt

Directions

Rinse drained spaghetti with cold water; drain again. Separate pieces of spaghetti and drop a few at a time into hot oil (375 degrees). Deep fry just enough at one time to cover bottom of fry basket. Fry about 3 minutes or until evenly and lightly browned. Spread on paper towels to drain. Sprinkle with onion salt. (Makes about 4 quarts, loosely packed.)

To get more unusual and tasty recipes using spaghetti WRITE FOR the free pamphlet *Think Spaghetti!* Send a postcard with your request to National Macaroni Institute, P.O. Box 336, Palatine, Ill. 60067.

FREE The Fourth R

In 1694 a ship was blown off its course and forced to land in Charleston, South Carolina. The people of Charleston were hospitable and kind, and before sailing away the captain expressed his gratitude by giving a handful of rough rice grains to the governor of the colony. This handful of rice was used for seed; America's extensive rice industry has grown from those first few grains. To learn more about rice and to receive a fun recipe for pizza made from rice WRITE TO Rice Council of America, P.O. Box 22802, Houston, Tex. 77027, for the free leaflet *The Fourth R.*

FREE Milk—from the Cow—by Truck—to You

This large folder is about the importance of the truck in getting milk to you. It describes the special refrigerator tank trucks that haul raw milk from the farms to the local plants that do the processing. For your free copy of the informative picture study *Milk—from the Cow—by Truck—to You* WRITE TO Educational Services, American Trucking Association, 1616 P St., NW, Washington, D.C. 20036.

My Friend the Cow

This is the story of milk and where it comes from, written and illustrated by one of the foremost authors of children's books. Lois Lenski, winner of the Newbery medal for children's stories, planned *My Friend the Cow* for youngsters of primary school age. There are 16 full-page color drawings. For your copy of this delightful little book SEND 20 cents to National Dairy Council, 111 N. Canal St., Chicago, Ill. 60606.

Let's Make Butter

This is a colorful activity sheet for young children that folds into a stage on which the child can use the accompanying puppet to tell how we make butter. Full instructions for the beginner are included, with an adult's guide to stimulate more artwork and activities. For your copy of *Let's Make Butter* SEND 12 cents (no stamps, please) to National Dairy Council, 111 N. Canal St., Chicago, Ill. 60606.

PUBLISHERS AND SUPPLIERS

Activity Resources Co.
P.O. Box 4875, Hayward, Calif. 94540

Addison-Wesley Publishing Co.
Reading, Mass. 01867

American Humane Education Society
180 Longwood Ave., Boston, Mass. 02115

Amsco School Publications
315 Hudson St., New York, N.Y. 10013

Animal Welfare Institute
P.O. Box 3650, Washington, D.C. 20036

Arc Books
219 Park Ave. S., New York, N.Y. 10003

Arco Publishing Co.
219 Park Ave. S., New York, N.Y. 10003

Art Education
Blauvelt, N.Y. 10913

ASPCA, Education Dept.
441 E. 92d St., New York, N.Y. 10028

Bantam Books
666 5th Ave., New York, N.Y. 10019

Bellerophon Books
153 Steuart St., San Francisco, Calif. 94105

Bobbs-Merrill Co.
4300 W. 62d St., Indianapolis, Ind. 46206

Boys' Life
North Brunswick, N.J. 08902

Milton Bradley Co.
Springfield, Mass. 01101

Camp Fire Girls
450 Ave. of the Americas, New York, N.Y. 10011

Candle Mill
East Arlington, Vt. 05252

Chatham Press
15 Wilmot Lane, Riverside, Conn. 06878

Chilton Book Co.
Chilton Way, Radnor, Pa. 19089

Coats & Clark
P.O. Box 1010, Toccoa, Ga. 30577

Collier Books
866 3d Ave., New York, N.Y. 10022

Countryside Press
230 W. Washington Sq., Philadelphia, Pa. 19105

Creative Publications
P.O. Box 10328, Palo Alto, Calif. 94303

Thomas Y. Crowell Co.
666 5th Ave., New York, N.Y. 10019

Crown Publishers
419 Park Ave. S., New York, N.Y. 10016

Daisy
Rogers, Ark. 72756

Dial Press
245 E. 47th St., New York, N.Y. 10017

Dodd, Mead & Co.
79 Madison Ave., New York, N.Y. 10016

Doubleday & Co.
Garden City, N.Y. 11530

Dover Publications
180 Varick St., New York, N.Y. 10014

E. P. Dutton & Co.
201 Park Ave. S., New York, N.Y. 10003

EDC Distribution Center
39 Chapel St., Newton, Mass. 02160

Edmund Scientific Co.
555 Edscorp Building, Barrington, N.J. 08007

Carolyn R. Fellman
313 S. Aurora St., Ithaca, N.Y. 14850

J. C. Ferguson Publishing Co.
6 N. Michigan Ave., Chicago, Ill. 60602

Flash Books
33 W. 60th St., New York, N.Y. 10023

Fleet Academic Editions
160 5th Ave., New York, N.Y. 10010

Four Winds Press
50 W. 44th St., New York, N.Y. 10036

Fun Publishing
P.O. Box 2049, Scottsdale, Ariz. 85252

Funk & Wagnalls
55 E. 77th St., New York, N.Y. 10021

GeoBooks
179 Oak St., San Francisco, Calif. 94102

Golden Books; Golden Fun & Games Books; Golden Guide Series; Golden Press
1220 Mound Ave., Racine, Wis. 53404

Grosset & Dunlap
51 Madison Ave., New York, N.Y. 10010

Harcourt Brace Jovanovich
757 3d Ave., New York, N.Y. 10017

Harper & Row
10 E. 53d St., New York, N.Y. 10022

Harvey House
Irvington-on-Hudson, N.Y. 10533

Kristin Hedberg Toys
Town Square, Nicasio, Calif. 94946

Holiday House
18 E. 56th St., New York, N.Y. 10022

Holt, Rinehart and Winston
383 Madison Ave., New York, N.Y. 10017

Houghton Mifflin Co.
2 Park St., Boston, Mass. 02107

House of Collectibles
17 Park Ave., New York, N.Y. 10016

Japan Publications Trading Co.
1255 Howard St., San Francisco, Calif. 94103

Kalmbach Books
1027 N. 7th St., Milwaukee, Wis. 53233

Alfred A. Knopf
201 E. 50th St., New York, N.Y. 10022

Learning Stuff
P.O. Box 4123, Modesto, Calif. 95352

J. B. Lippincott Co.
E. Washington Square, Philadelphia, Pa. 19105

Little, Brown and Co.
34 Beacon St., Boston, Mass. 02106

Lothrop, Lee & Shepard Co.
105 Madison Ave., New York, N.Y. 10016

Macmillan Publishing Co.
866 3d Ave., New York, N.Y. 10022

Macoy Publishing Co.
P.O. Box 9825, Richmond, Va. 23228

Mangelsen's
P.O. Box 3314, Omaha, Nebr. 68127

McGraw-Hill Book Co.
1221 Ave. of the Americas, New York, N.Y. 10020

David McKay Co.
750 3d Ave., New York, N.Y. 10017

William Morrow & Co.
105 Madison Ave., New York, N.Y. 10016

Museum of Fine Arts
Boston, Mass. 02115

Nabisco
P.O. Box 29, New York, N.Y. 10017

National Dairy Council
111 N. Canal St., Chicago, Ill. 60606

National Geographic Society
17th and M Sts., NW, Washington, D.C. 20036

Natural History Press
Central Park West and 79th St., New York, N.Y. 10024

Thomas Nelson
407 7th Ave., S., Nashville, Tenn. 37203

Nitty Gritty Productions
P.O. Box 457, Concord, Calif. 94522

Oak Publications
33 W. 60th St., New York, N.Y. 10023

101 Productions
834 Mission St., San Francisco, Calif. 94103

Open Door Enterprises
1249 Dell Ave., Campbell, Calif. 95008

Pantheon Books
201 E. 50th St., New York, N.Y. 10022

Parents' Magazine Press
52 Vanderbilt Ave., New York, N.Y. 10017

Pet Library
P.O. Box D, Harrison, N.J. 07029

Pocket Books
630 5th Ave., New York, N.Y. 10020

Polk's Hobby Department Store
314 5th Ave., New York, N.Y. 10001

Praeger Publishers
111 4th Ave., New York, N.Y. 10003

Prentice-Hall
Englewood Cliffs, N.J. 07632

Random House
201 E. 50th St., New York, N.Y. 10022

Saturday Evening Post
1100 Waterway Blvd., Indianapolis, Ind. 46202

Scholastic Book Services
50 W. 44th St., New York, N.Y. 10036

Scott Publishing Co.
530 5th Ave., New York, N.Y. 10036

Charles Scribner's Sons
597 5th Ave., New York, N.Y. 10017

Scrimshaw Press
149 9th St., San Francisco, Calif. 94103

Sentinel Books
17 E. 22d St., New York, N.Y. 10010

Signet
1301 Ave. of the Americas, New York, N.Y. 10019

Simon and Schuster
630 5th Ave., New York, N.Y. 10020

Stein & Day
Scarborough House, Briarcliff Manor, N.Y. 10510

Sterling Publishing Co.
419 Park Ave. S., New York, N.Y. 10016

Sun River Press
132 Old Post Road, N. Croton-on-Hudson, N.Y. 10520

Louis Tannen Magic Shop
1540 Broadway, New York, N.Y. 10036

Taplinger Publishing Co.
200 Park Ave. S., New York, N.Y. 10003

Tempo Books
51 Madison Ave., New York, N.Y. 10010

Ten Speed Press
P.O. Box 4310, Berkeley, Calif. 94704

Troubadour Press
126 Folsom St., San Francisco, Calif. 94105

Two Continents Publishing Group
30 E. 42d St., New York, N.Y. 10017

UNICEF
331 E. 38th St., New York, N.Y. 10016

United States Playing Card Co.
P.O. Box 12126, Cincinnati, Ohio 45212

University of California, Dept. of Landscape Architecture
202 Wurster Hall, Berkeley, Calif. 94702

Van Nostrand Reinhold Co.
450 W. 33d St., New York, N.Y. 10001

Viking Press
625 Madison Ave., New York, N.Y. 10022

Walker & Co.
720 5th Ave., New York, N.Y. 10019

Frederick Warne & Co.
101 5th Ave., New York, N.Y. 10003

Watson-Guptill Publications
1 Astor Plaza, New York, N.Y. 10036

Samuel Weiser
734 Broadway, New York, N.Y. 10003

Workman Publishing Co.
231 E. 51st St., New York, N.Y. 10022

Workshop for Learning Things
5 Bridge St., Watertown, Mass. 02172

ART AND PHOTO CREDITS

P. 3, top left: drawing by Ann Rees from *Modelling Is Easy.* By permission of Arco Publishing Co. P. 4, center: mouse drawing by Steve Madison from *Felt Craft.* By permission of Doubleday & Co.; right: "Stone Frog" by Margaret Hartelius from *Pebbles and Pods: A Book of Nature Crafts.* By permission of Scholastic Magazines. P. 7, left: "clothespin mortar" by Harvey Weiss from *The Gadget Book.* By permission of Thomas Y. Crowell Co.; bottom center: drawing by Gretchen Schields from *International Folk Crafts.* By permission of Troubador Press. P. 9, left: drawing by Tom Cooke; center, top: drawing by Arielle Mather; bottom: drawing by Tom Cooke; all from *Steven Caney's Play Book.* By permission of Workman Publishing Co.; right: 2 photos from *The Little Kid's Four Seasons Craft Book.* By permission of Taplinger Publishing Co. P. 11: photo by Carl Fischer, drawings by chas. b. slackman from *Trash Can Toys and Games.* By permission of Viking Press. P. 13, center: photo from *Woodstock Kid's Crafts.* By permission of Bobbs-Merrill Co.; right: "Tick-Tack-Toe" from *Best Rainy Day Book Ever.* By permission of Random House. P. 18, left: photo from *Step by Step Candlemaking.* By permission of Western Publishing Co.; center: photo by Barbara Weakley from *How to Make Candles;* right: 3 photos from *The Art of Thread Design.* By permission of Mark Jansen and Open Door Enterprises. P. 19: photo by True Kelly from *Finding One's Way with Clay.* By permission of Simon and Schuster. P. 20, top: photo from *Flower Making for Beginners.* By permission of Taplinger Publishing Co.; bottom: photo from *Origami for Displays.* P. 21, left: drawing by Loretta Trezzo from *Jewelry Craft for Beginners.* By permission of Bobbs-Merrill, Co.; center: photo by George C. Bradbury from *Tincraft.* By permission of Simon and Schuster. P. 26, left: illustrations from *A Christmas Stained Glass Coloring Book.* By permission of Dover Publications. P. 27, right: illustration from *Make Your Own World of Christmas.* By permission of Bobbs-Merrill Co. P. 32, left: from *Altair Design.* By permission of Pantheon Books, a division of Random House. P. 35, center: photo from *Art from Found Objects.* By permission of Lothrop, Lee & Shepard; right: from *Draw 50 Animals.* By permission of Doubleday & Co. P. 36, center: "Moses" from *Stained Glass Windows Coloring Book.* By permission of Dover Publications; right: drawing from *Creative Ways with Drawing.* By permission of Western Publishing Co. P. 37, right: "I" from *A Medieval Alphabet Coloring Book.* By permission of Bellerophon Books. P. 38: detail from "A Computer Portrait of Lincoln" by Leon Harmon and Ken Knowlton from *Optricks: A Book of Optical Illusions.* By permission of Troubador Press. P. 39, bottom right: from *Optricks: A Book of Optical Illusions.* By permission of Troubador Press. P. 44, left: from *Puzzlers* by John Hull. By permission of Troubador Press. P. 45, right: from *Maze Craze* by John Hull. By permission of Troubador Press. P. 47: from *Puzzlers* by John Hull. By permission of Troubador Press. P. 55: illustration from *Zodiac Coloring Book.* By permission of Troubador Press. Pp. 56–57: photos by David Attlie from *Making Puppets Come Alive: A Method of Learning and Teaching Hand Puppetry.* By permission of Taplinger Publishing Co. P. 60: photo by Alfonso Barrios from *Young Filmmakers.* By permission of E. P. Dutton & Co. Pp. 62–63: 2 photos courtesy Yellow Ball Workshop. Pp. 64–65: photo by Viki Holland from *How to Photograph Your World.* By permission of Charles Scribner's Sons. P. 77: Illustration from *Stone Soup,* Volume III, No. 2. P. 78: illustration from *Fun Projects for Dad and the Kids.* By permission of Arco Publishing Co. P. 79: illustrations by Jan Adkins from *How a House Happens.* By permission of Walker & Co. P. 80, top: photo © 1974 by Harvey Weiss from *Model Cars and Trucks and How to Build Them.* By permission of Thomas Y. Crowell Co. P. 81: 2 photos from *The Buffy-Porson: A Car You Can Build and Drive.* By permission of Charles Scribner's Sons. P. 82, left: drawing by Judith Lane from *Making Children's Furniture and Play Structures.* By permission of Workman Publishing Co. P. 84, left: drawings by Liz Green from *Step by Step Bargello.* By permission of Western Publishing Co. Pp. 84–85: 6 photos by George Ancona from *Step by Step Bargello.* By permission of Western Publishing Co. P. 87, left: 2 drawings by John Giannoni from *Step by Step Weaving.* By permission of Western Publishing Co.; right: 2 drawings by Dag Olsen from *Step by Step Rugmaking.* By permission of Western Publishing Co. P. 88: photo by Jerry Wainwright from *Native Funk & Flash.* By permission of Scrimshaw Press. P. 88, left and p. 89, bottom: 3 illustrations by Ava Morgan from *Sew It and Wear It.* By permission of Thomas Y. Crowell Co. P. 89, top right: illustration from *A Beginner's Book of Patchwork Appliqué & Quilting.* By permission of Dodd, Mead & Co. P. 90: "The Scholars" photo from *Creative Soft Toy Making.* By permission of Bobbs-Merrill Co. P. 96: illustration by "Wolo" from *Hippo Cook Book.* By permission of Nitty Gritty Productions. P. 98, left; p. 99, right; p. 100, bottom: 4 drawings by Marvin Rubin from *Mother Earth's Hassle-Free Indoor Plant Book.* By permission of J. P. Tarcher. Pp. 98–99: "Gaillardia" from *Garden Flowers Coloring Book.* By permission of Dover Publications. P. 100, top: illustration by Norman Erler Rahn from *Growing Up Green.* By permission of Workman Publishing Co. P. 101: "Phlox" from *Garden Flowers Coloring Book.* By permission of Dover Publications. P. 106: illustration by Winston Tong from *The Dinosaur Coloring Book.* By permission of Troubador Press. P. 107: illustration by Paul E. Kennedy from *Audubon's Birds of America Coloring Book.* By permission of Dover Publications. P. 108: "Iris" by Gompers Saijo from *North American Wildflowers Coloring Album.* By permission of Troubador Press. P. 109: "Sea Horse" by Gompers Saijo from *North American Sea Life Coloring Album.* By permission of Troubador Press. P. 111: "Wild Cat" by Gompers Saijo from *North American Wildlife Coloring Album.* By permission of Troubador Press. P. 113: photo by Dan Levin from *The Air We Breathe.* By permission of Doubleday & Co. P. 115, top: illustration by Charles Jakubowski from *Science Projects in Ecology.* By permission of Holiday House. P. 117, top: 2 illustrations by Susan Perl from *Sparrows Don't Drop Candy Wrappers.* By permission of Dodd, Mead & Co. P. 124, bottom: drawing from *Build Your Own Early American Village.* By permission of Pantheon Books, a division of Random House. Pp. 124–25, top: "Lexington Fight" from *Coloring Book of the American Revolution.* By permission of Bellerophon Books. P. 125, bottom: illustration from *Paper Soldiers of the American Revolution.* By permission of Bellerophon Books. P. 128: photo courtesy of Ocko Associates, Inc. Pp. 129–30: "Drummer" and "Pioneer" by Peter F. Copeland from *Uniforms of the American Revolution Coloring Book.* By permission of Dover Publications. P. 131: "Broom Seller" from *Everyday Dress of the American Revolution Coloring Book.* By permission of Dover Publications. P. 138, top: photo courtesy of Living World®; bottom: photo by Frederick Breda from *Hamsters: All About Them.* By permission of Lothrop, Lee & Shepard Co. P. 140, top right: illustration by Haris Petie from *Both Ends of a Leash.* By permission of Prentice-Hall. P. 147, top: illustration by Frank Mullins from *Track and Field for Young Champions.* By permission of McGraw-Hill Book Co. P. 149, left: illustration from *Junior Judo.* By permission of Sterling Publishing Co.; center: illustration by Frank Robbins from *Bowling Talk for Beginners.* By permission of Julian Messner, a division of Simon and Schuster. P. 150, top: illustration by Albert Nebeker from *Auto Racing.* By permission of Troubador Press. P. 154: photo by Don Miller, with permission of Grey Owl Indian Craft Manufacturing Co. Pp. 154–55: 2 illustrations from *Kachina Doll Coloring Book 2.* By permission of Fun Publishing Co. P. 156: "Plains Warrior" from *A Coloring Book of American Indians.* By permission of Bellerophon Books. P. 157, top: "Indian Symbols and Designs" from *The Complete How-to Book of Indiancraft.* By permission of Macmillan Publishing Co. P. 161: photo by Edward Kimball, Jr. from *Be a Frog, a Bird, or a Tree.* By permission of Doubleday & Co. P. 163, right: "Super Highriser" and "Extreme Automotive Style" from *Bikes.* By permission of Chatham Press. P. 164, left, and pp. 164–65, top: 2 illustrations by Robert Smith from *The American Biking Atlas & Touring Guide.* By permission of Workman Publishing Co. P. 168, center and right: 2 illustrations from *25 Kites That Fly.* By permission of Dover Publications. P. 169, left: illustration from *Boomerangs: How to Make and Throw Them.* By permission of Dover Publications. P. 172–73: photos courtesy Sig Mfg. Co. Pp. 175–76: illustrations from *Paper Airplanes.* By permission of Troubador Press. P. 177, top and left: illustrations by John Kaufmann from *Flying Hand-Launched Gliders.* By permission of William Morrow & Co.; right: illustration from *How to Make and Fly Paper Airplanes.* By permission of Four Winds Press. P. 178: illustrations from *How It Works Volume 2.* By permission of Grosset & Dunlap. P. 179, left: photo courtesy of Things of Science. P. 182: photo courtesy of Edcom Systems. P. 188, left: illustration by Peter Lippman from *Science Experiments You Can Eat.* By permission of J. B. Lippincott Co. Pp. 192, left, 193, top, and 197, left: 3 illustrations from *A Chaucer Coloring Book.* By permission of Bellerophon Books. Pp. 192–93: illustration from *A Coloring Book of Horses.* By permission of Bellerophon Books. P. 194, bottom: detail from "Isis" poster. By permission of Bellerophon Books. P. 197: photo by Roy Killeen. By permission of 101 Productions. P. 205: illustration from *A First Book of Space Form Making.* By permission of GeoBooks. Supplemental illustrations and decorative elements from the Dover Pictorial Archive Books: *Decorative Frames and Borders, 1800 Woodcuts by Thomas Bewick and His School, Catchpenny Prints, Curious Woodcuts of Fanciful and Real Beasts, Quaint Cuts in the Chap Book Style, Art Nouveau and Art Deco Type and Design, Picture Sourcebook for Collage and Decoupage, Handbook of Early Advertising Art, Original Art Deco Designs, Exotic Alphabets and Ornament, An Old Fashioned Christmas in Illustration and Decoration, Graphic Trade Symbols by German Designers.*